Topics in Contemporary Mathematical Analysis and Applications

Mathematics and its Applications
Modelling, Engineering, and Social Sciences

Series Editor:
Hemen Dutta
Department of Mathematics, Gauhati University

Tensor Calculus and Applications
Simplified Tools and Techniques
Bhaben Kalita

Discrete Mathematical Structures
A Succinct Foundation
Beri Venkatachalapathy Senthil Kumar and Hemen Dutta

Methods of Mathematical Modelling
Fractional Differential Equations
Edited by Harendra Singh, Devendra Kumar, and Dumitru Baleanu

Mathematical Methods in Engineering and Applied Sciences
Edited by Hemen Dutta

Sequence Spaces
Topics in Modern Summability Theory
Mohammad Mursaleen and Feyzi Başar

Fractional Calculus in Medical and Health Science
Devendra Kumar and Jagdev Singh

Topics in Contemporary Mathematical Analysis and Applications
Hemen Dutta

Sloshing in Upright Circular Containers
Theory, Analytical Solutions, and Applications
Alexander Timokha and Ihor Raynovskyy

ISSN (online): 2689-0224
ISSN (print): 2689-0232

For more information about this series, please visit:
https://www.routledge.com/Mathematics-and-its-applications/book-series/MES

Topics in Contemporary Mathematical Analysis and Applications

Edited by

Hemen Dutta

CRC Press
Taylor & Francis Group
Boca Raton London New York

CRC Press is an imprint of the
Taylor & Francis Group, an **informa** business

First edition published 2021
by CRC Press
6000 Broken Sound Parkway NW, Suite 300, Boca Raton, FL 33487-2742

and by CRC Press
4 Park Square, Milton Park, Abingdon, Oxon OX14 4RN

ISBN: 978-0-367-53266-6 (hbk)
ISBN: 978-1-003-08119-7 (ebk)

Typeset in Times
by codeMantra

Contents

Preface

This book covers certain aspects of contemporary mathematical analysis and its applications. It focuses on enriching the understanding of several methods, problems, and applications in the area of mathematical analysis having contemporary research and study significances. Each chapter of this book aims to offer readers the understanding of discussed research problems by presenting related developments in reasonable details. This book is expected to be a valuable resource for graduate students, researchers, educators, and teachers interested in contemporary mathematical analysis. There are 12 chapters in this book, and they are organized as follows.

Chapter "Certain Banach-Space Operators Acting on Free Poisson Elements Induced by Orthogonal Projections" investigates the free (weighted-) Poisson elements in the Banach $*$-probability space generated by integer-many mutually free, semicircular elements, induced by mutually orthogonal integer-many projections in a fixed C^*-probability space. It constructs corresponding $*$-homomorphisms acting on the Banach $*$-probability space in order to study how they affect the free (weighted-) Poisson distributions and the free distributions of free (weighted-) Poisson elements. It also constructs certain Banach-space operators generated by the $*$-homomorphisms to examine how such operators deform free probability and to characterize how the free (weighted-) Poisson distributions are deformed by the action of those operators.

Chapter "Linear positive operators involving orthogonal polynomials" aims to survey extensively on the available research work regarding approximations by the linear positive operators defined by using orthogonal polynomials. In the literature, several sequences of linear positive operators involving orthogonal polynomials have been defined, and their Durrmeyer and Kantorovich-type variants have been investigated. In recent years, there has been a significant increase in activities in the approximation of continuous functions on $[0, \infty)$ by the linear positive operators based on orthogonal polynomials.

Chapter "Approximation by Kantorovich variant of λ-Schurer operators and related numerical results" aims to construct Kantorovich variant of λ-Schurer operators using certain polynomials in order to study Voronovskaja-type theorems and obtain a global approximation formula for such operators. It presents a local direct estimate of the rate of convergence by Lipschitz-type function involving two parameters. The proposed operators reduce to the classical Schurer-Kantorovich operators as well as to the classical Bernstein-Kantorovich operators and λ-Bernstein-Kantorovich operators under certain special cases. Finally, it provides tables and graphs based on numerical experiments in order to justify the efficiency of the results in comparison with other available results.

Chapter "Characterizations of rough fractional-type integral operators on variable exponent vanishing Morrey-type spaces" aims to apply relevant properties of variable exponent to study on Adams and Spanne-type estimates for a class of fractional integral operators with variable orders on variable exponent vanishing generalized Morrey spaces. Finally, it claims to introduce the variable exponent-generalized Campanato spaces and then obtain the boundedness of the commutators of these operators on such spaces.

Chapter "Compact-like operators in vector lattices normed by locally solid lattices" investigates continuous, bounded, and compact operators on locally solid Riesz spaces with respect to a concept of convergence, which was introduced and investigated on locally solid Riesz spaces. This chapter describes the generalization of several known classes of operators on Riesz spaces and Banach lattices such as norm continuous, order continuous, order bounded, bounded, and compact operators. It also presents the relations among such operators and examines the basic properties of the operators.

Chapter "On indexed product summability of an infinite series" presents the findings based on the indexed product summability of an infinite series by using Nörlund mean. In summability theory, indexed summability and indexed product summability of an infinite series have significant importance and relevance.

Chapter "On some important inequalities" aims to present some useful mathematical inequalities such as the Jensen, Hermite-Hadamard, Rogers-Hölder, and Minkowski inequalities. The aforementioned four inequalities have been presented in discrete and integral forms, and their generalizations have also been discussed. The Jensen and Hermite-Hadamard inequalities have been considered in more detail. An expansion of the initial integral form of Jensen's inequality is promoted for convex functions of several variables. The Rogers-Hölder and Minkowski inequalities have been derived from the integral form of Jensen's inequality for convex functions of several variables. The Minkowski inequality is realized independently of the Rogers-Hölder inequality.

Chapter "Refinements of Young integral inequality via fundamental inequalities and mean value theorems for derivatives" reviews several refinements of Young's integral inequality via several mean value theorems, such as Lagrange's and Taylor's mean value theorems of Lagrange's and Cauchy's type remainders, and via several fundamental inequalities, such as Čebyšev's integral inequality, Hermite-Hadamard's type integral inequalities, Hölder's integral inequality, and Jensen's discrete and integral inequalities, in terms of higher-order derivatives and their norms. It also surveys several applications of several refinements of Young's integral inequality and further refines Young's integral inequality via Pólya's type integral inequalities.

Chapter "On the coefficient estimates for new subclasses of bi-univalent functions associated with subordination and Fibonacci numbers" investigates new subclasses of the bi-univalent function class associated with subordination and Fibonacci numbers in the open unit disc. It obtains estimates on the first two Taylor-Maclaurin coefficients, derives Fekete-Szegö inequalities for functions belonging to the newly defined classes, and also discusses relevant connections to some of the earlier known results.

Chapter "Fixed point problems of multivalued mappings" aims to construct the fixed-point results of multivalued mappings subject to the satisfaction of generalized cyclic F-contractive conditions in metric spaces. It obtains the fixed points of multi-valued mappings under cyclic simulation functions, and then establishes the stability of fixed-point sets of generalized cyclic F-contractive mappings and mappings under cyclic simulation function. The results have been obtained without using any form of continuity of multivalued mappings involved. Further, suitable examples have also been given to support the results.

Chapter "Significance and relevance of functional equations in various fields" aims to impart the significant role of functional equations in various fields. The study of functional equations is a growing, and an important, area in mathematics. It covers many other areas of mathematics, and recently, their role in science and engineering has found to be very attractive. It also portrays a few applications of functional equations in geometry, finance, information theory, wireless sensor networks and electric circuits with parallel resistances, physics, and electromagnetism.

Chapter "Unified type nondifferentiable second-order symmetric duality results over arbitrary cones" claims to formulate a new mixed-type second-order nondifferentiable symmetric duality in scalar-objective programming over arbitrary cones. It first discusses mixed-type primal-dual results based on available literature and constructs some numerical examples to justify the presented definitions. It derives the combined result with one model over arbitrary cones, and the duality theorems have been derived for some problems under bonvexity and pseudobonvexity with respect to η assumptions over arbitrary cones.

The editor sincerely acknowledges the cooperation and patience of contributors during the entire process of handling their works submitted for this book. Reviewers deserve deep gratitude for selflessly offering their help for successfully bringing out this book. The editor also thankfully acknowledges the support of editorial staff at Taylor & Francis. Encouragement from several colleagues and friends is the primary source of inspiration to work for such book projects, and the editor gratefully acknowledges their mental support.

Hemen Dutta
Guwahati, India
26th April, 2020

Editor

Hemen Dutta is a regular faculty member in Mathematics at Gauhati University, India. He did his Master of Science (MSc) in Mathematics, Post Graduate Diploma in Computer Application (PGDCA), Master of Philosophy (MPhil) in Mathematics, and Doctor of Philosophy (PhD) in Mathematics, respectively. His topics of primary research interest are in the area of mathematical analysis. He has contributed in authoring around 150 items so far as research papers for journals, chapters for books, and proceedings papers. He has already published 20 books as textbooks, reference books, edited books, and proceedings of conferences. He has organized many academic events for researchers and academicians, and also associated with several other such activities in different capacities. He has contributed several general and popular articles in newspaper, science magazines, science portals, popular books, etc. He has offered his service as a resource person/invited speaker in workshops, conferences, teachers' training program, faculty development programs, etc. He has also visited foreign institutions on invitation for research collaboration and delivering talks.

Contributors

Mujahid Abbas
Department of Mathematics
Government College University Lahore
Lahore, Pakistan

P. N. Agrawal
Department of Mathematics
IIT Roorkee
Roorkee, India

Şahsene Altınkaya
Department of Mathematics
Bursa Uludag University
Bursa, Turkey

Abdullah Aydın
Department of Mathematics
Muş Alparslan University
Muş, Turkey

P. Baliarsingh
Department of Mathematics
Gangadhar Meher University
Sambalpur, Odisha, India

Ruchi Chauhan
Department of Mathematics
IIT Roorkee
Roorkee, India

Ilwoo Cho
Department of Mathematics and Statistics
Saint Ambrose University
Davenport, Iowa, U.S.A.

Kamil Demirci
Department of Mathematics
Sinop University
Sinop, Turkey

Ramu Dubey
Department of Mathematics
J.C. Bose University of Science and Technology
Faridabad, India

Hemen Dutta
Department of Mathematics
Gauhati University
Guwahati, India

Shenghu Ding
School of Mathematics and Statistics
Ningxia University
Yinchuan, P. R. China

Ferit Gürbüz
Faculty of Education, Department of Mathematics Education
Hakkari University
Hakkari, Turkey

Bai-Ni Guo
School of Mathematics and Informatics
Henan Polytechnic University
Jiaozuo, China

Huili Han
School of Mathematics and Statistics
Ningxia University
Yinchuan, P. R. China

Wen-Hui Li
Department of Fundamental Courses
Zhengzhou University of Science and Technology
Zhengzhou, China

Pinhong Long
School of Mathematics and Statistics
Ningxia University
Yinchuan, P. R. China

Vishnu Narayan Mishra
Department of Mathematics
Indira Gandhi National Tribal University
Lalpur, Amarkantak, India

Talat Nazir
Department of Mathematical Sciences
University of South Africa
Johannesburg, South Africa
and
Department of Mathematics
COMSATS University Islamabad
Abbottabad, Pakistan

Faruk Özger
Department of Engineering Sciences
İzmir Kâtip Çelebi University
İzmir, Turkey

B. P. Padhy
School of Applied Sciences, Department of Mathematics
KIIT Deemed to be University
Bhubaneswar, India

Zlatko Pavić
Department of Mathematics
Mechanical Engineering Faculty in Slavonski Brod
University of Osijek
Osijek, Croatia

Feng Qi
Institute of Mathematics, Henan Polytechnic University, Jiaozuo 454010, Henan, China
College of Mathematics and Physics
Inner Mongolia University for Nationalities Tongliao, China
and
School of Mathematical Sciences
Tianjin Polytechnic University
Tianjin, China

B. V. Senthil Kumar
Department of Information Technology
Nizwa College of Technology
Nizwa, Oman

Guo-Sheng Wu
School of Computer Science
Sichuan Technology and Business University
Chengdu, China

Sevda Yıldız
Department of Mathematics
Sinop University
Sinop, Turkey

1 Certain Banach-Space Operators Acting on Free Poisson Elements Induced by Orthogonal Projections

Ilwoo Cho
Saint Ambrose University

CONTENTS

1.1 INTRODUCTION

Let (A, ψ) be a C^**-probability space* containing mutually orthogonal $|\mathbb{Z}|$-many *projections* $\{q_j\}_{j\in\mathbb{Z}}$. Such projections q_j's induce the corresponding *free weighted-semicircular family* $\{u_j\}_{j\in\mathbb{Z}}$ in a certain *Banach* $*$*-probability space* $(\mathfrak{L}_Q, \tau)$, and under additional conditions, this free family $\{u_j\}_{j\in\mathbb{Z}}$ generates the *free semicircular family* $\{U_j\}_{j\in\mathbb{Z}}$ in $(\mathfrak{L}_Q, \tau)$; see [6]. This free semicircular family $\{U_j\}_{j\in\mathbb{Z}}$ generates the Banach $*$-probabilistic substructure $(\mathbb{L}_Q, \tau)$ of $(\mathfrak{L}_Q, \tau)$. The free probability on $\mathbb{L}_Q$ was studied in [7,8,10,11].

There are ways to construct semicircular elements in earlier works, e.g., see [1,2,17,18,19,20,21,29,30]. However, our construction is different from those of them. Ours is motivated by the construction of the *weighted-semicircular elements* of [5,9,11]. In [5,9,11], (weighted-)semicircular elements are naturally constructed from analyses on the *p-adic number fields* $\mathbb{Q}_p$, for primes p (and their globalization, the analysis on the *finite Adelic ring* $A_{\mathbb{Q}}$), from $|\mathbb{Z}|$-many orthogonal projections induced by $|\mathbb{Z}|$-many measurable *p-adic characteristic functions* of spheres in the p-adic number fields $\mathbb{Q}_p$ (e.g., [26] and [27]). Motivated by these, we mimic the construction of [5,9,11], in the cases where a C^*-probability space (A, ψ) has mutually orthogonal $|\mathbb{Z}|$-many projections in [6] (also, see below). For more about details and applications, see [7,8,10] and [11].

The main purposes of this paper are (i) to construct *free (weighted-)Poisson elements* of the *Banach* $*$*-probability space* $(\mathbb{L}_Q, \tau)$, (ii) to consider the $*$-homomorphisms $\{\beta_{\pm}^{n}\}_{n\in\mathbb{N}_0}$ acting on $\mathbb{L}_Q$ induced by certain order-preserving bijective functions $h_{\pm}$ on the set $\mathbb{Z}$ of all integers, (iii) to study how these $*$-homomorphisms of (ii) affect the free probability on $\mathbb{L}_Q$, (iv) to consider how the *free (weighted-)Poisson distributions*, which are the free distributions of our free (weighted-)Poisson elements of $(\mathbb{L}_Q, \tau)$, are affected by the action of $\{\beta_{\pm}^{n}\}_{n\in\mathbb{N}_0}$, based on (iii), where

$$\mathbb{N}_0 = \mathbb{N} \cup \{0\},$$

and (v) to construct a certain Banach $*$-algebra $\mathfrak{B}$ generated by $\{\beta_{\pm}^{n}\}_{n\in\mathbb{N}_0}$, and to investigate how the operators of $\mathfrak{B}$ distort the free probability on $(\mathbb{L}_Q, \tau)$.

Our main results show that (I) typical types of free (weighted-)Poisson elements generated by (weighted-)semicircular elements are characterized by the construction of $\mathbb{L}_Q$, and the universality of the semicircular law; (II) our free weighted-Poisson distributions are completely characterized by free Poisson distributions; (III) the $*$-homomorphisms of (ii) are *free-homomorphisms*, i.e., $*$-homomorphisms of (ii) preserve the free probability on $\mathbb{L}_Q$, and hence, the free (weighted-)Poisson distributions of (i) are preserved by the action of these $*$-homomorphisms on $\mathbb{L}_Q$; and (IV) the distortion of free probability on $\mathbb{L}_Q$, under the action of the Banach algebra $\mathfrak{B}$ of (v), is characterized, and it shows how our free (weighted-)Poisson distributions are deformed by the action of $\mathfrak{B}$ on $\mathbb{L}_Q$.

In earlier works, free Poisson elements have been studied. Even though free Poisson elements and free Poisson distributions are well known (e.g., see [15,16] and [17]), our free "weighted-Poisson" elements are newly introduced here. Moreover, their free distributions, the *free weighted-Poisson distributions*, are completely characterized with respect to certain free Poisson distributions. In the long run, we show that free weighted-Poisson elements are factorized by certain operators and free Poisson elements. More interestingly, distorted free (weighted-)Poisson distributions are studied under the action of $\mathfrak{B}$ on $\mathbb{L}_Q$.

1.2 PRELIMINARIES

Free probability is the noncommutative operator-algebraic version of classical *measure theory* and *statistical analysis*. The operator-algebraic *freeness* plays like the classical functional *independence* by replacing measures on sets to linear functionals on noncommutative algebras (e.g., [17,22,23,28] and [30]). It is one of the main subjects not only in pure mathematics (e.g., [3,4,20,21,24,25] and [29]), but also in related fields (e.g., [5] through [12]). In particular, here, the combinatorial free probability is used (e.g., [17,22] and [23]).

In the text, without definitions and backgrounds, *free moments* and *free cumulants* of operators will be computed for analyzing free distributions of the operators. Also, we use *free product $*$-probability spaces*, without a detailed introduction.

By the *central limit theorem(s)*, studying semicircular elements whose free distributions are *the semicircular law*, is one of the main topics in free probability theory (e.g., [1,2,18,19,20,21] and [29]). As an application, *free Poisson elements* whose free distributions are the free Poisson distributions, have been considered (e.g., [15,16] and [17]).

1.3 SOME BANACH $*$-ALGEBRAS INDUCED BY PROJECTIONS

In this section, we establish backgrounds of our proceeding works. Let (B, φ) be a topological $*$-probability space (a *C^*-probability space*, or a *W^*-probability space*, or a *Banach $*$-probability space*, etc.), where B is a topological $*$-algebra (a C^*-algebra, resp., a W^*-algebra, resp., a Banach $*$-algebra, etc.), and φ is a bounded linear functional on B.

An operator a of B is said to be a *free random variable* whenever it is regarded as an element of (B, φ). As usual in *operator theory*, a free random variable a is said to be *self-adjoint* in (B, φ) if $a^* = a$ in B, where a^* is the *adjoint of a* (e.g., [14]).

Definition 1.1. *A self-adjoint free random variable a is said to be weighted-semicircular in (B, φ) with its weight $t_0 \in \mathbb{C}^\times = \mathbb{C} \setminus \{0\}$ (or, in short, t_0-semicircular), if a satisfies the free-cumulant computations,*

$$k_n(a, ..., a) = \begin{cases} k_2(a, a) = t_0 & \text{if } n = 2 \\ 0 & \text{otherwise,} \end{cases} \tag{1.1}$$

for all $n \in \mathbb{N}$, where $k_\bullet(...)$ is the free cumulant on B in terms of φ under the Möbius inversion of [17,22,23].

If $t_0 = 1$ in (1.1), the 1-semicircular element a is said to be semicircular in (B, φ), i.e., a is semicircular in (B, φ) if a satisfies

$$k_n(a, ..., a) = \begin{cases} 1 & \textit{if } n = 2 \\ 0 & \textit{otherwise,} \end{cases} \tag{1.2}$$

for all $n \in \mathbb{N}$.

By the Möbius inversion of [17,22] and [23], one can characterize the weighted-semicircularity (1.1) as follows: A self-adjoint operator a is t_0-semicircular in (B, φ) if and only if

$$\varphi(a^n) = \omega_n \left(t_0^{\frac{n}{2}} c_{\frac{n}{2}} \right), \tag{1.3}$$

where

$$\omega_n \overset{def}{=} \begin{cases} 1 & \text{if } n \text{ is even} \\ 0 & \text{if } n \text{ is odd,} \end{cases}$$

for all $n \in \mathbb{N}$, and c_k are the k-th *Catalan numbers*,

$$c_k \overset{def}{=} \frac{1}{k+1} \begin{pmatrix} 2k \\ k \end{pmatrix} = \frac{1}{k+1} \left(\frac{(2k)!}{k!(2k-k)!} \right) = \frac{(2k)!}{k!(k+1)!},$$

for all $k \in \mathbb{N}_0$.

Similarly, a self-adjoint free random variable a is semicircular in (B, φ) if and only if a is 1-semicircular in (B, φ), and if and only if

$$\varphi(a^n) = \omega_n c_{\frac{n}{2}}, \tag{1.4}$$

by (1.2) and (1.3), for all $n \in \mathbb{N}$, where ω_n are in the sense of (1.3).

So we use the t_0-semicircularity (1.1) (or, the semicircularity (1.2)) and its characterization (1.3) (resp., (1.4)) alternatively from below.

If a is a self-adjoint free random variable in (B, φ), then the sequences consisting of

$$\text{the } \textit{free moments } (\varphi(a^n))_{n=1}^{\infty},$$

and

$$\text{the } \textit{free cumulants } (k_n(a, ..., a))_{n=1}^{\infty}$$

provide equivalent free-distributional data of a in (B, φ), characterizing the *free distribution of* a (e.g., [17,22] and [23]).

In the rest of this paper, we fix a C^*-probability space (A, ψ) and assume that there are $|\mathbb{Z}|$-many projections $\{q_j\}_{j \in \mathbb{Z}}$ in the C^*-algebra A, i.e., the operators q_j satisfy

$$q_j^* = q_j = q_j^2 \text{ in } A,$$

for all $j \in \mathbb{Z}$ (e.g., [14]). Assume further that these projections $\{q_j\}_{j \in \mathbb{Z}}$ are *mutually orthogonal* in A, in the sense that:

$$q_i q_j = \delta_{i,j} q_j \text{ in } A, \text{ for all } i, j \in \mathbb{Z}, \tag{1.5}$$

where δ is the *Kronecker delta*. Now, we fix the family

$$\mathbf{Q} = \{q_j\}_{j\in\mathbb{Z}} \tag{1.6}$$

of mutually orthogonal projections (1.5) of A.

Remark 1.1. *One can have such a C^*-algebraic structure A containing a family $\mathbf{Q}$ of (1.6), naturally or artificially. Clearly, in the settings of [5,9,12], one can naturally take such structures.*

Suppose there is a C^*-algebra A_0 containing a family $\mathbf{Q}_N = \{q_1, ..., q_N\}$ of mutually orthogonal N-many projections $q_1, ..., q_N$, for $N \in \mathbb{N}_\infty = \mathbb{N} \cup \{\infty\}$. Then, under a suitable direct product, or tensor product, or free product of copies of A_0 under product topology, one can construct a C^*-algebra A containing a family $\mathbf{Q}$ with $|\mathbb{Z}|$-many mutually orthogonal projections, containing $\mathbf{Q}_N$ under unitarily equivalence. For example, see [10].

Let Q be the C^*-subalgebra of A generated by the family $\mathbf{Q}$ of (1.6),

$$Q \overset{def}{=} C^*(\mathbf{Q}) \subseteq A. \tag{1.7}$$

Proposition 1.1. *Let Q be a C^*-subalgebra (1.7) of a fixed C^*-algebra A. Then*

$$Q \overset{*\text{-iso}}{=} \underset{j\in\mathbb{Z}}{\oplus} (\mathbb{C}\cdot q_j) \overset{*\text{-iso}}{=} \mathbb{C}^{\oplus|\mathbb{Z}|}, \text{ in } (A,\ \psi). \tag{1.8}$$

Proof. The proof of (1.8) is done by the orthogonality (1.5) of $\mathbf{Q}$ in A. □

Define now linear functionals ψ_j on the C^*-algebra Q of (1.7) by

$$\psi_j(q_i) = \delta_{ij}\psi(q_j), \text{for all } i \in \mathbb{Z}, \tag{1.9}$$

for all $j \in \mathbb{Z}$, where ψ is the linear functional of the fixed C^*-probability space $(A,\ \psi)$. The linear functionals $\{\psi_j\}_{j\in\mathbb{Z}}$ of (1.9) are well defined on Q by (1.8).

Assumption Let $(A,\ \psi)$ be a fixed C^*-probability space, and let Q be the C^*-subalgebra (1.7) of A. In the rest of this paper, we assume that

$$\psi(q_j) \in \mathbb{C}^\times, \text{ for all } j \in \mathbb{Z}.$$

□

Definition 1.2. *The C^*-probability spaces $(Q,\ \psi_j)$ are called the j-th filters of Q in a given C^*-probability space $(A,\ \psi)$, where Q is in the sense of (1.7) and ψ_j are the linear functionals of (1.9), for all $j \in \mathbb{Z}$.*

Now, let's define the bounded linear transformations $\mathbf{c}$ and $\mathbf{a}$ acting on the C^*-algebra Q, by linear morphisms satisfying

$$\mathbf{c}(q_j) = q_{j+1}, \text{and } \mathbf{a}(q_j) = q_{j-1}, \tag{1.10}$$

for all $j \in \mathbb{Z}$. Then, $\mathbf{c}$ and $\mathbf{a}$ are the well-defined bounded linear operators "on Q." One can understand that they are *Banach-space operators* in the *operator space* $B(Q)$, consisting of all bounded linear transformations acting on Q, by regarding Q as a *Banach space* equipped with its C^*-*norm* (e.g., [13]).

Definition 1.3. *We call these Banach-space operators* **c** *and* **a** *of (1.10) the* creation *and, respectively, the* annihilation *on* Q.

Define now a new Banach-space operator **l** on Q by

$$\mathbf{l} = \mathbf{c} + \mathbf{a} \in B(Q). \tag{1.11}$$

Definition 1.4. *The Banach-space operator* $\mathbf{l} \in B(Q)$ *of (1.11) is called the* radial operator on Q.

Note that if **l** is in the sense of (1.11), then the powers $\mathbf{l}^n$ are also contained in $B(Q)$, for all $n \in \mathbb{N}_0$, with axiomatization:

$$\mathbf{l}^0 = 1_Q, \text{ the identity operator of } B(Q).$$

Now, define a closed subspace $\mathfrak{L}$ of the operator space $B(Q)$ by

$$\mathfrak{L} \overset{def}{=} \overline{\mathbb{C}[\{\mathbf{l}\}]}^{\|.\|} = \overline{span_{\mathbb{C}}\{\mathbf{l}^n : n \in \mathbb{N}_0\}}^{\|.\|}, \tag{1.12}$$

generated by the radial operator **l**, where $\|.\|$ is the operator norm on $B(Q)$, defined to be

$$\|T\| = \sup\{\|Tq\|_Q : \|q\|_Q = 1\},$$

for all $T \in B(Q)$, where $\|.\|_Q$ is the C^**-norm on* Q, and $\overline{X}^{\|.\|}$ are the *operator-norm closures* of subsets X of the *operator space* $B(Q)$ (e.g., [13]). It is not difficult to check that this subspace $\mathfrak{L}$ forms an algebra in the vector space $B(Q)$, by (1.12). So, it forms a Banach algebra in the topological vector space $B(Q)$.

On this Banach algebra $\mathfrak{L}$ of (1.12), define a unary operation $(*)$ by

$$\left(\sum_{n=0}^{\infty} t_n \mathbf{l}^n\right)^* = \sum_{n=0}^{\infty} \overline{t_n}\, \mathbf{l}^n \text{ in } \mathfrak{L}, \tag{1.13}$$

where $\overline{z}$ are the *conjugates of* $z \in \mathbb{C}$.

Then, this operation (1.13) is a well-defined *adjoint on* the Banach algebra $\mathfrak{L}$ (e.g., [6] and [12]), and hence, every element of $\mathfrak{L}$ is *adjointable* in $B(Q)$ in the sense of [13]. So the algebra $\mathfrak{L}$ forms a *Banach* $*$*-algebra* in $B(Q)$ with (1.13).

Now, let $\mathfrak{L}$ be the Banach $*$-algebra (1.12). Define the *tensor product Banach* $*$*-algebra* $\mathfrak{L}_Q$ by

$$\mathfrak{L}_Q = \mathfrak{L} \otimes_{\mathbb{C}} Q, \tag{1.14}$$

where $\otimes_{\mathbb{C}}$ is the tensor product of Banach $*$-algebras.

Definition 1.5. *We call the Banach* $*$*-algebra* $\mathfrak{L}_Q$ *of (1.14) the radial projection (Banach* $*$*-)algebra on* Q.

1.4 WEIGHTED-SEMICIRCULAR ELEMENTS INDUCED BY Q

We here construct the weighted-semicircular elements induced by the family **Q** of mutually orthogonal projections in a fixed C^*-probability space (A, ψ), inducing the radial projection algebra $\mathfrak{L}_Q$ of (1.14). Let (Q, ψ_j) be the j-th filters of Q in (A, ψ), where ψ_j are the linear functionals (1.9), for all $j \in \mathbb{Z}$.

Remark that if u_j are the operators of $\mathfrak{L}_Q$,

$$u_j \overset{def}{=} \mathbf{1} \otimes q_j \in \mathfrak{L}_Q, \text{ for all } j \in \mathbb{Z}, \tag{1.15}$$

then

$$u_j^n = (\mathbf{1} \otimes q_j)^n = \mathbf{l}^n \otimes q_j, \text{ for all } n \in \mathbb{N},$$

with axiomatization:

$$u_j^0 \overset{\text{axiom}}{=} \mathbf{l}^0 \otimes q_j = 1_Q \otimes q_j,$$

for all $n \in \mathbb{N}_0$, for $j \in \mathbb{Z}$. That is, the operators $\{u_j\}_{j \in \mathbb{Z}}$ of (1.15) generate $\mathfrak{L}_Q$, by (1.8) and (1.12).

One can construct a linear functional φ_j on $\mathfrak{L}_Q$ by a linear morphism satisfying that

$$\varphi_j(u_i^n) = \varphi_j(\mathbf{l}^n \otimes q_i) \overset{def}{=} \psi_j(\mathbf{l}^n(q_i)), \tag{1.16}$$

for all $n \in \mathbb{N}_0$, for all $i, j \in \mathbb{Z}$.

These linear functionals $\{\varphi_j\}_{j \in \mathbb{Z}}$ of (1.16) are well defined by (1.8), (1.12), and (1.14).

Definition 1.6. *We call the Banach $*$-probability spaces*

$$(\mathfrak{L}_Q, \varphi_j), \textit{ for all } j \in \mathbb{Z}, \tag{1.17}$$

*the j-th (*Banach-$*$-probability*) spaces on Q.*

Observe that if $\mathbf{c}$ and $\mathbf{a}$ are the creation, and, respectively, the annihilation, then

$$\mathbf{ca} = 1_Q = \mathbf{ac} \text{ on } Q.$$

More generally, one has

$$\mathbf{c}^n\mathbf{a}^n = (\mathbf{ca})^n = 1_Q = (\mathbf{ac})^n = \mathbf{a}^n\mathbf{c}^n, \forall n \in \mathbb{N},$$

and hence, (1.18)

$$\mathbf{c}^{n_1}\mathbf{a}^{n_2} = \mathbf{a}^{n_2}\mathbf{c}^{n_1}, \forall n_1, n_2 \in \mathbb{N}.$$

Thus, one obtains that

$$\mathbf{l}^n = (\mathbf{c}+\mathbf{a})^n = \Sigma_{k=0}^n \begin{pmatrix} n \\ k \end{pmatrix} \mathbf{c}^k\mathbf{a}^{n-k}, \tag{1.19}$$

for all $n \in \mathbb{N}$, by (1.18), where

$$\begin{pmatrix} n \\ k \end{pmatrix} = \frac{n!}{k!(n-k)!}, \forall k \leq n \in \mathbb{N}_0.$$

Note that, for any $n \in \mathbb{N}$,

$$\mathbf{l}^{2n-1} = \Sigma_{k=0}^{2n-1} \begin{pmatrix} 2n-1 \\ k \end{pmatrix} \mathbf{c}^k \mathbf{a}^{n-k}, \tag{1.20}$$

by (1.19). So formula (1.20) does not contain 1_Q-terms by (1.18).

Note also that for any $n \in \mathbb{N}$, one has

$$\mathbf{l}^{2n} = \Sigma_{k=0}^{2n} \begin{pmatrix} 2n \\ k \end{pmatrix} \mathbf{c}^k \mathbf{a}^{n-k} = \begin{pmatrix} 2n \\ n \end{pmatrix} \mathbf{c}^n \mathbf{a}^n + [\text{Rest terms}], \tag{1.21}$$

by (1.19). So $\mathbf{l}^{2n}$ contains $\begin{pmatrix} 2n \\ n \end{pmatrix}$-many 1_Q-terms by (1.18) and (1.21).

Proposition 1.2. *Let* $\mathbf{l}$ *be the radial operator on* Q. *Then for any* $n \in \mathbb{N}$,

$$\mathbf{l}^{2n-1} \text{ does not contain } 1_Q\text{-terms in } \mathfrak{L}, \tag{1.22}$$

$$\mathbf{l}^{2n} \text{ contains } \begin{pmatrix} 2n \\ n \end{pmatrix} \cdot 1_Q \text{ in } \mathfrak{L}. \tag{1.23}$$

Proof. Statements (1.22) and (1.23) are proven by (1.20), respectively, by (1.21). □

Remark that one has

$$\varphi_j\left(u_j^{2n-1}\right) = \psi_j\left(\mathbf{l}^{2n-1}(q_j)\right) = 0, \tag{1.24}$$

for all $n \in \mathbb{N}$, by (1.9) and (1.22).

Similarly, we have

$$\varphi_j\left(u_j^{2n}\right) = \psi_j\left(\mathbf{l}^{2n}(q_j)\right) = \psi_j\left(\begin{pmatrix} 2n \\ n \end{pmatrix} q_j + [\text{Rest terms}](q_j)\right)$$

by (1.21)

$$= \begin{pmatrix} 2n \\ n \end{pmatrix} \psi_j(q_j) = \begin{pmatrix} 2n \\ n \end{pmatrix} \psi(q_j),$$

by (1.9) and (1.23). That is,

$$\varphi_j\left(u_j^{2n}\right) = \begin{pmatrix} 2n \\ n \end{pmatrix} \psi(q_j), \tag{1.25}$$

for all $n \in \mathbb{N}$.

Proposition 1.3. *Fix* $j \in \mathbb{Z}$, *and let* $u_k = \mathbf{l} \otimes q_k$ *be the k-th generating operators of the j-th space* $(\mathfrak{L}_Q, \varphi_j)$, *for all* $k \in \mathbb{Z}$. *Then,*

$$\varphi_j(u_k^n) = \delta_{j,k}\omega_n\left(\left(\frac{n}{2}+1\right)\psi(q_j)\right)c_{\frac{n}{2}}, \tag{1.26}$$

where ω_n *are in the sense of (1.3) for all* $n \in \mathbb{N}$, *and* c_k *are the k-th Catalan numbers for all* $k \in \mathbb{N}_0$.

Proof. First, take the j-th generating operator u_j in $(\mathfrak{L}_Q,\ \varphi_j)$. By (1.24) and (1.25), one can get that:

$$\varphi_j\left(u_j^{2n-1}\right) = 0,$$

and

$$\varphi_j\left(u_j^{2n}\right) = \left(\begin{array}{c} 2n \\ n \end{array}\right)\psi(q_j) = \left(\tfrac{n+1}{n+1}\right)\left(\begin{array}{c} 2n \\ n \end{array}\right)\psi(q_j)$$

$$= ((n+1)\psi(q_j))\,c_n,$$

for all $n \in \mathbb{N}$, where c_n are the n-th Catalan numbers.

Assume now that $k \neq j$ in $\mathbb{Z}$. Then, by (1.9), (1.16), and (1.23),

$$\varphi_j\left(u_k^n\right) = 0, \text{ for all } n \in \mathbb{N}.$$

Therefore, formula (1.26) holds. □

Motivated by (1.26), we define a linear morphism,

$$E_{j,Q} : \mathfrak{L}_Q \to \mathfrak{L}_Q$$

by a bounded linear transformation satisfying

$$E_{j,Q}\left(u_i^n\right) \overset{def}{=} \begin{cases} \frac{\psi(q_j)^{n-1}}{\left(\left[\frac{n}{2}\right]+1\right)}\, u_j^n & \text{if } i = j \\ \\ 0_{\mathfrak{L}_Q}, \text{ the zero operator of } \mathfrak{L}_Q & \text{otherwise,} \end{cases} \tag{1.27}$$

for all $n \in \mathbb{N}$, $i,\ j \in \mathbb{Z}$, where $\left[\frac{n}{2}\right]$ mean the *minimal integers* greater than or equal to $\frac{n}{2}$, e.g.,

$$\left[\tfrac{3}{2}\right] = 2 = \left[\tfrac{4}{2}\right].$$

The linear transformations $E_{j,Q}$ of (1.27) are the well-defined bounded linear transformations on $\mathfrak{L}_Q$, because of the cyclicity (1.12) of the tensor factor $\mathfrak{L}$ of $\mathfrak{L}_Q$, and the structure theorem (1.8) of the other tensor factor Q of $\mathfrak{L}_Q$, for all $j \in \mathbb{Z}$.

Define now the new linear functionals τ_j on $\mathfrak{L}_Q$ by

$$\tau_j \overset{def}{=} \varphi_j \circ E_{j,Q} \text{ on } \mathfrak{L}_Q, \text{ for all } j \in \mathbb{Z}, \tag{1.28}$$

where φ_j are in the sense of (1.16), and $E_{j,Q}$ are in the sense of (1.27).

Definition 1.7. *The Banach $*$-probability spaces*

$$\mathfrak{L}_Q(j) \overset{denote}{=} (\mathfrak{L}_Q,\ \tau_j) \tag{1.29}$$

are called the j-th filtered (Banach-$*$-probability) spaces of $\mathfrak{L}_Q$, where τ_j are the linear functionals (1.28) on the radial projection algebra $\mathfrak{L}_Q$, for all $j \in \mathbb{Z}$.

On the j-th filtered space $\mathfrak{L}_Q(j)$ of (1.29), one can get that

$$\tau_j\left(u_j^n\right) = \varphi_j\left(E_{j,Q}\left(u_j^n\right)\right)$$

$$= \varphi_j\left(\frac{\psi(q_j)^{n-1}}{([\frac{n}{2}]+1)}\left(u_j^n\right)\right) = \frac{\psi(q_j)^{n-1}}{([\frac{n}{2}]+1)}\varphi_j\left(u_j^n\right)$$

$$= \frac{\psi(q_j)^{n-1}}{([\frac{n}{2}]+1)}\omega_n\left(\left(\tfrac{n}{2}+1\right)\psi(q_j)\right)c_{\frac{n}{2}},$$

by (1.26), i.e.,

$$\tau_j\left(u_j^n\right) = \omega_n\psi(q_j)^n c_{\frac{n}{2}}, \tag{1.30}$$

for all $n \in \mathbb{N}$, for $j \in \mathbb{Z}$, where ω_n are in the sense of (1.3).

Lemma 1.1. *Let $\mathfrak{L}_Q(j) = (\mathfrak{L}_Q,\ \tau_j)$ be the j-th filtered space of $\mathfrak{L}_Q$, for $j \in \mathbb{Z}$. Then*

$$\tau_j\left(u_i^n\right) = \delta_{j,i}\left(\omega_n\psi(q_j)^n c_{\frac{n}{2}}\right), \tag{1.31}$$

for all $n \in \mathbb{N}$, for all $i \in \mathbb{Z}$.

Proof. If $i = j$ in $\mathbb{Z}$, then formula (1.31) holds by (1.30), for all $n \in \mathbb{N}$. If $i \neq j$ in $\mathbb{Z}$, then, by (1.16) and (1.27),

$$\tau_j\left(u_i^n\right) = 0, \text{ for all } n \in \mathbb{N}.$$

Therefore, the free-distributional data (1.31) holds true for all $i \in \mathbb{Z}$. □

The following theorem is proven by the aforementioned free-distributional data (1.31).

Theorem 1.1. *Let $\mathfrak{L}_Q(j)$ be the j-th filtered space $(\mathfrak{L}_Q,\ \tau_j)$ of $\mathfrak{L}_Q$ for $j \in \mathbb{Z}$. Then, the "j-th" generating operator u_j is $\psi(q_j)^2$-semicircular in $\mathfrak{L}_Q(j)$. Meanwhile, for all $i \neq j \in \mathbb{Z}$, the i-th generating operators u_i of $\mathfrak{L}_Q$ have the zero free distribution.*

Proof. First of all, the generating operators u_i are self-adjoint in $\mathfrak{L}_Q$, for all $i \in \mathbb{Z}$, since

$$u_i^* = (\mathbf{1} \otimes q_i)^* = \mathbf{1} \otimes q_i = u_i \text{ in } \mathfrak{L}_Q,$$

for all $i \in \mathbb{Z}$, by (1.13).

Let's fix $j \in \mathbb{Z}$, and let $u_j = \mathbf{1} \otimes q_j$ be the j-th generating operator (1.15) of $\mathfrak{L}_Q(j)$. Then, by (1.31), we have that

$$\tau_j\left(u_j^n\right) = \omega_n\left(\psi(q_j)^2\right)^{\frac{n}{2}} c_{\frac{n}{2}},$$

for all $n \in \mathbb{N}$, where c_k are the k-th Catalan numbers, for all $k \in \mathbb{N}_0$. Therefore, this self-adjoint element u_j is $\psi(q_j)^2$-semicircular in $\mathfrak{L}_Q(j)$, by (1.3).

Consider now the i-th generating operators u_i of $\mathfrak{L}_Q(j)$, for any $i \neq j$ in $\mathbb{Z}$. Since u_i are self-adjoint in $\mathfrak{L}_Q$, the free distributions of u_i are completely characterized by the free-moment sequences,

$$\left(\tau_j(u_i^n)\right)_{n=1}^{\infty} = (0,\ 0,\ 0,\ ...),$$

the zero sequence, by (1.31). So the free distributions of $u_i \in \mathfrak{L}_Q(j)$ are the zero free distribution, whenever $i \neq j$ in $\mathbb{Z}$. □

The above theorem characterizes the free-probabilistic information of the generators $\{u_i\}_{i\in\mathbb{Z}}$ in the j-th filtered space $\mathfrak{L}_Q(j)$, for $j \in \mathbb{Z}$.

Note that, by the Möbius inversion of [17], if u_i are the i-th generating operators of the j-th filtered space $\mathfrak{L}_Q(j)$, then

$$k_n^j(u_i,\ ...,\ u_i) = \begin{cases} \delta_{j,i}\psi(q_j)^2 & \text{if } n = 2 \\ 0 & \text{otherwise,} \end{cases} \tag{1.32}$$

for all $n \in \mathbb{N}$, and $i \in \mathbb{Z}$, by (1.31), where $k_\bullet^j(...)$ is the free cumulant on $\mathfrak{L}_Q$ with respect to the linear functional τ_j, for $j \in \mathbb{Z}$.

1.5 SEMICIRCULAR ELEMENTS INDUCED BY Q

As in Section 1.4, let $\mathfrak{L}_Q(j)$ be the j-th filtered space (1.29) of Q for $j \in \mathbb{Z}$. Then, the j-th generating operator u_j is $\psi(q_j)^2$-semicircular in $\mathfrak{L}_Q(j)$, satisfying that

$$\tau_j\left(u_j^n\right) = \omega_n \psi(q_j)^n c_{\frac{n}{2}},$$

equivalently, (1.33)

$$k_n^j(u_j,\ ...,\ u_j) = \begin{cases} \psi(q_j)^2 & \text{if } n = 2 \\ 0 & \text{otherwise,} \end{cases}$$

for all $n \in \mathbb{N}$, by (1.31) and (1.32).

By the weighted-semicircularity (1.33), one may/can obtain the following semicircular element U_j of $\mathfrak{L}_Q(j)$ (under an additional condition),

$$U_j \overset{def}{=} \frac{1}{\psi(q_j)}\, u_j \in \mathfrak{L}_Q(j), \tag{1.34}$$

for $j \in \mathbb{Z}$. Recall that we assumed $\psi(q_k) \in \mathbb{C}^\times$, for all $k \in \mathbb{Z}$, and hence, the above operator U_j of (1.34) is well defined in $\mathfrak{L}_Q(j)$.

Theorem 1.2. *Let $U_j = \frac{1}{\psi(q_j)}\, u_j$ be a free random variable (1.34) of $\mathfrak{L}_Q(j)$, for $j \in \mathbb{Z}$, where u_j is the j-th generating operator of $\mathfrak{L}_Q$. If*

$$\psi(q_j) \in \mathbb{R}^\times = \mathbb{R} \setminus \{0\} \text{ in } \mathbb{C}^\times,$$

then U_j is semicircular in $\mathfrak{L}_Q(j)$.

Proof. Fix $j \in \mathbb{Z}$, and assume $\psi(q_j) \in \mathbb{R}^{\times}$ in $\mathbb{C}^{\times}$. Then,

$$U_j^* = \left(\frac{1}{\psi(q_j)}u_j\right)^* = U_j,$$

by the self-adjointness of u_j in $\mathfrak{L}_Q$. Observe that

$$\begin{aligned}\tau_j\left(U_j^n\right) &= \left(\frac{1}{\psi(q_j)}\right)^n \tau_j\left(u_j^n\right) \\ &= \left(\frac{1}{\psi(q_j)^n}\right)\left(\omega_n \psi(q_j)^n c_{\frac{n}{2}}\right) = \omega_n c_{\frac{n}{2}},\end{aligned} \tag{1.35}$$

for all $n \in \mathbb{N}$. So this operator U_j is semicircular in $\mathfrak{L}_Q(j)$, whenever $\psi(q_j) \in \mathbb{R}^{\times}$, by (1.35) and (1.4). □

Assumption 1.1 (in short, **A 1.1**, from below) We further assume that

$$\psi(q_j) \in \mathbb{R}^{\times} \text{ in } \mathbb{C}, \text{ for } q_j \in \mathbf{Q},$$

for all $j \in \mathbb{Z}$. □

1.6 THE SEMICIRCULAR FILTERIZATION ($\mathbb{L}_Q$, τ)

Let (A, ψ) be a fixed C^*-probability space containing a family $\mathbf{Q} = \{q_k\}_{k\in\mathbb{Z}}$ of mutually orthogonal projections satisfying

$$\psi(q_k) \in \mathbb{R}^{\times}, \text{ for all } k \in \mathbb{Z},$$

(under **A 1.1**). For the system

$$\{\mathfrak{L}_Q(j) : j \in \mathbb{Z}\}$$

of j-th filtered spaces (1.29), define the *free product Banach $*$-probability space* $\mathfrak{L}_Q(\mathbb{Z})$ by

$$\begin{aligned}\mathfrak{L}_Q(\mathbb{Z}) &\overset{denote}{=} (\mathfrak{L}_Q(\mathbb{Z}), \tau) \\ &\overset{def}{=} \underset{j\in\mathbb{Z}}{\star} \mathfrak{L}_Q(j) = \left(\underset{j\in\mathbb{Z}}{\star} \mathfrak{L}_{Q,j}, \underset{j\in\mathbb{Z}}{\star} \tau_j\right).\end{aligned} \tag{1.36}$$

That is, our j-th filtered spaces $\mathfrak{L}_Q(j)$ form the *free blocks of* $\mathfrak{L}_Q(\mathbb{Z})$, for all $j \in \mathbb{Z}$ (e.g., [17] and [30]).

Definition 1.8. *Let $\mathfrak{L}_Q(\mathbb{Z})$ be the free product Banach $*$-probability space (1.36) of the filtered spaces $\{\mathfrak{L}_Q(j)\}_{j\in\mathbb{Z}}$. Then, it is said to be the free filterization of Q.*

Now, construct two subsets $\mathcal{X}$ and $\mathcal{S}$ of $\mathfrak{L}_Q(\mathbb{Z})$,

$$\mathcal{X} = \{u_j \in \mathfrak{L}_Q(j) : j \in \mathbb{Z}\},$$

and (1.37)

$$\mathcal{S} = \{U_j \in \mathfrak{L}_Q(j) : j \in \mathbb{Z}\},$$

where u_j are the j-th generating operators (1.15) of $\mathfrak{L}_Q(j)$, and $U_j = \frac{1}{\psi(q_j)} u_j$ are the operators (1.34) in $\mathfrak{L}_Q(j)$ (under **A 1.1**), for all $j \in \mathbb{Z}$.

Recall that a subset $\mathcal{Y}$ of an arbitrary topological $*$-probability space (B, φ) is said to be a *free family* if all elements of $\mathcal{Y}$ are mutually free from each other in (B, φ). Also, a free family $\mathcal{Y}$ is called a *free (weighted-)semicircular family* in (B, φ) if this family $\mathcal{Y}$ is not only a free family in (B, φ), but also a subset of B whose elements are (weighted-)semicircular in (B, φ). (e.g., [6] and [30]).

Theorem 1.3. *Let $\mathcal{X}$ and $\mathcal{S}$ be the subsets (1.37) of the free filterization $\mathfrak{L}_Q(\mathbb{Z})$.*
(1.38) The family $\mathcal{X}$ is a free weighted-semicircular family in $\mathfrak{L}_Q(\mathbb{Z})$.
(1.39) The family $\mathcal{S}$ is a free semicircular family in $\mathfrak{L}_Q(\mathbb{Z})$.

Proof. Let $\mathcal{X}$ be in the sense of (1.37) in $\mathfrak{L}_Q(\mathbb{Z})$. All elements u_j of $\mathcal{X}$ are taken from mutually distinct free blocks $\mathfrak{L}_Q(j)$ of $\mathfrak{L}_Q(\mathbb{Z})$, for all $j \in \mathbb{Z}$, and hence, they are free from each other in $\mathfrak{L}_Q(\mathbb{Z})$. Thus, this family $\mathcal{X}$ is a free family in $\mathfrak{L}_Q(\mathbb{Z})$. Moreover, the n-th powers u_j^n of $u_j \in \mathcal{X}$ are again contained in the free block $\mathfrak{L}_Q(j)$ as free reduced words with their lengths-1, for all $n \in \mathbb{N}$ and for $j \in \mathbb{Z}$. Thus,

$$\tau\left(u_j^n\right) = \tau_j\left(u_j^n\right) = \omega_n \psi(q_j)^n c_{\frac{n}{2}},$$

by (1.33), for all $n \in \mathbb{N}$ and for all $j \in \mathbb{Z}$. It shows every element $u_j \in \mathcal{X}$ is $\psi(q_j)^2$-semicircular in $\mathfrak{L}_Q(\mathbb{Z})$, for all $j \in \mathbb{Z}$. Therefore, the family $\mathcal{X}$ is a free weighted-semicircular family in $\mathfrak{L}_Q(\mathbb{Z})$. Equivalently, statement (1.38) holds.

Similarly, one can verify that the family $\mathcal{S}$ of (1.37) is a free semicircular family in $\mathfrak{L}_Q(\mathbb{Z})$ (under **A 1.1**). That is, statement (1.39) holds. □

By (1.31), (1.32), (1.36), (1.38), and (1.39), the only "j-th" generating operators u_j of the free blocks $\mathfrak{L}_Q(j)$ provide possible nonzero free distributions on $\mathfrak{L}_Q(\mathbb{Z})$. More precisely, all free reduced words of $\mathfrak{L}_Q(\mathbb{Z})$ in

$$\underset{j \in \mathbb{Z}}{\cup} \{u_i \in \mathfrak{L}_Q(j) : i \in \mathbb{Z}\},$$

are the free reduced words in $\mathcal{X}$, having their nonzero free distributions in $\mathfrak{L}_Q(\mathbb{Z})$. So we now restrict our interests to the Banach $*$-subalgebra $\mathbb{L}_Q$ of the free filterization $\mathfrak{L}_Q(\mathbb{Z})$, whose elements have possible nonzero free distributions in $\mathfrak{L}_Q(\mathbb{Z})$.

Definition 1.9. *Let $\mathfrak{L}_Q(\mathbb{Z})$ be the free filterization of Q. Define a Banach $*$-subalgebra $\mathbb{L}_Q$ of $\mathfrak{L}_Q(\mathbb{Z})$ by*

$$\mathbb{L}_Q \overset{def}{=} \overline{\mathbb{C}[\mathcal{X}]}, \tag{1.40}$$

where $\mathcal{X}$ is the free weighted-semicircular family (1.38) in $\mathfrak{L}_Q(\mathbb{Z})$, and $\overline{Y}$ are the Banach-topology closures of subsets Y of $\mathfrak{L}_Q(\mathbb{Z})$. Construct the Banach $*$-probability space,

$$\mathbb{L}_Q \overset{denote}{=} \left(\mathbb{L}_Q,\ \tau = \tau\mid_{\mathbb{L}_Q}\right), \tag{1.41}$$

as a free-probabilistic substructure of $\mathfrak{L}_Q(\mathbb{Z}) = (\mathfrak{L}_Q(\mathbb{Z}),\ \tau)$.

We call the Banach $*$-algebra $\mathbb{L}_Q$ of (1.40), or the Banach $*$-probability space $\mathbb{L}_Q$ of (1.41), the semicircular (free-sub-)filterization of $\mathfrak{L}_Q(\mathbb{Z})$.

The semicircular filterization $\mathbb{L}_Q$ satisfies the following structure theorem.

Theorem 1.4. *Let $\mathbb{L}_Q$ be the semicircular filterization (1.40) of $\mathfrak{L}_Q(\mathbb{Z})$. Then,*

$$\begin{aligned} \mathbb{L}_Q &\overset{def}{=} \overline{\mathbb{C}[\mathcal{X}]} = \overline{\mathbb{C}[\mathcal{S}]} \\ &\overset{*\text{-}iso}{=} \underset{j\in\mathbb{Z}}{\star}\, \overline{\mathbb{C}[\{u_j\}]} \overset{*\text{-}iso}{=} \overline{\mathbb{C}\left[\underset{j\in\mathbb{Z}}{\star}\{u_j\}\right]}, \end{aligned} \tag{1.42}$$

in $\mathfrak{L}_Q(\mathbb{Z})$, where "$\overset{*\text{-}iso}{=}$" means "being Banach-$*$-isomorphic," and where ($\star$) in the first $*$-isomorphic relation of (1.42) is the free-probabilistic free product of [17,30], and ($\star$) in the second $*$-isomorphic relation of (1.42) is the pure-algebraic free product inducing noncommutative free words in $\mathcal{X}$.

Proof. The free weighted-semicircular family $\mathcal{X}$ of (1.38) can be rewritten as

$$\mathcal{X} = \{\psi(q_j)U_j \in \mathfrak{L}_Q(j) : U_j \in \mathcal{S}\}$$

in the free filterization $\mathfrak{L}_Q(\mathbb{Z})$ of Q, where $\mathcal{S}$ is the free semicircular family (1.39). Therefore,

$$\overline{\mathbb{C}[\mathcal{X}]} = \overline{\mathbb{C}[\mathcal{S}]} \text{ in } \mathfrak{L}_Q(\mathbb{Z}),$$

by (1.40). It shows that the set equality (=) of (1.42) holds.

By definition (1.40) of $\mathbb{L}_Q$, it is generated by the free family $\mathcal{X}$ by (1.38), and hence, the first $*$-isomorphic relation of (1.42) holds in the free filterization $\mathfrak{L}_Q(\mathbb{Z})$ by (1.36), because

$$\overline{\mathbb{C}[\{u_j\}]} \subset \mathfrak{L}_Q(j) \text{ in } \mathfrak{L}_Q(\mathbb{Z}), \text{ for all } j \in \mathbb{Z}.$$

Since

$$\mathbb{L}_Q \overset{*\text{-iso}}{=} \underset{j\in\mathbb{Z}}{\star}\, \overline{\mathbb{C}[\{u_j\}]} \text{ in } \mathfrak{L}_Q(\mathbb{Z}),$$

every element T of $\mathbb{L}_Q$ is a limit of linear combinations of free reduced words in $\mathcal{X}$. Also, all (pure-algebraic) free words in $\mathcal{X}$ have their unique free-reduced-word forms under operator-multiplication on $\mathfrak{L}_Q(\mathbb{Z})$. Therefore, the second $*$-isomorphic relation of (1.42) holds, too. □

1.7 FREE POISSON ELEMENTS OF $\mathbb{L}_Q$

Let (A, ψ) be a fixed C^*-probability space, containing a family $\mathbf{Q} = \{q_j\}_{j \in \mathbb{Z}}$ of mutually orthogonal projections, and let $\mathbb{L}_Q$ be the semicircular filterization (1.41) of the free filterization $\mathfrak{L}_Q(\mathbb{Z})$ of the C^*-subalgebra $Q = C^*(\mathbf{Q})$ of A. Throughout this section, we also assume **A 1.1**, and hence, the family $\mathcal{S}$ of (1.37) is a well-determined free semicircular family in $\mathfrak{L}_Q(\mathbb{Z})$, generating $\mathbb{L}_Q$ by (1.39) and (1.42). From the generating free family $\mathcal{S}$ of $\mathbb{L}_Q$, we here study free Poisson elements.

1.7.1 FREE POISSON ELEMENTS

Let (B, φ) be a topological $*$-probability space. If $x \in (B, \varphi)$ is a free random variable, then the free distribution of x is characterized by the *joint free moments of* $\{x, x^*\}$,

$$\varphi\left(x^{r_1} x^{r_2} ... x^{r_n}\right),$$

or, by the *joint free cumulants of* $\{x, x^*\}$,

$$k_n^B\left(x^{r_1}, x^{r_2}, ..., x^{r_n}\right),$$

for all $(r_1, ..., r_n) \in \{1, *\}^n$, and for all $n \in \mathbb{N}$, where $k_n^B(..)$ is the free cumulant on B in terms of φ under the Möbius inversion of [17,22,23]. And they provide equivalent free-distributional data of $x \in (B, \varphi)$, representing its free distribution.

Thus, if x is a self-adjoint free random variable of (B, φ_B), satisfying $x = x^*$ in B, then the free distribution of x is characterized by the free-moment sequence

$$\left(\varphi(x^n)\right)_{n=1}^{\infty} = \left(\varphi(x), \varphi(x^2), \varphi(x^3), ...\right),$$

or, by the free-cumulant sequence

$$\left(k_n^B(x, ..., x)\right)_{n=1}^{\infty} = \left(k_1^B(x) = \varphi(x), k_2^B(x, x), ...\right).$$

For example, every semicircular element $s \in (B, \varphi)$ has its free distribution, the semicircular law, characterized by the free-moment sequence

$$(0, c_1, 0, c_2, 0, c_3, ...),$$

or, by the free-cumulant sequence

$$(0, 1, 0, 0, 0, ...),$$

where c_k are the k-th Catalan numbers for all $k \in \mathbb{N}$.

Notation From below, we will write " a free random variable $x \in (B, \varphi)$ has its free distribution $\left(\varphi(x^n)\right)_{n=1}^{\infty}$," and equivalently, "$x \in (B, \varphi)$ has its free distribution $\left(k_n^B(x, ..., x)\right)_{n=1}^{\infty}$," if (i) x is self-adjoint in B and (ii) the free distribution of x is characterized by

$$\left(\varphi(x^n)\right)_{n=1}^{\infty}, \text{ or, } \left(k_n^B(x, ..., x)\right)_{n=1}^{\infty}.$$

□

Definition 1.10. *Let $s \in (B, \varphi)$ be a semicircular element, and let a be a self-adjoint free random variable of (B, φ), having its free distribution $\left(\varphi(a^n)\right)_{n=1}^{\infty}$. Assume that a and s are free in (B, φ). Then, a new free random variable*

$$W_s^a = sas \in (B, \varphi) \tag{1.43}$$

is called the free Poisson element generated by s and a.

By definition (1.43), a free Poisson element W_s^a is a free reduced word with its length-3 in (B, φ). Also, by (1.43),

$$(W_s^a)^* = (sas)^* = s^*a^*s^* = sas = W_s^a \tag{1.44}$$

in B, and hence, it is a self-adjoint free random variable of (B, φ), too.

Let $\Omega = \{e_1, ..., e_n\}$ be a finite set with its cardinality $n \in \mathbb{N}$. Then, the *lattice* $NC(\Omega)$ *of noncrossing partitions of* Ω is well defined with its partial ordering $\leq$,

$$\theta_1 \leq \theta_2 \Longleftrightarrow \forall V_1 \in \theta_1, \exists V_2 \in \theta_2, \text{ s.t., } V_1 \subseteq V_2,$$

where $\subseteq$ is the usual set inclusion, where "$V \in \theta$" means "V is a block of θ."

For example, if $\Omega_5 = \{1, 2, ..., 5\}$, and if

$$\theta_1 = \{(1, 4), (2, 3), (5)\},$$

and

$$\theta_2 = \{(1, 2, 3, 4), (5)\}$$

in $NC(\Omega_5)$, then $\theta_1 \leq \theta_2$.

Notation From below, if a given finite set Ω is a subset $\{1, 2, ..., n\}$ of $\mathbb{N}$, for some $n \in \mathbb{N}$, then we denote Ω by Ω_n. □

Under $(\leq)$, the lattice $NC(\Omega)$ has its maximal element,

$$1_\Omega = \{(e_1, ..., e_n)\}, \text{ the 1-block partition,}$$

and its minimal element,

$$0_\Omega = \{(e_1), (e_2), ..., (e_n)\}, \text{ the } n\text{-block partition.}$$

Suppose

$$\Omega = \Omega^1 \sqcup \Omega^2, \text{ with } \Omega^l \subset \Omega, \forall l = 1, 2.$$

where $\sqcup$ is the disjoint union, and let $NC\left(\Omega^l\right)$ be the noncrossing partition lattices for Ω^l, for $l = 1, 2$.

Then, for $\theta_l \in NC(\Omega^l)$, for $l = 1, 2$, one can construct a noncrossing partition $\theta \in NC(\Omega)$ as the join of θ_1 and θ_2,

$$\theta = \theta_1 \vee \theta_2 \in NC(\Omega),$$

as in [17,22,23]. For example, if

$$\theta_1 = \{(2, 5), (3)\} \in NC(\{2, 3, 5\}),$$

and

$$\theta_2 = \{(4), (1, 6, 7)\} \in NC(\{1, 4, 6, 7\}),$$

then, we obtain

$$\theta_1 \vee \theta_2 = \{(1,6,7), (2, 5), (3), (4)\},$$

in $NC\left(\Omega_7\right)$.

Now, fix $n \in \mathbb{N}$, and let $NC(\Omega_{3n})$ be the noncrossing partition lattice, and let

$$\Omega_{3n}^1 = \{1, 3, 4, 6, 7, 9, 10, ..., 3n-3, 3n-2, 3n\},$$

and (1.45)

$$\Omega_{3n}^2 = \Omega_{3n} \setminus \Omega_{3n}^1 = \{2, 5, 8, 11, ..., 3n-1\},$$

satisfying

$$\Omega_{3n} = \Omega_{3n}^1 \sqcup \Omega_{3n}^2.$$

And then take $\theta_o \in NC\left(\Omega_{3n}^1\right)$,

$$\theta_o = \{(1, 3n), (3,4), (6,7), ..., (3n-3, 3n-2)\}, \tag{1.46}$$

where Ω_{3n}^1 is in the sense of (1.45).

The following lemma is already shown (e.g., in the page 207 of [17]), but we provide a sketch of the proof for our future works.

Lemma 1.2. *Let $a \in (B, \varphi)$ be a self-adjoint free random variable having its free distribution $\left(\varphi(a^n)\right)_{n=1}^{\infty}$, and let $s \in (B, \varphi)$ be a semicircular element, free from a in (B, φ). If $W_s^a = sas$ is a free Poisson element (1.43), then*

$$k_n^B\left(\underbrace{W_s^a, W_s^a, ..., W_s^a}_{n\text{-times}}\right) = \varphi(a^n), \tag{1.47}$$

for all $n \in \mathbb{N}$.

Proof. By the semicircularity (1.2) of s,

$$k_n^B(s, ..., s) = \begin{cases} 1 & \text{if } n = 2 \\ 0 & \text{otherwise,} \end{cases} \tag{1.48}$$

for all $n \in \mathbb{N}$. So,

$$k_n^B(W_s^a, ..., W_s^a) = k_n^B(sas, sas, ..., sas)$$

$$= \sum_{\theta \in NC(\Omega_{3n}^2),\ \theta \vee \pi_0 \leq 1_{3n}} k_\theta^B(sas, ..., sas)$$

where Ω_{3n}^2 is the set (1.45), and $k_\theta^B(...)$ are the block-depending free cumulants of [27], and 1_{3n} is the maximal 1-block partition of $NC(\Omega_{3n})$, and

$$\pi_0 = \{(1, 2, 3), (4, 5, 6), ..., (3n-2, 3n-1, 3n)\}$$

(see [17,22] and [23] for details), and hence, it goes to

$$= \sum_{\theta \in NC(\Omega_{3n}^2),\ \theta_o \vee \theta \in NC(\Omega_{3n})} k_{\theta_o \vee \theta}^B(sas, ..., sas)$$

by (1.48), where Ω_{3n}^2 is in the sense of (1.45), and θ_o is the partition (1.46)

$$= \sum_{\theta \in NC(\Omega_{3n}^2)} k_{\theta_o}^B\left(\underbrace{s, s, s, ..., s}_{2n\text{-times}}\right) k_\theta^B\left(\underbrace{a, a, ..., a}_{n\text{-times}}\right)$$

since all "mixed" free cumulants of s and a vanish by the freeness of s and a

$$= \sum_{\theta \in NC(\Omega_{3n}^2)} \left((k_2^B(s, s))^n\right) k_\theta^B\left(\underbrace{a, a, ..., a}_{n\text{-times}}\right)$$

$$= \sum_{\theta \in NC(\Omega_{3n}^2)} k_\theta^B\left(\underbrace{a, a, ..., a}_{n\text{-times}}\right)$$

by (1.2)

$$= \sum_{\pi \in NC(\Omega_n)} k_\pi^B(a, ..., a)$$

since the sublattice $NC(\Omega_{3n}^2)$ of $NC(\Omega_{3n})$ is equivalent to the lattice $NC(\Omega_n)$

$$= \varphi(a^n)$$

by the Möbius inversion of [17], for all $n \in \mathbb{N}$. Therefore, the free-distributional data (1.47) is obtained. □

The free-cumulant formula (1.47) shows that a *free Poisson distribution*, and the free distribution of a free Poisson element W_s^a, is characterized by the free distribution $(\varphi(a^n))_{n=1}^{\infty}$ of a fixed self-adjoint free random variable $a \in (B, \varphi)$.

Theorem 1.5. *Let $W_s^a = sas \in (B, \varphi)$ be a free Poisson element (1.43). Then,*

$$\varphi\left((W_s^a)^n\right) = k_n^B\left(\underbrace{a, a, ..., a}_{n\text{-times}}\right), \tag{1.49}$$

for all $n \in \mathbb{N}$, where $|V|$ are the cardinalities of blocks V.

Proof. Observe that

$$\varphi\left((W_s^a)^n\right) = \sum_{\theta \in NC(\Omega_n)} k_\theta^B\left(W_s^a, ..., W_s^a\right)$$

by the Möbius inversion

$$= \sum_{\theta \in NC(\Omega_n)} \left(\prod_{V \in \theta} k_{|V|}^B\left(\underbrace{W_s^a, ..., W_s^a}_{|V|\text{-times}}\right)\right)$$

$$= \sum_{\theta \in NC(\Omega_n)} \left(\prod_{V \in \theta} \varphi\left(a^{|V|}\right)\right)$$

by (1.47)

$$= k_n^B(a, ..., a),$$

by the Möbius inversion, for all $n \in \mathbb{N}$. So formula (1.49) holds. □

1.7.2 CERTAIN FREE POISSON ELEMENTS INDUCED BY $\mathcal{S}$

Let $\mathbb{L}_Q = (\mathbb{L}_Q, \tau)$ be the semicircular filterization (1.41). In this section, we fix a semicircular element $U_j \in \mathcal{S}$ in $\mathbb{L}_Q$, for a fixed $j \in \mathbb{Z}$. By Section 1.7.1, if $T \in \mathbb{L}_Q$ is a self-adjoint operator, and if T is free from U_j in $\mathbb{L}_Q$, then one obtains the corresponding free Poisson element

$$W_j^T = U_j T U_j \text{ in } \mathbb{L}_Q,$$

satisfying that

$$k_n\left(W_j^T, ..., W_j^T\right) = \tau\left(T^n\right),$$

and (1.50)

$$\tau\left(\left(W_j^T\right)^n\right) = k_n\left(T, ..., T\right),$$

by (1.47) and (1.49), for all $n \in \mathbb{N}$, where $k_\bullet(...)$ is the free cumulant on $\mathbb{L}_Q$ in terms of the linear functional τ of (1.41).

Define free Poisson elements $W_j^k = W_j^{U_k}$ of (1.50) by

$$W_j^k \overset{def}{=} U_j U_k U_j \in \mathbb{L}_Q, \tag{1.51}$$

where $j \neq k$ in $\mathbb{Z}$. Then, one has that

$$k_n\left(W_j^k, ..., W_j^k\right) = \tau\left(U_k^n\right) = \omega_n c_{\frac{n}{2}},$$

and (1.52)

$$\tau\left(\left(W_j^k\right)^n\right) = k_n\left(U_k, ..., U_k\right) = \delta_{n,2}$$

by (1.50) and (1.2), for all $n \in \mathbb{N}$, where δ is the Kronecker delta.

More generally, for $k \neq j \in \mathbb{Z}$, define a free Poisson element,

$$W_j^{k,N} = U_j U_k^N U_j \in \mathbb{L}_Q, \text{ for } N \in \mathbb{N}. \tag{1.53}$$

Theorem 1.6. *Let $W_j^{k,N}$ be a free Poisson element (1.53) for $N \in \mathbb{N}$. Then,*

$$k_n\left(W_j^{k,N}, ..., W_j^{k,N}\right) = \omega_{nN} c_{\frac{nN}{2}},$$

and *(1.54)*

$$\tau\left(\left(W_j^{k,N}\right)^n\right) = \begin{cases} \sum\limits_{\pi \in NC(\Omega_n)} \left(\prod\limits_{B \in \theta} c_{\frac{N|B|}{2}}\right) & \text{if } N \text{ is even} \\ \omega_n \left(\sum\limits_{\theta \in NC_e(\Omega_n)} \left(\prod\limits_{V \in \theta} c_{\frac{N|V|}{2}}\right)\right) & \text{if } N \text{ is odd,} \end{cases}$$

for all $n \in \mathbb{N}$, where

$$NC_e(\Omega_n) = \{\pi \in NC(\Omega_n) : |V| \text{ is even}, \forall V \in \pi\}.$$

Proof. Observe first that

$$\begin{aligned} k_n\left(W_j^{k,N}, ..., W_j^{k,N}\right) &= \tau\left(\left(U_k^N\right)^n\right) \\ &= \tau\left(U_k^{nN}\right) = \omega_{nN} c_{\frac{nN}{2}}, \end{aligned}$$

by (1.47), for all $n \in \mathbb{N}$. Thus, the free-cumulant formula of (1.54) holds.

Consider now that

$$\tau\left(\left(W_j^{k,N}\right)^n\right) = k_n\left(U^N, ..., U^N\right)$$

by (1.50)

$$= \sum_{\pi \in NC(\Omega_n)} \left(\prod_{V \in \pi} \left(\tau \left(U_k^{N|V|} \right) \right) \right)$$

by the Möbius inversion

$$= \sum_{\pi \in NC(\Omega_n)} \left(\prod_{V \in \pi} \left(\omega_{N|V|} c_{\frac{N|V|}{2}} \right) \right)$$

by the semicircularity (1.4) of $U_k \in \mathcal{S}$ in $\mathbb{L}_Q$

$$= \begin{cases} \sum_{\pi \in NC(\Omega_n)} \left(\prod_{B \in \theta} c_{\frac{N|B|}{2}} \right) & \text{if } N \text{ is even} \\ \omega_n \left(\sum_{\theta \in NC_e(\Omega_n)} \left(\prod_{V \in \theta} c_{\frac{N|V|}{2}} \right) \right) & \text{if } N \text{ is odd,} \end{cases}$$

because

if N is even, then lN is even for all $l \in \mathbb{N}$,

respectively,

if N is odd, then lN is even if and only if l is even in $\mathbb{N}$.

Therefore, the free-moment formula of (1.54) holds true. □

Trivially, if $N = 1$ in (1.54), then relation (1.52) is automatically obtained.

1.7.3 SOME FREE POISSON ELEMENTS INDUCED BY $\mathcal{S} \cup \mathcal{X}$

As in Section 1.7.2, let's fix a semicircular element $U_j \in \mathcal{S}$, and let $\mathcal{X}$ be the free weighted-semicircular family (1.38) in the semicircular filterization $\mathbb{L}_Q$. For any $k \neq j$ in $\mathbb{Z}$, we have the corresponding free Poisson elements

$$Y_j^{k,N} = U_j u_k^N U_j \text{ in } \mathbb{L}_Q, \tag{1.55}$$

where $u_k \in \mathcal{X}$ is a $\psi(q_j)^2$-semicircular element of $\mathbb{L}_Q$, for all $N \in \mathbb{N}$.

Since

$$u_k = \psi(q_k) U_k \in \mathcal{X}, \text{ in } \mathbb{L}_Q, \tag{1.56}$$

by (1.55)

$$Y_j^{k,N} = \psi(q_k)^N U_j U_k^N U_{p,j} = \psi(q_k)^N W_j^{k,N}, \tag{1.57}$$

in $\mathbb{L}_Q$, by (1.56), where $W_j^{k,N}$ is the free Poisson element (1.55).

Theorem 1.7. *Let $Y_j^{k,N}$ be a free Poisson element (1.55) in $\mathbb{L}_Q$, for $N \in \mathbb{N}$. Then*

$$k_n \left(Y_j^{k,N}, \ldots, Y_j^{k,N} \right) = \omega_{nN} \psi(q_k)^{nN} c_{\frac{nN}{2}},$$

and *(1.58)*

$$\tau\left(\left(Y_j^{k,N}\right)^n\right) = \begin{cases} \psi(q_k)^{nN}\left(\sum\limits_{\pi\in NC(\Omega_n)}\left(\prod\limits_{B\in\theta} c_{\frac{N|B|}{2}}\right)\right) & \textit{if } N \textit{ is even} \\ \omega_n\psi(q_k)^{nN}\left(\sum\limits_{\theta\in NC_e(\Omega_n)}\left(\prod\limits_{V\in\theta} c_{\frac{N|V|}{2}}\right)\right) & \textit{if } N \textit{ is odd,} \end{cases}$$

for all $n \in \mathbb{N}$.

Proof. Consider that

$$k_n\left(Y_j^{k,N}, \ldots, Y_j^{k,N}\right) = k_n\left(\psi(q_k)^N W_j^{k,N}, \ldots, \psi(q_k)^N W_j^{k,N}\right)$$

by (1.57)

$$= \psi(q_k)^{nN} k_n\left(W_j^{k,N}, \ldots, W_j^{k,N}\right)$$

by the bimodule-map property of free cumulants (e.g., [17,22] and [23])

$$= \psi(q_k)^{nN}\ \tau\left(U_k^{nN}\right) = \psi(q_k)^{nN}\left(\omega_{nN} c_{\frac{nN}{2}}\right)$$

$$= \omega_{nN}\psi(q_k)^{nN} c_{\frac{nN}{2}} = \tau\left(u_k^{nN}\right),$$

for all $n \in \mathbb{N}$, by (1.54).

Similarly, for any $n \in \mathbb{N}$, one obtains that

$$\tau\left(\left(Y_j^{k,N}\right)^n\right) = \tau\left(\psi(q_k)^{nN}\left(W_j^{k,N}\right)^n\right)$$

by (1.57)

$$= \psi(q_k)^{nN}\tau\left((W_j^{k,N})^n\right)$$

$$= \begin{cases} \psi(q_k)^{nN}\left(\sum\limits_{\pi\in NC(\Omega_n)}\left(\prod\limits_{B\in\theta} c_{\frac{N|B|}{2}}\right)\right) & \text{if } N \text{ is even} \\ \omega_n\psi(q_k)^{nN}\left(\sum\limits_{\theta\in NC_e(\Omega_n)}\left(\prod\limits_{V\in\theta} c_{\frac{N|V|}{2}}\right)\right) & \text{if } N \text{ is odd,} \end{cases}$$

by (1.54).

Therefore, the free-distributional data (1.58) hold in $\mathbb{L}_Q$. □

The above theorem shows that the free Poisson elements $Y_j^{k,N}$ of (1.55) induced by the free weighted-semicircular family $\mathcal{X}$ of (1.38) have their free Poisson distributions, affected by the weights of fixed weighted-semicircular elements of $\mathcal{X}$ in $\mathbb{L}_Q$.

Corollary 1.1. *Let $Y_j^{k,1} = U_j u_k U_j$ be a free Poisson element (1.55) in $\mathbb{L}_Q$. Then,*

$$k_n\left(Y_j^{k,1}, ..., Y_j^{k,1}\right) = \omega_n \psi(q_k)^n c_{\frac{n}{2}},$$

and *(1.59)*

$$\tau\left(\left(Y_j^{k,1}\right)^n\right) = \delta_{n,2}\psi(q_k)^2,$$

for all $n \in \mathbb{N}$.

Proof. The free-distributional data (1.59) is directly obtained by (1.58), by replacing N to 1. □

1.8 FREE WEIGHTED-POISSON ELEMENTS OF $\mathbb{L}_Q$

In this section, we are interested in free-Poisson-like free random variables in our semicircular filterization $\mathbb{L}_Q$.

1.8.1 FREE WEIGHTED-POISSON ELEMENTS

Let (B, φ) be an arbitrary topological $*$-probability space, and let $x \in (B, \varphi)$ be a t_0-semicircular element for some $t_0 \in \mathbb{C}^{\times}$, satisfying

$$\varphi(x^n) = \omega_n \, t_0^{\frac{n}{2}} c_{\frac{n}{2}},$$

and (1.60)

$$k_n^B(x, ..., x) = \delta_{n,2} t_0$$

for all $n \in \mathbb{N}$, where $k_n^B(...)$ is the free cumulant on B in terms of φ.

Definition 1.11. *Let $x \in (B, \varphi)$ be a t_0-semicircular element (1.60). Suppose $a \in (B, \varphi)$ is a self-adjoint free random variable, which is free from x in (B, φ). A free random variable*

$$T_x^a = xax \in (B, \varphi) \tag{1.61}$$

is called a free weighted-Poisson element with its weight t_0 (in short, free t_0-Poisson element) of (B, φ).

Let's consider free-distributional data of a free t_0-Poisson element T_x^a of (1.61).

Theorem 1.8. *Let $T_x^a = xax$ be a free t_0-Poisson element (1.61), where x is a fixed t_0-semicircular element (1.60) in (B, φ), and a has its free distribution $(\varphi(a^n))_{n=1}^{\infty}$. Then,*

$$k_n^B\left(\underbrace{T_x^a,\ T_x^a,\ ...,\ T_x^a}_{n\text{-}times}\right) = t_0^n\left(\varphi(a^n)\right), \tag{1.62}$$

for all $n \in \mathbb{N}$.

Proof. Observe that

$$\begin{aligned} k_n^B\left(T_x^a,\ ...,\ T_x^a\right) &= k_n^B\left(xax,\ xax, ...,\ xax\right) \\ &= \sum_{\theta\in NC(\Omega_{3n}),\ \theta\vee\pi_0\leq 1_{3n}} k_\theta^B\left(xax,\ ...,xax\right) \end{aligned}$$

where Ω_{3n} is in the sense of (1.45), and $k_\theta^B(...)$ are the block-depending free cumulants of [17], and

$$\pi_0 = \{(1,2,3),(4,5,6),...,(3n-2,3n-1,3n)\},$$

and hence, it goes to

$$= \sum_{\theta\in NC\left(\Omega_{3n}^2\right),\ \theta_o\vee\theta\in NC(\Omega_{3n})} k_{\theta_o\vee\theta}^B\left(xax,\ ...,\ xax\right)$$

by (1.48), where θ_o is in the sense of (1.46)

$$= \sum_{\theta\in NC\left(\Omega_{3n}^2\right)} k_{\theta_o}^B\left(\underbrace{x,\ x,\ x,\ ...,\ x}_{2n\text{-times}}\right) k_\theta^B\left(\underbrace{a,\ a,\ ...,\ a}_{n\text{-times}}\right)$$

$$= \sum_{\theta\in NC\left(\Omega_{3n}^2\right)} \left(\left(k_2^B(x,\ x)\right)^n\right) k_\theta^B\left(a,\ a,\ ...,\ a\right)$$

$$= t_0^n\left(\sum_{\theta\in NC\left(\Omega_{3n}^2\right)} k_\theta^B\left(a,\ a,\ ...,\ a\right)\right)$$

by (1.48), because x is a t_0-semicircular element satisfying (1.60)

$$= t_0^n\left(\sum_{\pi\in NC(\Omega_n)} k_\pi^B\left(a,\ ...,\ a\right)\right)$$

since the sublattice $NC\left(\Omega_{3n}^2\right)$ of $NC\left(\Omega_{3n}\right)$ is equivalent to the lattice $NC(\Omega_n)$

$$= t_0^n\left(\varphi(a^n)\right),$$

by the Möbius inversion, for all $n \in \mathbb{N}$. Therefore, the free-distributional data (1.62) is obtained. $\square$

The above theorem illustrates that the free distributions of our free weighted-Poisson elements T_x^a of (1.61) are depending not only on the free distributions of a, but also on the weights of fixed weighted-semicircular elements x in (B, φ), by (1.62). By the Möbius inversion, we obtain the following equivalent result.

Theorem 1.9. *Let T_x^a be a free t_0-Poisson element (1.61) in (B, φ). Then,*

$$\varphi\left((T_x^a)^n\right) = t_0^n \left(k_n^B \left(\underbrace{a,\ a,\ \ldots\ldots,\ a}_{n\text{-times}} \right) \right), \tag{1.63}$$

for all $n \in \mathbb{N}$.

Proof. Let $T_x^a \in (B, \varphi)$ be a free t_0-Poisson element (1.61). Then,

$$\varphi\left((T_x^a)^n\right) = \sum_{\pi \in NC(\Omega_n)} \left(\prod_{V \in \pi} k_{|V|}^B \left(T_x^a, \ \ldots, \ T_x^a\right) \right)$$

by the Möbius inversion

$$= \sum_{\pi \in NC(\Omega_n)} \left(\prod_{V \in \pi} \left(t_0^{|V|} \varphi\left(a^{|V|}\right) \right) \right)$$

by (1.62)

$$= \sum_{\pi \in NC(\Omega_n)} \left(\prod_{V \in \pi} t_0^{|V|} \right) \left(\prod_{V \in \pi} \varphi\left(a^{|V|}\right) \right)$$

$$= \sum_{\pi \in NC(\Omega_n)} \left(t_0^{\sum_{V \in \pi} |V|} \right) \left(\prod_{V \in \pi} \varphi\left(a^{|V|}\right) \right)$$

$$= \sum_{\pi \in NC(\Omega_n)} \left(t_0^n\right) \left(\prod_{V \in \pi} \varphi\left(a^{|V|}\right) \right)$$

since $\sum_{V \in \pi} |V| = |\Omega_n| = n$, for all $\pi \in NC(\Omega_n)$

$$= \left(t_0^n\right) \left(\sum_{\pi \in NC(\Omega_n)} \left(\prod_{V \in \pi} \varphi\left(a^{|V|}\right) \right) \right),$$

for all $n \in \mathbb{N}$. Therefore, the free-momental data (1.63) is obtained by the Möbius inversion. $\square$

The aforementioned free-distributional data (1.62) and (1.63) provide the following free-distributional information for the free weighted-Poisson elements.

Theorem 1.10. *Let T_x^a be a free t_0-Poisson element (1.61) in (B, φ), and let W_s^a be a free Poisson element (1.43) of (B, φ), where s is an arbitrary semicircular element of (B, φ), which is free from a fixed self-adjoint free random variable a of (B, φ). Then,*

$$k_n^B(T_x^a, ..., T_x^a) = t_0^n k_n^B(W_s^a, ..., W_s^a) = t_0^n \varphi(a^n),$$

and *(1.64)*

$$\varphi((T_x^a)^n) = t_0^n \ \varphi((W_s^a)^n) = t_0^n k_n \left(\underbrace{a, ..., a}_{n\text{-times}} \right),$$

for all $n \in \mathbb{N}$.

Proof. Under hypothesis, one has that

$$k_n^B(T_x^a, ..., T_x^a) = t_0^n \ \varphi(a^n) = t_0^n \ k_n^B(W_s^a, ..., W_s^a),$$

for all $n \in \mathbb{N}$, by (1.47) and (1.62).

Similarly, we have

$$\varphi((T_x^a)^n) = t_0^n (k_n(a, ..., a)) = t_0^n \ \varphi((W_s^a)^n),$$

for all $n \in \mathbb{N}$, by (1.49) and (1.63).

Therefore, relation (1.64) holds in (B, φ). □

The above theorem characterizes the difference between the free Poisson distributions and the free weighted-Poisson distributions by (1.64). It also shows how our weights act free-probabilistically.

1.8.2 FREE WEIGHTED-POISSON ELEMENTS INDUCED BY $\mathcal{S} \cup \mathcal{X}$

Let $\mathbb{L}_Q$ be the semicircular filterization, and let $\mathcal{S}$ be the free semicircular family (1.39) in $\mathbb{L}_Q$. Let's fix a $\psi(q_j)^2$-semicircular element $u_j \in \mathcal{X}$ in $\mathbb{L}_Q$, where $\mathcal{X}$ is the free weighted-semicircular family (1.38) of $\mathbb{L}_Q$. Then, as in (1.61), one can define the free $\psi(q_j)^2$-Poisson elements

$$T_j^{k,N} = u_j U_k^N u_j \in \mathbb{L}_Q, \tag{1.65}$$

for $U_k \in \mathcal{S}$, any $N \in \mathbb{N}$, whenever $k \neq j$ in $\mathbb{Z}$.

By (1.62), (1.63), and (1.64), we obtain the following free-distributional data of free $\psi(q_j)^2$-Poisson elements (1.65) in $\mathbb{L}_Q$.

Theorem 1.11. *Let $T_j^{k,N} = u_j U_k^N u_j$ be a free $\psi(q_j)^2$-Poisson element (1.65) in $\mathbb{L}_Q$, for $N \in \mathbb{N}$. If $W_j^{k,N} = U_j U_k^N U_j$ is a free Poisson element (1.53) in $\mathbb{L}_Q$, then*

$$k_n\left(T_j^{k,N}, \dots, T_j^{k,N}\right) = \psi(q_j)^{2n} k_n\left(W_j^{l,N}, \dots, W_j^{l,N}\right)$$

$$= \psi(q_j)^{2n}\left(\omega_{nN} c_{\frac{nN}{2}}\right),$$

and *(1.66)*

$$\tau\left(\left(T_j^{k,N}\right)^n\right) = \psi(q_j)^{2n}\tau\left(\left(W_j^{l,N}\right)^n\right)$$

$$= \begin{cases} \psi(q_j)^{2n}\left(\sum\limits_{\pi\in NC(\Omega_n)}\left(\prod\limits_{V\in\pi} c_{\frac{N|V|}{2}}\right)\right) & \text{if } N \text{ is even} \\ \\ \omega_n\psi(q_j)^{2n}\left(\sum\limits_{\theta\in NC_e(\Omega_n)}\left(\prod\limits_{V\in\theta} c_{\frac{N|V|}{2}}\right)\right) & \text{if } N \text{ is odd,} \end{cases}$$

for all $n \in \mathbb{N}$, where $NC_e(\Omega_n)$ are the subsets (1.54) of the lattices $NC(\Omega_n)$.

Proof. The proof of (1.66) is done by (1.54) and (1.64). □

The following corollary is obtained immediately by (1.66).

Corollary 1.2. *Let $T_j^{k,1}$ be a $\psi(q_j)^2$-free Poisson element (1.65) with $N = 1$. Then*

$$k_n\left(T_j^{k,1}, \dots, T_j^{k,1}\right) = \omega_n\psi(q_j)^{2n} c_{\frac{n}{2}},$$

and *(1.67)*

$$\tau\left(\left(T_j^{k,1}\right)^n\right) = \delta_{n,2}\psi(q_j)^2,$$

for all $n \in \mathbb{N}$.

Proof. The free-distributional data (1.67) of $T_j^{k,1}$ holds by (1.52) and (1.66). □

The above corollary shows the connections between our $\psi(q_j)^2$-semicircular laws, and the $\psi(q_j)^2$-free Poisson distributions on $\mathbb{L}_Q$, by (1.67).

1.8.3 FREE WEIGHTED-POISSON ELEMENTS INDUCED BY $\mathcal{X}$

Let $\mathbb{L}_Q$ be the semicircular Adelic filterization, and $\mathcal{X}$, the free weighted-semicircular family (1.38) of $\mathbb{L}_Q$, and let's fix $u_j \in \mathcal{X}$ in $\mathbb{L}_Q$. For any $k \neq j$ in $\mathbb{Z}$, define the corresponding free $\psi(q_j)^2$-Poisson elements

$$X_j^{k,N} = u_j u_k^N u_j \in \mathbb{L}_Q, \tag{1.68}$$

for $N \in \mathbb{N}$.

Since

$$U_j = \frac{1}{\psi(q_j)} u_j \text{ in } \mathcal{S} \Longleftrightarrow u_j = \psi(q_j)U_j \text{ in } \mathcal{X},$$

in $\mathbb{L}_Q$, for all $j \in \mathbb{Z}$, our free $\psi(q_j)^2$-Poisson elements $X_j^{k,N}$ of (1.68) are also understood as

$$\begin{aligned} X_j^{k,N} &= \psi(q_j)^2 \psi(q_k)^N U_j U_k^N U_j \\ &= \left(\psi(q_j)^2 \psi(q_k)^N\right) W_j^{k,N}, \end{aligned} \tag{1.69}$$

in $\mathbb{L}_Q$, where $W_j^{k,N}$ are the free Poisson element (1.53) of $\mathbb{L}_Q$, for all $N \in \mathbb{N}$.

Theorem 1.12. *Let $X_j^{k,N} = u_j u_k^N u_j$ be a free $\psi(q_j)^2$-Poisson element (1.68) of $\mathbb{L}_Q$, for $N \in \mathbb{N}$. Then,*

$$\begin{aligned} k_n\left(X_j^{k,N}, \ldots, X_j^{k,N}\right) &= \beta_n k_n\left(W_j^{l,N}, \ldots, W_j^{l,N}\right) \\ &= \beta_n\left(\omega_{nN} c_{\frac{nN}{2}}\right), \end{aligned}$$

and *(1.70)*

$$\tau\left(\left(X_j^{k,N}\right)^n\right) = \beta_n \tau\left(\left(W_j^{l,N}\right)^n\right)$$

$$= \begin{cases} \beta_n\left(\sum\limits_{\pi \in NC(\Omega_n)} \left(\prod\limits_{V \in \pi} c_{\frac{N|V|}{2}}\right)\right) & \text{if } N \text{ is even} \\ \omega_n \beta_n\left(\sum\limits_{\theta \in NC_e(\Omega_n)} \left(\prod\limits_{V \in \theta} c_{\frac{N|V|}{2}}\right)\right) & \text{if } N \text{ is odd,} \end{cases}$$

with

$$\beta_n = \left(\psi(q_j)^2 \psi(q_k)^N\right)^n = \psi(q_j)^{2n} \psi(q_k)^{nN},$$

in $\mathbb{R}^{\times}$, for all $n \in \mathbb{N}$.

Proof. The proof of the free-distributional data (1.70) is done by (1.58) and (1.69). □

The above theorem shows that the free weighted-Poisson distributions induced by the free weighted-semicircular family $\mathcal{X}$ are characterized by the free Poisson distributions induced by the free semicircular family $\mathcal{S}$ in the semicircular filterization $\mathbb{L}_Q$ up to the weights $\{\psi(q_j)^2\}_{j \in \mathbb{Z}}$ from $\mathcal{X}$, where the quantities $\{\psi(q_j)\}_{j \in \mathbb{Z}}$ represent the free distributions of our mutually orthogonal projections $\{q_j\}_{j\in\mathbb{Z}}$ in the fixed C^*-probability space $(A,\ \psi)$. That is, the study of our free weighted-Poisson elements of $\mathbb{L}_Q$ is to investigate the free Poisson elements of $\mathbb{L}_Q$ up to certain scalar multiples, characterized by (1.70). Therefore, from below, we concentrate on studying the free Poisson elements of $\mathbb{L}_Q$.

1.9 SHIFTS ON $\mathbb{Z}$ AND INTEGER-SHIFTS ON $\mathbb{L}_Q$

As before, let $(A,\ \psi)$ be a fixed C^*-probability space containing a family $\mathbf{Q} = \{q_j\}_{j\in\mathbb{Z}}$ of mutually orthogonal projections q_j's having

$$\psi(q_j) \in \mathbb{R}^{\times}, \text{ for all } j \in \mathbb{Z},$$

(under **A 1.1**), and let $\mathbb{L}_Q$ be the semicircular filterization.

1.9.1 $(\pm)$-SHIFTS ON $\mathbb{Z}$

Define bijections h_+ and h_- on the set $\mathbb{Z}$ of all integers by

$$h_{\pm}(j) = j \pm 1, \tag{1.71}$$

for all $j \in \mathbb{Z}$. By definition (1.71), indeed, these functions are bijective on $\mathbb{Z}$, since $h_+^{-1} = h_-$, where f^{-1} means the inverses of bijections f.

Then, for these bijections $h_{\pm}$ of (1.71), one can construct the following bijections $h_{\pm}^{(n)}$ on $\mathbb{Z}$,

$$h_{\pm}^{(n)} = \underbrace{h_{\pm} \circ h_{\pm} \circ \cdots \circ h_{\pm}}_{n\text{-times}}, \tag{1.72}$$

for all $n \in \mathbb{N}$, with identities, $h_{\pm}^{(1)} = h_{\pm}$, where $(\circ)$ is the composition. Then,

$$h_{\pm}^{(n)}(j) = j \pm n, \text{ for all } j \in \mathbb{Z},$$

satisfying $\left(h_{+}^{(n)}\right)^{-1} = h_{-}^{(n)}$ on $\mathbb{Z}$, for all $n \in \mathbb{N}_0$, with axiomatization:

$$h_{\pm}^{(0)} = id_{\mathbb{Z}}, \text{ the identity function on } \mathbb{Z}.$$

Definition 1.12. *Let $h_{\pm}^{(n)}$ be the bijections (1.72) on $\mathbb{Z}$, for all $n \in \mathbb{N}_0$. Then, we call $h_{\pm}^{(n)}$, the n-$(\pm)$-shifts on $\mathbb{Z}$.*

1.9.2 INTEGER-SHIFTS ON $\mathbb{L}_Q$

Let $h_{\pm}^{(n)}$ be n-$(\pm)$-shifts (1.72) on $\mathbb{Z}$, for $n \in \mathbb{N}_0$. In this section, by using $h_{\pm}^{(n)}$, certain $*$-isomorphisms $\beta_{\pm}^{(n)}$ on $\mathbb{L}_Q$ are constructed, and we study how these $*$-isomorphisms act on $\mathbb{L}_Q$, for $n \in \mathbb{N}_0$.

Define "multiplicative" bounded linear transformations $\beta_{\pm}$ on $\mathbb{L}_Q$ by the morphisms satisfying that:

$$\beta_{\pm}(U_j) = U_{h_{\pm}(j)} = U_{j\pm 1}, \tag{1.73}$$

for $U_j \in \mathcal{S}$, and for all $j \in \mathbb{Z}$, where $\mathcal{S}$ is the free semicircular family (1.39).

By the structure theorem (1.42), the aforementioned multiplicative linear transformations $\beta_{\pm}$ of (1.73) are well defined on $\mathbb{L}_Q$. More precisely, one obtains the following computations.

Lemma 1.3. *Let $Y = \prod_{l=1}^{N} U_{j_l}^{n_l}$ be a free reduced word of $\mathbb{L}_Q$ with its length-N, for $U_{j_1}, ..., U_{j_N} \in \mathcal{S}$, and $n_1, ..., n_N \in \mathbb{N}$, where $(j_1, ..., j_N)$ is alternating in $\mathbb{Z}^N$, in the sense that:*

$$j_1 \neq j_2,\ j_2 \neq j_3,\ ...,\ j_{N-1} \neq j_N \text{ in } \mathbb{Z},$$

for $N \in \mathbb{N}$. Then, $\beta_{\pm}(Y)$ become new free reduced words of $\mathbb{L}_Q$ with their lengths-N in $\mathcal{S}$,

$$\beta_{\pm}(Y) = \prod_{l=1}^{N} U_{j_l \pm 1}^{n_l}, \text{ in } \mathbb{L}_Q. \tag{1.74}$$

Proof. Let Y be given as above in $\mathbb{L}_Q$. Then, by the multiplicativity of the linear transformations $\beta_{\pm}$ of (1.73), one has that

$$\beta_{\pm}(Y) = \prod_{l=1}^{N} \beta_{\pm}\left(U_{j_l}^{n_l}\right) = \prod_{l=1}^{N} \left(\beta_{\pm}(U_{j_l})\right)^{n_l} = \prod_{l=1}^{N} U_{h_{\pm}(j_l)}^{n_l}.$$

Therefore, formula (1.74) holds.

Moreover, it is easily checked that if $(j_1, ..., j_N)$ is an alternating N-tuple of $\mathbb{Z}^N$, then the N-tuples

$$(j_1 \pm 1,\ j_2 \pm 1,\ ...,\ j_N \pm 1)$$

are alternating in $\mathbb{Z}^N$, too. Therefore, the images $\beta_{\pm}(Y)$ form new free reduced words of $\mathbb{L}_Q$ with their lengths-N in $\mathcal{S}$. □

As one can see in (1.74), the morphisms $\beta_{\pm}$ of (1.73) preserve the freeness on $\mathbb{L}_Q$, by (1.42).

Lemma 1.4. *The morphisms $\beta_{\pm}$ of (1.73) are $*$-isomorphisms on $\mathbb{L}_Q$.*

Proof. By (1.40), (1.41), and (1.42), all elements of the semicircular filterization $\mathbb{L}_Q$ are the limits of linear combinations of free reduced words in the free semicircular family $\mathcal{S}$ of (1.39). So, let's focus on the free reduced words of $\mathbb{L}_Q$ in $\mathcal{S}$.

Let $(j_1, ..., j_N)$ be an alternating N-tuple of $\mathbb{Z}^N$ for $N \in \mathbb{N}$, and

$$Y = \prod_{l=1}^{N} U_{j_l}^{n_l}, \text{ for } n_1, ..., n_N \in \mathbb{N}.$$

Then, by (1.74),

$$\beta_{\pm}(Y) = \prod_{l=1}^{N} U_{h_{\pm}(j_l)}^{n_l}, \tag{1.75}$$

where $h_{\pm}$ are the $(\pm)$-shifts (1.71) on $\mathbb{Z}$.

By the bijectivity of $h_{\pm}$, relation (1.75) implies the bijectivity of $\beta_{\pm}$ on $\mathbb{L}_Q$. That is, these multiplicative linear transformations $\beta_{\pm}$ of (1.73) are generator-preserving (bijective), and bounded on $\mathbb{L}_Q$.

Consider now that if Y is as above, then

$$\beta_{\pm}(Y^*) = \beta_{\pm}\left(\prod_{l=1}^{N} U_{j_{N-l+1}}^{n_{N-l+1}}\right) = \prod_{l=1}^{N} U_{h_{\pm}(j_{N-l+1})}^{n_{N-l+1}}$$

by (1.74)

$$= \left(\prod_{l=1}^{N} U_{h_{\pm}(j_l)}^{n_l}\right)^* = \left(\beta_{\pm}(Y)\right)^*. \tag{1.76}$$

So

$$\beta_{\pm}(S^*) = \left(\beta_{\pm}(S)\right)^*, \text{ for all } S \in \mathbb{L}_Q,$$

by (1.76). Therefore, the bounded, bijective, multiplicative linear transformations $\beta_{\pm}$ are adjoint-preserving on $\mathbb{L}_Q$, by (1.77). That is, they are $*$-isomorphisms on $\mathbb{L}_Q$. □

The above lemma shows that the $(\pm)$-shifts $h_{\pm}$ on $\mathbb{Z}$ induce the corresponding $*$-isomorphisms $\beta_{\pm}$ on $\mathbb{L}_Q$.

Let $\beta_{\pm}$ be the $*$-isomorphisms (1.73). Then, one can construct $*$-isomorphisms,

$$\beta_{\pm}^n = \underbrace{\beta_{\pm}\beta_{\pm}\cdots\cdots\beta_{\pm}}_{n\text{-times}} \text{ on } \mathbb{L}_Q, \tag{1.77}$$

for all $n \in \mathbb{N}_0$, with identity: $\beta_{\pm}^0 = 1_{\mathbb{L}_Q}$, the identity map on $\mathbb{L}_Q$.

Since $\beta_{\pm}$ and $1_{\mathbb{L}_Q}$ are $*$-isomorphisms, the morphisms $\beta_{\pm}^n$ of (1.77) are well-defined $*$-isomorphisms on $\mathbb{L}_Q$, too, for all $n \in \mathbb{N}_0$.

Definition 1.13. *The $*$-isomorphisms $\beta_{\pm}^n$ of (1.77) are called the n-$(\pm)$-(integer-)shifts on $\mathbb{L}_Q$, for all $n \in \mathbb{N}_0$.*

These $*$-isomorphisms $\{\beta_{\pm}^n\}_{n\in\mathbb{N}_0}$ satisfy the following identity relation on $\mathbb{L}_Q$.

Proposition 1.4. *Let $\beta_{\pm}^n$ be the n-$(\pm)$-integer-shifts (1.77) on $\mathbb{L}_Q$. Then,*

$$(\beta_+\beta_-)^n = \beta_+^n\beta_-^n = 1_{\mathbb{L}_Q} = \beta_-^n\beta_+^n = (\beta_-\beta_+)^n \text{ on } \mathbb{L}_Q, \tag{1.78}$$

for all $n \in \mathbb{N}_0$. Moreover,

$$\beta_+^{n_1}\beta_-^{n_2} = \beta_-^{n_2}\beta_+^{n_1} = \begin{cases} 1_{\mathbb{L}_Q} & \text{if } n_1 = n_2 \\ \beta_+^{n_1-n_2} & \text{if } n_1 > n_2 \\ \beta_-^{n_2-n_1} & \text{if } n_1 < n_2, \end{cases} \tag{1.79}$$

on $\mathbb{L}_Q$, for all $n_1, n_2 \in \mathbb{N}_0$.

Proof. As we discussed above, it suffices to consider the cases where we have the free reduced words

$$Y = \prod_{l=1}^{N} U_{j_l}^{n_l} \text{ of } \mathbb{L}_Q, \text{ for } n_1, ..., n_N \in \mathbb{N},$$

for $N \in \mathbb{N}$, by (1.74) and (1.42). Observe that

$$\begin{aligned}
\beta_+\beta_-(Y) &= \beta_+\left(\prod_{l=1}^{N} U_{j_l-1}^{n_l}\right) = \prod_{l=1}^{N} U_{(j_l-1)+1}^{n_l} \\
&= Y = \prod_{l=1}^{N} U_{(j_l+1)-1}^{n_l} \\
&= \beta_-\left(\prod_{l=1}^{N} U_{j_l+1}^{n_l}\right) = \beta_-\beta_+(Y).
\end{aligned}$$

Therefore,

$$\beta_+\beta_- = 1_{\mathbb{L}_Q} = \beta_-\beta_+ \text{ on } \mathbb{L}_Q. \tag{1.80}$$

By (1.80), one can get that

$$\beta_+^n\beta_-^n = (\beta_+\beta_-)^n = 1_{\mathbb{L}_Q} = (\beta_-\beta_+)^n = \beta_-^n\beta_+^n,$$

on $\mathbb{L}_Q$, for all $n \in \mathbb{N}_0$. Therefore, formula (1.78) holds.

By (1.78), one has that if $n_1 > n_2$ in $\mathbb{N}_0$, then

$$\beta_+^{n_1}\beta_-^{n_2} = \beta_+^{n_1-n_2}\beta_+^{n_2}\beta_-^{n_2} = \beta_+^{n_1-n_2} \text{ on } \mathbb{L}_Q,$$

and similarly, if $n_1 < n_2$ in $\mathbb{N}_0$, then

$$\beta_+^{n_1}\beta_-^{n_2} = \beta_+^{n_1}\beta_-^{n_1}\beta_-^{n_2-n_1} = \beta_-^{n_2-n_1}.$$

Therefore, formula (1.79) holds. □

The above relations (1.78) and (1.79) can be re-expressed simply by

$$\beta_{e_1}^{n_1}\beta_{e_2}^{n_2} = \beta_{e_2}^{n_2}\beta_{e_1}^{n_1} = \beta_{sgn(e_1n_1+e_2n_2)}^{|e_1n_1+e_2n_2|} \text{ on } \mathbb{L}_Q, \tag{1.81}$$

for all $e_1, e_2 \in \{\pm\}$, and $n_1, n_2 \in \mathbb{N}_0$, where $sgn(\cdot)$ is the sign-map on $\mathbb{Z}$,

$$sgn(j) = \begin{cases} + & \text{if } j \geq 0 \\ - & \text{if } j < 0 \end{cases}$$

for all $j \in \mathbb{Z}$, and where $|.|$ is the absolute value on $\mathbb{Z}$.

Now, consider the system $\mathfrak{B}$ of all n-($\pm$)-shifts $\beta_\pm^n$ on $\mathbb{L}_Q$, i.e.,

$$\mathfrak{B} = \{\beta_\pm^n\}_{n\in\mathbb{N}_0}. \tag{1.82}$$

Let $Aut(\mathbb{L}_Q)$ be the *automorphism group*

$$Aut\,(\mathbb{L}_Q) = \left(\left\{\alpha : \mathbb{L}_Q \to \mathbb{L}_Q \;\middle|\; \begin{array}{c}\alpha \text{ is a} \\ *\text{-isomorphism} \\ \text{on } \mathbb{L}_Q\end{array}\right\}, \cdot\right) \tag{1.83}$$

where the operation $(\cdot)$ is the product of $*$-isomorphisms.

By definition (1.82), the system $\mathfrak{B}$ is contained in the automorphism group $Aut(\mathbb{L}_Q)$ of (1.83). Note that the operation $(\cdot)$ is closed on $\mathfrak{B}$, in the sense that:

$$\left(\beta_{e_1}^{n_1},\ \beta_{e_2}^{n_2}\right) \in \mathfrak{B} \times \mathfrak{B} \longmapsto \beta_{e_1}^{n_1} \beta_{e_2}^{n_2} \in \mathfrak{B}, \tag{1.84}$$

for all $e_1, e_2 \in \{\pm\}$, and $n_1, n_2 \in \mathbb{N}_0$, by (1.81).

Clearly, by (1.84), one can get that

$$\left(\beta_e^{n_1} \beta_e^{n_2}\right) \beta_e^{n_3} = \beta_e^{n_1+n_2+n_3} = \beta_e^{n_1} \left(\beta_e^{n_2} \beta_e^{n_3}\right), \tag{1.85}$$

for all $e \in \{\pm\}$, and $n_1, n_2, n_3 \in \mathbb{N}_0$.

Observe now that

$$\left(\beta_{e_1}^{n_1} \beta_{e_2}^{n_2}\right) \beta_{e_3}^{n_3} = \beta_{sgn(e_1 n_1 e_2 n_2)}^{|e_1 n_1 e_2 n_2|} \beta_{e_3}^{n_3} = \beta_{sgn(e_1 n_1 e_2 n_2 e_3 n_3)}^{|e_1 n_1 e_2 n_2 e_3 n_3|},$$

and (1.86)

$$\beta_{e_1}^{n_1} \left(\beta_{e_2}^{n_2} \beta_{e_3}^{n_3}\right) = \beta_{e_1}^{n_1} \beta_{sgn(e_2 n_2 e_3 n_3))}^{|e_2 n_2 e_3 n_3|} = \beta_{sgn(e_1 n_1 e_2 n_2 e_3 n_3))}^{|e_1 n_1 e_2 n_2 e_3 n_3)|}.$$

Thus, by (1.85) and (1.86),

$$\left(\beta_{e_1}^{n_1} \beta_{e_2}^{n_2}\right) \beta_{e_3}^{n_3} = \beta_{e_1}^{n_1} \left(\beta_{e_2}^{n_2} \beta_{e_3}^{n_3}\right) \text{ on } \mathbb{L}_Q, \tag{1.87}$$

for all $e_1, e_2, e_3 \in \{\pm\}$, and $n_1, n_2, n_3 \in \mathbb{N}_0$.

Theorem 1.13. *If $\mathfrak{B}$ is the set (1.82), then $\mathfrak{B}$ is an abelian subgroup of $Aut(\mathbb{L}_Q)$.*

Proof. By (1.81), the operation $(\cdot)$ is closed on $\mathfrak{B}$. So the algebraic pair $\mathfrak{B} = (\mathfrak{B}, \cdot)$ is a well-defined algebraic substructure of $Aut(\mathbb{L}_Q)$. By (1.87), this operation $(\cdot)$ is associative on $\mathfrak{B}$, and hence, the pair forms a semigroup. Since

$$\beta_+^0 = 1_{\mathbb{L}_Q} = \beta_-^0 \in \mathfrak{B},$$

and since

$$\beta_e^n \cdot 1_{\mathbb{L}_Q} = \beta_e^n = 1_{\mathbb{L}_Q} \cdot \beta_e^n \text{ on } \mathbb{L}_Q,$$

for all $e \in \{\pm\}$, and $n \in \mathbb{N}_0$, the semigroup $\mathfrak{B}$ contains its $(\cdot)$-identity $1_{\mathbb{L}_Q}$, and hence, it forms a monoid.

By (1.78), all elements $\beta_\pm^n \in \mathfrak{B}$ have their unique $(\cdot)$-inverses $\beta_\mp^n \in \mathfrak{B}$, for all $n \in \mathbb{N}_0$. So this monoid $\mathfrak{B}$ is a group.

The commutativity on $\mathfrak{B}$ is clear by (1.79). Therefore, the system $\mathfrak{B}$ is an abelian subgroup of the automorphism group $Aut(\mathbb{L}_Q)$. □

The above theorem characterizes the system $\mathfrak{B}$ of (1.82) as an abelian subgroup of $Aut(\mathbb{L}_Q)$. As a group, $\mathfrak{B}$ satisfies the following group-property.

Theorem 1.14. *Let $\mathfrak{B}$ be the abelian group (1.82). Then,*

$$\mathfrak{B} \overset{Group}{=} (\mathbb{Z}, +), \tag{1.88}$$

where "$\overset{Group}{=}$" means "being group-isomorphic to."

Proof. Define now a function $\Phi : \mathbb{Z} \to \mathfrak{B}$ by

$$\Phi : j \in \mathbb{Z} \longmapsto \beta_{sgn(j)}^{|j|} \in \mathfrak{B}, \tag{1.89}$$

with identity: $\Phi(0) = 1_{\mathbb{L}_Q}$ in $\mathfrak{B}$. By (1.79), this map Φ is a well-defined bijection from $\mathbb{Z}$ onto $\mathfrak{B}$. Observe that

$$\begin{aligned}\Phi(j_1 + j_2) &= \beta_{sgn(j_1+j_2)}^{|j_1+j_2|} = \beta_{sgn(j_1)}^{|j_1|}\beta_{sgn(j_2)}^{|j_2|} \\ &= \Phi(j_1)\Phi(j_2),\end{aligned} \tag{1.90}$$

in $\mathfrak{B}$, by (1.81), for all $j_1, j_2 \in \mathbb{Z}$.

So the bijection Φ of (1.89) is a group-homomorphism by (1.90). Therefore, the group-isomorphic relation (1.88) holds true. □

The above theorem characterizes the group-structure of the subgroup $\mathfrak{B} = \{\beta_{\pm}^n\}_{n\in\mathbb{N}_0}$ in $Aut(\mathbb{L}_Q)$. That is, $\mathfrak{B}$ is an infinite abelian cyclic group $\langle\beta_+\rangle = \langle\beta_-\rangle$.

Definition 1.14. *Let $\mathfrak{B}$ be the abelian group (1.82). We call $\mathfrak{B}$ the integer-shift (sub)group (of the automorphism group $Aut(\mathbb{L}_Q)$ acting) on $\mathbb{L}_Q$.*

1.9.3 FREE PROBABILITY ON $\mathbb{L}_Q$ UNDER THE GROUP-ACTION OF $\mathfrak{B}$

Let $\mathfrak{B}$ be the integer-shift group (1.84) acting on the semicircular filterization $\mathbb{L}_Q$, which is an infinite abelian cyclic subgroup of the automorphism group $Aut(\mathbb{L}_Q)$, by (1.88). In this section, we consider how our $*$-isomorphisms $\beta_{\pm}^n \in \mathfrak{B}$ affect the free probability on $\mathbb{L}_Q$. Throughout this section, let's fix $n_0 \in \mathbb{N}_0$ and $\beta_{\pm}^{n_0} \in \mathfrak{B}$.

Lemma 1.5. *Let $u_j \in \mathcal{X}$ be the $\psi(q_j)^2$-semicircular element, and $U_j \in \mathcal{S}$, a semicircular element of $\mathbb{L}_Q$. Then,*

$$\tau\left(\left(\beta_{\pm}^{n_0}(u_j)\right)^n\right) = \omega_n \psi(q_j)^n c_{\frac{n}{2}} = \tau\left(u_j^n\right),$$

and *(1.91)*

$$\tau\left(\left(\beta_{\pm}^{n_0}(U_j)\right)^n\right) = \omega_n c_{\frac{n}{2}} = \tau\left(U_j^n\right),$$

for all $n \in \mathbb{N}$.

Proof. Observe that, for any $n \in \mathbb{N}$, one has that

$$\tau\left(\left(\beta_{\pm}^{n_0}(u_j)\right)^n\right) = \tau\left(\left(\psi(q_j)\beta_{\pm}^{n_0}(U_j)\right)^n\right)$$

since $\beta_{\pm}^{n_0}$ are linear on $\mathbb{L}_Q$ and $u_j = \psi(q_j)U_j$ in $\mathbb{L}_Q$

$$= \psi(q_j)^n\, \tau\left(U_{j\pm n_0}^n\right) = \psi(q_j)^n\left(\omega_n c_{\frac{n}{2}}\right) \tag{1.92}$$

by the semicircularity (1.39) of $U_{j \pm n_0} \in \mathcal{S}$

$$= \omega_n \left(\psi(q_j)^2\right)^{\frac{n}{2}} c_{\frac{n}{2}} = \tau\left(u_j^n\right), \tag{1.93}$$

by (1.38). Therefore, the first free-distributional data of (1.91) hold by (1.93).

As one can see in (1.92), we have that

$$\tau\left(\left(\beta_{\pm}^{n_0}(U_j)\right)^n\right) = \omega_n c_{\frac{n}{2}} = \tau\left(U_j^n\right), \text{ for all } n \in \mathbb{N},$$

by (1.39), for all $j \in \mathbb{Z}$. So the second free-distributional data of (1.91) hold. □

The above lemma shows how the original free-distributional data on $\mathbb{L}_Q$ is affected by the group-action of $\mathfrak{B}$. That is, the action preserves the free distributions of generating operators of $\mathbb{L}_Q$.

Definition 1.15. *Let (B_1, φ_1) and (B_2, φ_2) be arbitrary topological $*$-probability spaces. We say that they are free-($*$-)isomorphic if (i) B_1 and B_2 are $*$-isomorphic via a $*$-isomorphism $\Omega : B_1 \to B_2$, and (ii)*

$$\varphi_2\left(\Omega(a)\right) = \varphi_1(a), \forall a \in (B_1, \varphi_1).$$

In such a case, we call the $*$-isomorphism Ω a free-isomorphism.

By (1.91), one obtains the following theorem.

Theorem 1.15. *All integer-shifts of $\mathfrak{B}$ are free-isomorphisms on $\mathbb{L}_Q$.*

Proof. Let $\beta_e^n \in \mathfrak{B}$, for $e \in \{\pm\}$ and $n \in \mathbb{N}_0$. Since $\mathfrak{B}$ is a subgroup of $Aut(\mathbb{L}_Q)$, a group-element $\beta_e^n \in \mathfrak{B}$ is a $*$-isomorphisms on $\mathbb{L}_Q$. Moreover, by (1.42) and (1.91),

$$\tau\left(\beta_e^n(T)\right) = \tau(T), \text{ for all } T \in \mathbb{L}_Q.$$

Therefore, the $*$-isomorphism $\beta_e^n \in \mathfrak{B}$ is a free-isomorphism. Since β_e^n is arbitrary in $\mathfrak{B}$, all integer-shifts of $\mathfrak{B}$ are free-isomorphisms on $\mathbb{L}_Q$. □

The above theorem illustrates that the free probability on the semicircular filterization $\mathbb{L}_Q$ is preserved by the action of the integer-shift group $\mathfrak{B}$.

1.10 BANACH-SPACE OPERATORS ON $\mathbb{L}_Q$ GENERATED BY $\mathfrak{B}$

Throughout this section, let $\mathbb{L}_Q$ be the semicircular filterization, and let $\mathfrak{B}$ be the integer-shift group acting on $\mathbb{L}_Q$. In this section, we consider certain Banach-space operators acting "on $\mathbb{L}_Q$," generated by $\mathfrak{B}$, by regarding $\mathbb{L}_Q$ as a Banach space (e.g., [13]).

Let $B(\mathbb{L}_Q)$ be the operator space consisting of all Banach-space operators on the Banach space $\mathbb{L}_Q$ equipped with its operator norm,

$$\|T\| = \sup \left\{ \|Tv\|_{\mathbb{L}_Q} \middle| \begin{array}{c} v \in \mathbb{L}_Q, \text{ and} \\ \|v\|_{\mathbb{L}_Q} = 1 \end{array} \right\}, \tag{1.94}$$

for all $T \in B(\mathbb{L}_Q)$, where $\|.\|_{\mathbb{L}_Q}$ is the norm on the Banach $*$-algebra $\mathbb{L}_Q$ (e.g., [13] and [14]).

Define now a (closed) subspace $\mathfrak{A}$ of the vector space $B(\mathbb{L}_Q)$ by

$$\mathfrak{A} \overset{def}{=} \overline{span_{\mathbb{C}}(\mathfrak{B})} = \overline{\mathbb{C}[\mathfrak{B}]}^{\|\cdot\|}, \tag{1.95}$$

where $\mathfrak{B}$ is the integer-shift group, and $\overline{Y}^{\|\cdot\|}$ are the operator-norm topology closures of the subsets Y of $B(\mathbb{L}_Q)$, where $\|.\|$ is in the sense of (1.94). Note that since $\mathfrak{B}$ is a group, the set equality of (1.95) holds. That is, $\mathfrak{A}$ is not only a closed subspace, but also an algebra of $B(\mathbb{L}_Q)$.

On this Banach algebra $\mathfrak{A}$, define a unary operation $(*)$ by

$$\left(\sum_{(e,n)\in\{\pm\}\times\mathbb{N}_0} t_{(e,n)} \beta_e^n \right)^* \overset{def}{=} \sum_{(e,n)\in\{\pm\}\times\mathbb{N}_0} \overline{t_{(e,n)}}\, \beta_{-e}^n, \tag{1.96}$$

on $\mathfrak{A}$, for all $t_{(e,n)} \in \mathbb{C}$ with their conjugates $\overline{t_{(e,n)}} \in \mathbb{C}$.

Proposition 1.5. *The subspace $\mathfrak{A}$ of (1.95) forms a Banach $*$-algebra in $B(\mathbb{L}_Q)$.*

Proof. First of all, for any $\beta_e^n \in \mathfrak{B} \subset \mathfrak{A}$, there always exists $(\beta_e^n)^* = \beta_{-e}^n \in \mathfrak{B} \subset \mathfrak{A}$, for all $e \in \{\pm\}$ and $n \in \mathbb{N}_0$. So for any $A \in \mathfrak{A}$, there exists a unique A^* in $\mathfrak{A}$. So the operation $(*)$ of (1.96) is closed on $\mathfrak{A}$. Observe that

$$((\beta_e^n)^*)^* = (\beta_{-e}^n)^* = \beta_{-(-e)}^n = \beta_e^n \text{ in } \mathfrak{A},$$

by (1.96), for all $e \in \{\pm\}$, and $n \in \mathbb{N}_0$. So if $A \in \mathfrak{A}$, then

$$A^{**} = A, \text{ in } \mathfrak{A}. \tag{1.97}$$

By the very definition (1.96), one immediately obtain that

$$(zA)^* = \overline{z}\, A^*, \; \forall z \in \mathbb{C}, \text{ and } A \in \mathfrak{A}. \tag{1.98}$$

Now, observe that

$$\left(\beta_{e_1}^{n_1} + \beta_{e_2}^{n_2}\right)^* = \beta_{-e_1}^{n_1} + \beta_{-e_2}^{n_2} = \left(\beta_{e_1}^{n_1}\right)^* + \left(\beta_{e_2}^{n_2}\right)^*,$$

in $\mathfrak{A}$, by (1.96), for all $e_1, e_2 \in \{\pm\}$, and $n_1, n_2 \in \mathbb{N}_0$. Thus, if $A_1, A_2 \in \mathfrak{A}$,

$$(A_1 + A_2)^* = A_1^* + A_2^* \text{ in } \mathfrak{A}, \tag{1.99}$$

by (1.98). Also, one can get that

$$\left(\beta_{e_1}^{n_1} \beta_{e_2}^{n_2}\right)^* = (\beta_e^n)^* = \beta_{-e}^n,$$

and

$$\left(\beta_{e_2}^{n_2}\right)^* \left(\beta_{e_1}^{n_1}\right)^* = \beta_{-e_2}^{n_2} \beta_{-e_1}^{n_1} = \beta_{-e_1}^{n_1} \beta_{-e_2}^{n_2} = \beta_{-e}^n,$$

where $e = sgn(e_1n_1 + e_2n_2) \in \{\pm\}$, and $n = |e_1n_1 + e_2n_2| \in \mathbb{N}_0$.
And hence,

$$\left(\beta_{e_1}^{n_1}\beta_{e_2}^{n_2}\right)^* = \left(\beta_{e_2}^{n_2}\right)^*\left(\beta_{e_1}^{n_1}\right)^*,$$

in $\mathfrak{A}$, for all $e_1, e_2 \in \{\pm\}$, and $n_1, n_2 \in \mathbb{N}_0$. So

$$(A_1A_2)^* = A_2^*A_1^* \text{ in } \mathfrak{A}, \tag{1.100}$$

for all $A_1, A_2 \in \mathfrak{A}$.

Therefore, the operation $(*)$ of (1.96) is a well-defined adjoint on $\mathfrak{A}$, by (1.97), (1.98), (1.99), and (1.100). So the Banach algebra $\mathfrak{A}$ forms a Banach $*$-algebra. □

The above proposition shows that every element of $\mathfrak{A}$ is an adjointable Banach-space operator on $\mathbb{L}_Q$ in the sense of [13].

Definition 1.16. *Let $\mathfrak{A}$ be the Banach $*$-algebra (1.95) embedded in the operator space $B(\mathbb{L}_Q)$. Then, we call $\mathfrak{A}$ the integer-shift operator algebra (on $\mathbb{L}_Q$). All elements of $\mathfrak{A}$ are said to be integer-shift operators on $\mathbb{L}_Q$.*

1.10.1 DEFORMED FREE PROBABILITY OF $\mathbb{L}_Q$ BY $\mathfrak{A}$

Let $\mathbb{L}_Q$ be the semicircular filterization, and $\mathfrak{B}$, the integer-shift group acting on $\mathbb{L}_Q$, and let $\mathfrak{A}$ be the integer-shift-operator algebra (1.95) generated by $\mathfrak{B}$ in the operator space $B(\mathbb{L}_Q)$. In this section, we act operators of $\mathfrak{A}$ on $\mathbb{L}_Q$ and study how the original free-distributional data on $\mathbb{L}_Q$ are deformed by the action. By (1.40) and (1.42), we focus on how our (weighted-)semicircularity induced by $(\mathcal{X}\cup)\mathcal{S}$ is affected by the action of $\mathfrak{A}$ on $\mathbb{L}_Q$.

Since we already considered how the integer-shift group $\mathfrak{B}$, the generator set of $\mathfrak{A}$, affects the free probability on $\mathbb{L}_Q$ in Section 1.9, we here are interested in the distorted (weighted-)semicircular law(s) by the action of nongenerator operators A of $\mathfrak{A}$, formed by

$$A = \sum_{(e,n)\in\mathbb{N}_0^{\pm}} t_{(e,n)}\beta_e^n \in \mathfrak{A}. \tag{1.101}$$

Note again that if $A \in \mathfrak{B} \subset \mathfrak{A}$, then such an integer-shift operator A is an integer-shift, which is free-isomorphism on $\mathbb{L}_Q$, and hence, it preserves the free probability on $\mathbb{L}_Q$. However, if $A \in \mathfrak{A} \setminus \mathfrak{B}$, then we cannot guarantee that A preserves the free-distributional data on $\mathbb{L}_Q$.

Notation From below, for convenience, we let

$$\mathbb{N}_0^{\pm} \overset{denote}{=} \{\pm\} \times \mathbb{N}_0.$$

□

Lemma 1.6. *Let $A = t\beta_e^k \in \mathfrak{A}$, for $t \in \mathbb{C}^{\times}$, and $(e,k) \in \mathbb{N}_0^{\pm}$, and let $u_j \in \mathcal{X}$ be a $\psi(q_j)^2$-semicircular element in $\mathbb{L}_Q$, for $j \in \mathbb{Z}$. Then,*

$$\tau\left((A(u_j))^n\right) = \omega_n\left(t\psi(q_j)\right)^n c_{\frac{n}{2}},$$

and (1.102)

$$\tau((A(u_j)^*)^n) = \omega_n (\bar{t}\psi(q_j))^n c_{\frac{n}{2}},$$

for all $n \in \mathbb{N}$.

Proof. Let $A = t\beta_e^k \in \mathfrak{A}$, for $(e, k) \in \mathbb{N}_0^{\pm}$, and let $t \in \mathbb{C}^{\times}$. Then,

$$\begin{aligned}\tau(A(u_j)^n) &= \tau\left(\left(t\beta_e^k(u_j)\right)^n\right) = \tau\left(\left(t\psi(q_j)U_{jek}\right)^n\right)\\ &= t^n\psi(q_j)^n\tau\left(U_{jek}^n\right) = \omega_n t^n\psi(q_j)^n c_{\frac{n}{2}},\end{aligned}$$

and

$$\begin{aligned}\tau((A(u_j)^*)^n) &= \tau\left(\left(\left(t\beta_e^k(u_j)\right)^*\right)^n\right)\\ &= \omega_n\bar{t}^n\psi(q_j)^n c_{\frac{n}{2}},\end{aligned}$$

for all $n \in \mathbb{N}$, and $j \in \mathbb{Z}$, implying the formulas of (1.104). □

By (1.102), we obtain the following result.

Theorem 1.16. *Let $A = t\beta_e^n \in \mathfrak{A}$, for $t \in \mathbb{C}^{\times}$, and $(e, n) \in \mathbb{N}_0^{\pm}$. Then,*

if $t \in \mathbb{R}^{\times}$, then $t\beta_e^n(u_j) \in \mathbb{L}_Q$ are $(t\psi(q_j))^2$-semicircular, (1.103)

in $\mathbb{L}_Q$, for all $(e, n) \in \mathbb{N}_0^{\pm}$ and for all $j \in \mathbb{Z}$.

Proof. Let $t \in \mathbb{R}^{\times}$ in $\mathbb{C}$, and let $A = t\beta_e^n$, for $(e, n) \in \mathbb{N}_0^{\pm}$, then, for any $j \in \mathbb{Z}$, one has that

$$\tau\left((A(u_j))^k\right) = \omega_k(t\psi(q_j))^k c_{\frac{k}{2}} = \tau\left((A(u_j)^*)^k\right),$$

for all $k \in \mathbb{N}$, by (1.102).

Indeed, since $t \in \mathbb{R}^{\times}$, we have

$$(A(u_j))^* = \left(t\psi(q_j)U_{jek}\right)^* = t\psi(q_j)U_{jek} = A(u_j),$$

in $\mathbb{L}_Q$ (under **A 1.1**); equivalently, $A(u_j)$ is self-adjoint in $\mathbb{L}_Q$, for all $j \in \mathbb{Z}$.

Therefore, $t\beta_e^n(u_j)$ is $(t\psi(q_j))^2$-semicircular in $\mathbb{L}_Q$, for all $j \in \mathbb{Z}$, whenever $t \in \mathbb{R}^{\times}$, for all $(e,n) \in \mathbb{N}_0^{\pm}$. □

The above theorem shows that the integer-shift operators $t\beta_e^n \in \mathfrak{A}$ distort the original $\psi(q_j)^2$-semicircular laws to the $(t\psi(q_j))^2$-semicircular laws on $\mathbb{L}_Q$, for all $(e, n) \in \mathbb{N}_0^{\pm}$, and $j \in \mathbb{Z}$, whenever $t \in \mathbb{R}^{\times} \setminus \{1\}$, by (1.103).

Corollary 1.3. *Let $A = t\beta_e^n \in \mathfrak{A}$, for $t \in \mathbb{R}^{\times}$, and $(e, n) \in \mathbb{N}_0^{\pm}$, and let $U_j \in \mathcal{S}$ be a semicircular element in $\mathbb{L}_Q$, for $j \in \mathbb{Z}$. Then,*

if $t \in \mathbb{R}^{\times}$, then $t\beta_e^n(U_j)$ is t^2-semicircular in $\mathbb{L}_Q$, (1.104)

for all $(e, n) \in \mathbb{N}_0^{\pm}$ and for all $j \in \mathbb{Z}$.

Proof. Statement (1.104) is shown by the proof of (1.103). □

Now, let $A \in \mathfrak{A}$ be in the sense of (1.101), and assume further that A has more than one summand. Then, for any $u_j \in \mathcal{X}$ and $U_j \in \mathcal{S}$, one can have that

$$A\left(U_j^n\right) = \sum_{(e,k)\in\mathbb{N}_0} t_{(e,k)} U_{jek}^n \text{ in } \mathbb{L}_Q,$$

and hence, (1.105)

$$A\left(u_j^n\right) = \psi(q_j)^n \left(\sum_{(e,k)\in\mathbb{N}_0^{\pm}} t_{(e,k)} U_{jek}^n \right) \text{ in } \mathbb{L}_Q,$$

for all $n \in \mathbb{N}$.

Theorem 1.17. *Let $A \in \mathfrak{A}$ be an integer-shift operator (1.101), and $u_j \in \mathcal{X}$, $U_j \in \mathcal{S}$ in $\mathbb{L}_Q$. Then,*

$$\tau\left(A\left(u_j^n\right)\right) = \left(\omega_n \psi(q_j)^n c_{\frac{n}{2}}\right) \left(\sum_{(e,k)\in\mathbb{N}_0^{\pm}} t_{(e,k)} \right)$$
$$= \tau\left(u_j^n\right) \left(\sum_{(e,k)\in\mathbb{N}_0^{\pm}} t_{(e,k)} \right),$$

and (1.106)

$$\tau\left(A\left(U_j^n\right)\right) = \left(\omega_n c_{\frac{n}{2}}\right) \left(\sum_{(e,k)\in\mathbb{N}_0^{\pm}} t_{(e,k)} \right)$$
$$= \tau\left(U_j^n\right) \left(\sum_{(e,k)\in\mathbb{N}_0^{\pm}} t_{(e,k)} \right),$$

for all $n \in \mathbb{N}$.

Proof. Remark that if $A \in \mathfrak{A}$ is in the sense of (1.101), then

$$A\left(u_j^n\right) = \sum_{(e,k)\in\mathbb{N}_0^{\pm}} t_{(e,k)} \beta_e^k \left(u_j^n\right)$$
$$= \sum_{(e,k)\in\mathbb{N}_0^{\pm}} t_{(e,k)} \left(\psi(q_j)^n U_{jek}^n\right),$$

by (1.105). So

$$\tau\left(A\left(u_j^n\right)\right) = \psi(q_j)^n \left(\sum_{(e,k)\in\mathbb{N}_0^{\pm}} t_{(e,k)} \tau\left(U_{jek}^n\right) \right)$$

$$= \psi(q_j)^n \left(\sum_{(e,k)\in \mathbb{N}_0^{\pm}} t_{(e,k)} \left(\omega_n c_{\frac{n}{2}} \right) \right) \tag{1.107}$$

$$= \left(\omega_n \psi(q_j)^n c_{\frac{n}{2}} \right) \left(\sum_{(e,k)\in \mathbb{N}_0^{\pm}} t_{(e,k)} \right)$$

$$= \left(\tau \left(u_j^n \right) \right) \left(\sum_{(e,k)\in \mathbb{N}_0^{\pm}} t_{(e,k)} \right), \tag{1.108}$$

for all $n \in \mathbb{N}$.

By (1.107), one obtains that

$$\tau \left(A \left(U_j^n \right) \right) = \left(\omega_n c_{\frac{n}{2}} \right) \left(\sum_{(e,k)\in \mathbb{N}_0^{\pm}} t_{(e,k)} \right)$$

$$= \tau \left(U_j^n \right) \left(\sum_{(e,k)\in \mathbb{N}_0^{\pm}} t_{(e,k)} \right),$$

for all $n \in \mathbb{N}$.

Therefore, by (1.107) and (1.109), the free-distributional data (1.106) hold. □

In the formulas of (1.106), the quantity

$$t_A \overset{def}{=} \sum_{(e,k)\in \mathbb{N}_0^{\pm}} t_{(e,k)} \in \mathbb{C} \tag{1.109}$$

is the sum of all coefficients $\{t_{(e,k)}\}_{(e,k)\in \mathbb{N}_0^{\pm}}$ of the operator $A \in \mathfrak{A}$. So

$$\tau \left(A(U_j^n) \right) = t_A \left(\omega_n c_{\frac{n}{2}} \right), \forall n \in \mathbb{N}, \tag{1.110}$$

for all $j \in \mathbb{Z}$, by (1.106) and (1.109). And, similar to (1.110),

$$\tau \left(A(u_j^n) \right) = (t_A \psi(q_j)^n) \left(\omega_n c_{\frac{n}{2}} \right), \forall n \in \mathbb{N}, \tag{1.111}$$

for all $j \in \mathbb{Z}$, where t_A is in the sense of (1.109).

Formulas (1.110) and (1.111) show how the integer-shift-operator algebra $\mathfrak{A}$ deforms the (weighted-)semicircular law(s) on $\mathbb{L}_Q$.

1.10.2 DEFORMED SEMICIRCULAR LAWS ON $\mathbb{L}_Q$ BY $\mathfrak{A}$

We, here, concentrate on the distortions of the semicircular law on $\mathbb{L}_Q$ by acting the integer-shift-operator algebra $\mathfrak{A}$. Now, observe that if A is an integer-shift operator,

$$A = \sum_{(e,k)\in \mathbb{N}_0^{\pm}} t_{(e,k)} \beta_e^k \in \mathfrak{A}, \tag{1.112}$$

then

$$A^2 = \sum_{((e_1,k_1),(e_2,k_2))\in \mathbb{N}_0^{\pm}\times \mathbb{N}_0^{\pm}} t_{(e_1,k_1)}t_{(e_2,k_2)}\beta_{e_1}^{k_1}\beta_{e_2}^{k_2}$$

$$= \sum_{((e_1,k_1),(e_2,k_2))\in \mathbb{N}_0^{\pm}\times \mathbb{N}_0^{\pm}} t_{(e_1,k_1)}t_{(e_2,k_2)}\beta_{sgn(e_1k_1+e_2k_2)}^{|e_1k_1+e_2k_2|},$$

and hence,

$$A^2 = \sum_{(e,k)\in \mathbb{N}_0^{\pm}} \left(\sum_{((e_1,k_1),(e_2,k_2))\in [e,k]_2} t_{(e_1,k_1)}t_{(e_2,k_2)} \right) \beta_e^k,$$

where (1.113)

$$[e,k]_2 = \left\{ ((e_1,k_1),(e_2,k_2)) \;\middle|\; \begin{array}{c} (e_1,k_1),\ (e_2,k_2) \in \mathbb{N}_0^{\pm}, \\ e = sgn(e_1k_1+e_2k_2), \\ \text{and } k = |e_1k_1+e_2k_2| \end{array} \right\},$$

for all $(e,\ k) \in \mathbb{N}_0^{\pm}$, because the integer-shift group $\mathfrak{B}$, the generator set of $\mathfrak{A}$, is abelian.

Lemma 1.7. *Let $A \in \mathfrak{A}$ be an integer-shift operator (1.112). Then,*

$$A^n = \sum_{(e,k)\in \mathbb{N}_0^{\pm}} \left(\sum_{((e_l,k_l))_{l=1}^n \in [e,k]_n} \left(\prod_{l=1}^{n} t_{(e_l,k_l)} \right) \right) \beta_e^k \in \mathfrak{A},$$

with *(1.114)*

$$[e,k]_n = \left\{ ((e_l,k_l))_{l=1}^n \;\middle|\; \begin{array}{c} (e_l,k_l) \in \mathbb{N}_0^{\pm},\ \forall l = 1,\dots,n, \\ e = sgn\left(\sum_{l=1}^n e_lk_l\right) \in \{\pm\}, \\ \textit{and } k = \left|\sum_{l=1}^n e_lk_l\right| \in \mathbb{N}_0 \end{array} \right\},$$

for all $n \in \mathbb{N}$.

Proof. The operator equality (1.114) is obtained by the induction on (1.113). □

Notation Let A^n be as in (1.114) for an integer-shift operator $A \in \mathfrak{A}$ of (1.112), for $n \in \mathbb{N}$. Then, we denote the quantities

$$\sum_{((e_l,k_l))_{l=1}^n \in [e,k]_n} \left(\prod_{l=1}^{n} t_{(e_l,k_l)} \right) \overset{denote}{=} t_{[e,k]_n}, \tag{1.115}$$

for all $(e,\ k) \in \mathbb{N}_0^{\pm}$, where $[e,\ k]_n$ are in the sense of (1.114). □

By notation (1.115), one can re-express (1.115) by

$$A^n = \sum_{(e,k)\in \mathbb{N}_0^{\pm}} t_{[e,k]_n}\, \beta_e^k \text{ in } \mathfrak{A},$$

for all $n \in \mathbb{N}$.

Theorem 1.18. *Let $U_j \in \mathcal{S}$ in $\mathbb{L}_Q$, and let $A \in \mathfrak{A}$ be an integer-shift operator (1.112). Then,*

$$\tau\left(A^{n_1}\left(U_j^{n_2}\right)\right) = \left(\omega_{n_2} c_{\frac{n_2}{2}}\right)\left(\sum_{(e,k)\in\mathbb{N}_0^{\pm}} t_{[e,k]_{n_1}}\right) \tag{1.116}$$

where $[e, k]_{n_1}$ are in the sense of (1.115), for all $(e, k) \in \mathbb{N}_0^{\pm}$, and $n_1, n_2 \in \mathbb{N}$.

Proof. Now, let $n_1, n_2 \in \mathbb{N}$, and $j \in \mathbb{Z}$. Then,

$$\tau\left(A^{n_1}\left(U_j^{n_2}\right)\right) = \tau\left(\sum_{(e,k)\in\mathbb{N}_0^{\pm}} t_{[e,k]_{n_1}} \beta_e^k\left(U_j^{n_2}\right)\right)$$

by (1.115)

$$= \sum_{(e,k)\in\mathbb{N}_0^{\pm}} t_{[e,k]_{n_1}} \tau\left(U_{jek}^{n_2}\right)$$

where $[e,k]_{n_1}$ are in the sense of (1.115) for all $(e,k) \in \mathbb{N}_0^{\pm}$

$$= \sum_{(e,k)\in\mathbb{N}_0^{\pm}} t_{[e,k]_{n_1}} \left(\omega_{n_2} c_{\frac{n_2}{2}}\right)$$

by the semicircularity of $U_{jek} \in \mathcal{S}$ in $\mathbb{L}_Q$, for all $j \in \mathbb{Z}$, and $(e, k) \in \mathbb{N}_0^{\pm}$

$$= \left(\omega_{n_2} c_{\frac{n_2}{2}}\right)\left(\sum_{(e,k)\in\mathbb{N}_0^{\pm}} t_{[e,k]_{n_1}}\right).$$

Therefore, formula (1.116) holds. □

Similar to (1.116), we also have the following free-distributional data on $\mathbb{L}_Q$.

Theorem 1.19. *Let $u_j \in \mathcal{X}$ in $\mathbb{L}_Q$, and let A be an integer-shift operator (1.112) in $\mathfrak{A}$. Then,*

$$\tau\left(A^{n_1}(u_j^{n_2})\right) = \left(\omega_{n_2}\psi(q_j)^{n_2} c_{\frac{n_2}{2}}\right)\left(\sum_{(e,k)\in\mathbb{N}_0^{\pm}} t_{[e,k]_{n_1}}\right), \tag{1.117}$$

for all $n_1, n_2 \in \mathbb{N}$.

Proof. The proof of (1.117) is similarly done by (1.114) and (1.116). □

The above two theorems generalize the distorted free-distributional data (1.106) of the semicircular law on $\mathbb{L}_Q$. Similar to the notation t_A of (1.109), one can define the following quantities.

Notation Let's denote the quantity

$$\sum_{(e,k)\in\mathbb{N}_0^{\pm}} t_{[e,k]_n} \overset{denote}{=} t_A^{(n)} \text{ in } \mathbb{C}, \tag{1.118}$$

where $\{t_{[e,k]_n}\}_{(e,k)\in\mathbb{N}_0^{\pm}}$ are the coefficients (1.115) of $A^n \in \mathfrak{A}$ of the integer-shift operator A of (1.112). Remark that, by (1.118),

$$t_A \text{ of } (1.109) = t_A^{(1)} \text{ of } (1.118).$$

□

By using the above new notation (1.118), formulas (1.116) and (1.117) can be re-expressed simply by

$$\tau\left(A^{n_1}\left(U_j^{n_2}\right)\right) = t_A^{(n_1)}\left(\omega_{n_2} c_{\frac{n_2}{2}}\right) = t_A^{(n_1)}\tau\left(U_j^{n_2}\right),$$

and (1.119)

$$\tau\left(A^{n_1}\left(u_j^{n_2}\right)\right) = t_A^{(n_1)}\left(\omega_{n_2}\psi(q_j)^{n_2} c_{\frac{n_2}{2}}\right) = t_A^{(n_1)}\tau\left(u_j^{n_2}\right),$$

for all $n_1, n_1 \in \mathbb{N}$, and for all $j \in \mathbb{Z}$.

1.11 DEFORMED FREE POISSON DISTRIBUTIONS ON $\mathbb{L}_Q$ BY $\mathfrak{A}$

In this section, as an application of Section 1.10, we study the deformed free (weighted-)Poisson distributions on $\mathbb{L}_Q$ under the action of the integer-shift-operator algebra $\mathfrak{A}$.

Throughout this section, we fix a free Poisson element

$$W_j^{m,N} = U_j U_m^N U_j \in \mathbb{L}_Q, \tag{1.120}$$

for $U_j, U_m \in \mathcal{S}$, and a free $\psi(q_j)^2$-Poisson element

$$Y_j^{m,N} = u_j U_m^N u_j \in \mathbb{L}_Q, \tag{1.121}$$

for $u_j \in \mathcal{X}$, where $j \neq m$ in $\mathbb{Z}$, for $N \in \mathbb{N}$.

By (1.54) and (1.66), we have the following free Poisson distribution of $W_j^{m,N}$ and the free $\psi(q_j)^2$-Poisson distribution of $Y_j^{m,N}$:

$$\tau\left(\left(W_j^{m,N}\right)^n\right) = \begin{cases} \sum\limits_{\pi\in NC(\Omega_n)} \left(\prod\limits_{V\in\pi} c_{\frac{N|V|}{2}}\right) & \text{if } N \text{ is even} \\ \omega_n \sum\limits_{\pi\in NC_e(\Omega_n)} \left(\prod\limits_{V\in\pi} c_{\frac{N|V|}{2}}\right) & \text{if } N \text{ is odd,} \end{cases} \tag{1.122}$$

and

$$\tau\left(\left(Y_j^{m,N}\right)^n\right) = \psi(q_j)^{2n}\, \tau\left(\left(W_j^{m,N}\right)^n\right), \tag{1.123}$$

for all $n \in \mathbb{N}$.

Corollary 1.4. *Let $W_j^{m,N}$ be the free Poisson element (1.120) of $\mathbb{L}_Q$. If $A \in \mathfrak{A}$, an integer-shift operator (1.112), then*

$$\tau\left(A^{n_1}\left(\left(W_j^{m,N}\right)^{n_2}\right)\right) = \begin{cases} t_A^{(n_1)}\left(\sum\limits_{\pi \in NC(\Omega_{n_2})}\left(\prod\limits_{V \in \pi} c_{\frac{n_2|V|}{2}}\right)\right) & \text{if } N \text{ is even} \\ \omega_n t_A^{(n_1)}\left(\sum\limits_{\pi \in NC_e(\Omega_{n_2})}\left(\prod\limits_{V \in \pi} c_{\frac{n_2|V|}{2}}\right)\right) & \text{if } N \text{ is odd,} \end{cases} \tag{1.124}$$

with

$$t_A^{(n_1)} = \sum_{(e,k)\in\mathbb{N}_0^{\pm}}\left(\sum_{((e_l,k_l))_{l=1}^{n}\in[e,k]_{n_1}}\left(\prod_{l=1}^{n} t_{(e_l,k_l)}\right)\right), \tag{1.125}$$

for all $n_1, n_2 \in \mathbb{N}$, where

$$[e,k]_{n_1} = \left\{ ((e_l,k_l))_{l=1}^{n_1} \;\middle|\; \begin{array}{c} (e_l,k_l) \in \mathbb{N}_0^{\pm},\ \forall l = 1,\ldots,n_1, \\ e = sgn\left(\sum_{l=1}^{n} e_l n_l\right) \in \{\pm\}, \\ \text{and } k = \left|\sum_{l=1}^{n} e_l n_l\right| \in \mathbb{N}_0 \end{array} \right\}.$$

Proof. Fix $n_1, n_2 \in \mathbb{N}$, and let

$$T_{n_1,n_2} = A^{n_1}\left(\left(W_j^{m,N}\right)^{n_2}\right) \in \mathbb{L}_Q.$$

Then

$$\begin{aligned} T_{n_1,n_2} &= A^{n_1}\left(\left(U_j U_m^N U_j\right)^{n_2}\right) \\ &= \sum_{(e,k)\in\mathbb{N}_0^{\pm}} t_{[e,k]_{n_1}} \beta_e^k\left(\left(U_j U_m^N U_j\right)^{n_2}\right) \end{aligned}$$

with

$$t_{[e,k]_{n_1}} = \sum_{((e_l,k_l))_{l=1}^{n_1}\in[e,k]_{n_1}}\left(\prod_{l=1}^{n_1} t_{(e_l,k_l)}\right)$$

by (1.115), where

$$[e,\,k]_{n_1} = \left\{ ((e_l,k_l))_{l=1}^{n_1} \;\middle|\; \begin{array}{c} (e_l,k_l) \in \mathbb{N}_0^{\pm},\ \forall l = 1,\ldots,n_1, \\ e = sgn\left(\sum_{l=1}^{n} e_l n_l\right) \in \{\pm\}, \\ \text{and } k = \left|\sum_{l=1}^{n} e_l n_l\right| \in \mathbb{N}_0 \end{array} \right\},$$

by (1.114), for all $(e,\,k) \in \mathbb{N}_0^{\pm}$, and hence, it goes to

$$= \sum_{(e,k)\in\mathbb{N}_0^{\pm}} t_{[e,k]_{n_1}}\left(U_{jek} U_{mek}^N U_{jek}\right)^{n_2}, \tag{1.126}$$

in $\mathbb{L}_Q$.

Note that in (1.126), the summands $U_{jek}U_{mek}^{N}U_{jek}$ are the free Poisson elements of $\mathbb{L}_Q$, because the freeness of semicircular elements U_{jek} and U_{mek} is preserved from the freeness of U_j and U_m in $\mathbb{L}_Q$ (by the action of the integer-shift group $\mathfrak{B}$), for all $j \neq m \in \mathbb{Z}$. Thus,

$$\tau(T_{n_1,n_2}) = \sum_{(e,k)\in\mathbb{N}_0^{\pm}} t_{[e,k]_{n_1}}\ \tau\left(\left(U_{jek}U_{mek}^{N}U_{jek}\right)^{n_2}\right)$$

$$= \sum_{(e,k)\in\mathbb{N}_0^{\pm}} t_{[e,k]_{n_1}}\ \tau\left(\left(U_{jek}U_{mek}^{N}U_{jek}\right)^{n_2}\right)$$

$$= \sum_{(e,k)\in\mathbb{N}_0^{\pm}} t_{[e,k]_{n_1}}\ \tau\left(\left(U_{j}U_{m}^{N}U_{j}\right)^{n_2}\right)$$

by Theorem 1.15

$$= \tau\left(\left(U_{j}U_{m}^{N}U_{j}\right)^{n_2}\right)\left(\sum_{(e,k)\in\mathbb{N}_0^{\pm}} t_{[e,k]_{n_1}}\right)$$

$$= t_A^{(n_1)}\ \tau\left(\left(W_j^{m,N}\right)^{n_2}\right), \tag{1.127}$$

where $t_A^{(n_1)}$ is in the sense of (1.125). Therefore, by (1.54) and (1.127), the free-distributional data (1.124) hold true. □

The aforementioned free-distributional data (1.124) shows how our integer-shift operators $A \in \mathfrak{A}$ of (1.112) distort the free Poisson distribution of $W_j^{m,N}$ on $\mathbb{L}_Q$. As a special case of (1.124), one obtains the following fact.

Corollary 1.5. *Let $W_j^{m,1}$ be the free Poisson element (1.120) of $\mathbb{L}_Q$. If $A \in \mathfrak{A}$, an integer-shift operator (1.112), then*

$$\tau\left(A^{n_1}\left(\left(W_j^{m,1}\right)^{n_2}\right)\right) = \delta_{n_2,2}\ t_A^{(n_1)}$$

with

$$t_A^{(n_1)} = \sum_{(e,k)\in\mathbb{N}_0^{\pm}}\left(\sum_{((e_l,k_l))_{l=1}^{n}\in[e,k]_{n_1}}\left(\prod_{l=1}^{n} t_{(e_l,k_l)}\right)\right),$$

for all $n_1, n_2 \in \mathbb{N}$. □

Also, we obtain the following corollary.

Corollary 1.6. *Let $Y_j^{m,N}$ be the free $\psi(q_j)^2$-Poisson element (1.121) of $\mathbb{L}_Q$, and let $A \in \mathfrak{A}$ be an integer-shift operator (1.112). Then,*

$$\tau\left(A^{n_1}\left(Y_j^{m,N}\right)^{n_2}\right)=\begin{cases} t_A^{(n_1)}\psi(q_j)^{2n_2}\left(\sum\limits_{\pi\in NC(\Omega_n)}\left(\prod\limits_{V\in\pi}c_{\frac{N|V|}{2}}\right)\right) & \text{if } N \text{ is even} \\ \omega_n t_A^{(n_1)}\psi(q_j)^{2n_2}\left(\sum\limits_{\pi\in NC_e(\Omega_n)}\left(\prod\limits_{V\in\pi}c_{\frac{N|V|}{2}}\right)\right) & \text{if } N \text{ is odd,} \end{cases} \tag{1.128}$$

for all $n_1, n_2 \in \mathbb{N}$, where $t_A^{(n_1)}$ are in the sense of (1.125).

Proof. The free-distributional data (1.128) is proven by (1.66), (1.126), and (1.127). □

The free-distributional data (1.128) illustrates how the integer-shift operator $A \in \mathfrak{A}$ distorts the free $\psi(q_j)^2$-Poisson distribution of $Y_j^{m,N}$ on $\mathbb{L}_Q$, in general. As a special case of (1.128), one has the following corollary.

Corollary 1.7. *Let $Y_j^{m,1}$ be the free $\psi(q_j)^2$-Poisson element (1.121) of $\mathbb{L}_Q$, and let $A \in \mathfrak{A}$ be an integer-shift operator (1.112). Then,*

$$\tau\left(A^{n_1}\left(Y_j^{m,1}\right)^{n_2}\right)=\delta_{n_2,2}\left(t_A^{(n_1)}\psi(q_j)^4\right),$$

for all $n_1, n_2 \in \mathbb{N}$, where $t_A^{(n_1)}$ are in the sense of (1.125). □

REFERENCES

[1] M. Ahsanullah, Some Inferences on Semicircular Distribution, *J. Stat. Theo. Appl.,* 15, no. 3, (2016) 207–213.

[2] H. Bercovici, and D. Voiculescu, Superconvergence to the Central Limit and Failure of the Cramer Theorem for Free Random Variables, *Probab. Theo. Related Fields,* 103, no. 2, (1995) 215–222.

[3] M. Bozejko, W. Ejsmont, and T. Hasebe, Noncommutative Probability of Type *D*, *Internat. J. Math.,* 28, no. 2, (2017) 1750010, 30.

[4] M. Bozheuiko, E. V. Litvinov, and I. V. Rodionova, An Extended Anyon Fock Space and Non-commutative Meixner-Type Orthogonal Polynomials in the Infinite-Dimensional Case, *Uspekhi Math. Nauk.,* 70, no. 5, (2015) 75–120.

[5] I. Cho, Semicircular Families in Free Product Banach $*$-Algebras Induced by p-Adic Number Fields over Primes p, *Compl. Anal. Oper. Theo.,* 11, no. 3, (2017) 507–565.

[6] I. Cho, Semicircular-Like Laws and the Semicircular Law Induced by Orthogonal Projections, *Compl. Anal. Oper. Theo.,* 12, no.1, DOI:10.1007/s11785-018-0781-x, (2018).

[7] I. Cho, Banach-Space Operators Acting on Semicircular Elements Induced by Orthogonal Projections, *Compl. Anal. Oper. Theo.,* 13, Issue 8, (2019) 4065–4115.

[8] I. Cho, Acting Semicircular Elements Induced by Orthogonal Projections on von Neumann Algebras, *Mathematics,* 5, 74, DOI:10.3390/math5040074, (2017).

[9] I. Cho, Semicircular-Like and Semicircular Laws on Banach $*$-Probability Spaces Induced by Dynamical Systems of the Finite Adele Ring, Adv. Oper. Theo., Special Issue: Trends in Operators on Banach Spaces (dedicated to Prof. S. Banach), *Adv. Oper. Theo.,* 4, no. 1, (2019) 24–70.

[10] I. Cho, and P. E. T. Jorgensen, *Semicircular Elements Induced by Projections on Separable Hilbert Spaces,* Monograph Ser., Operator Theory: Advances & Applications, Published by Birkhauser, Basel, (2019) To Appear.

[11] I. Cho, and P. E. T. Jorgensen, Banach $*$-Algebras Generated by Semicircular Elements Induced by Certain Orthogonal Projections, Opuscula Math., 38, no. 4, (2018) 501–535.

[12] I. Cho, and P. E. T. Jorgensen, Semicircular Elements Induced by p-Adic Number Fields, *Opuscula Math.,* 35, no. 5, (2017) 665–703.

[13] A. Connes, *Noncommutative Geometry,* ISBN: 0-12-185860-X, (1994) Academic Press (San Diego, CA).

[14] P. R. Halmos, *Hilbert Space Problem Books,* Grad. Texts in Math., 19, ISBN: 978-0387906850, (1982) Published by Springer.

[15] I. Kaygorodov, and I. Shestakov, Free Generic Poisson Fields and Algebras, *Comm. Alg.,* 46, issue 4, DOI:10.1080/00927872.2017.1358269, (2018).

[16] L. Makar-Limanov, and I. Shestakov, Polynomials and Poisson Dependence in Free Poisson Algebras and Free Poisson Fields, *J. Alg.,* vol. 349, issue 1, (2012) 372–379.

[17] A. Nica, and R. Speicher, *Lectures on the Combinatorics of Free Probability* (1-st Ed.), London Math. Soc. Lecture Note Ser., 335, ISBN-13:978-0521858526, (2006) Cambridge Univ. Press.

[18] I. Nourdin, G. Peccati, and R. Speicher, Multi-Dimensional Semicircular Limits on the Free Wigner Chaos, *Progr. Probab.,* 67, (2013) 211–221.

[19] V. Pata, The Central Limit Theorem for Free Additive Convolution, *J. Funct. Anal.,* 140, no. 2, (1996) 359–380.

[20] F. Radulescu, Random Matrices, Amalgamated Free Products and Subfactors of the C^*-Algebra of a Free Group of Nonsingular Index, *Invent. Math.,* 115, (1994) 347–389.

[21] F. Radulescu, Free Group Factors and Hecke Operators, notes taken by N. Ozawa, Proceed. 24-th Conference in Oper. Theo., Theta Advanced Series in Math., (2014) Theta Foundation.

[22] R. Speicher, Combinatorial Theory of the Free Product with Amalgamation and Operator-Valued Free Probability Theory, *Amer. Math. Soc. Mem.,* vol 132, no. 627, (1998).

[23] R. Speicher A Conceptual Proof of a Basic Result in the Combinatorial Approach to Freeness, Infinit. Dimention. *Anal. Quant. Prob. & Related Topics,* 3, (2000) 213–222.

[24] R. Speicher, and T. Kemp, Strong Haagerup Inequalities for Free R-Diagonal Elements, *J. Funct. Anal.,* 251, Issue 1, (2007) 141–173.

[25] R. Speicher, and U. Haagerup, Brown's Spectrial Distribution Measure for R-Diagonal Elements in Finite Von Neumann Algebras, *J. Funct. Anal.,* 176, Issue 2, (2000) 331–367.

[26] V. S. Vladimirov, p-Adic Quantum Mechanics, *Comm. Math. Phy.,* 123, no. 4, (1989) 659–676.

[27] V. S. Vladimirov, I. V. Volovich, and E. I. Zelenov, *p-Adic Analysis and Mathematical Physics,* Ser. Soviet & East European Math., vol 1, ISBN: 978-981-02-0880-6, (1994) World Scientific.

[28] D. Voiculescu, Aspects of Free Analysis, *Jpn. J. Math.,* 3, no. 2, (2008) 163–183.

[29] D. Voiculescu, Free Probability and the Von Neumann Algebras of Free Groups, *Rep. Math. Phy.,* 55, no. 1, (2005) 127–133.

[30] D. Voiculescu, K. Dykemma, and A. Nica, Free Random Variables, CRM Monograph Series, vol 1., ISBN-13: 978-0821811405, (1992) Published by Amer. Math. Soc.

2 Linear Positive Operators Involving Orthogonal Polynomials

P. N. Agrawal and Ruchi Chauhan
IIT Roorkee

CONTENTS

2.1 OPERATORS BASED ON ORTHOGONAL POLYNOMIALS

In recent years, there has been a significant increase in activities in approximation of continuous functions on the semi-real axis by the linear positive operators based on orthogonal polynomials. Jakimovski and Leviatan [21] initiated the study in this direction by defining the generalization of Szász operators based on Appell polynomials. Ismail [19] generalized the Szász operators by means of Sheffer polynomials.

Varma and Tasdelen [42] proposed Szász-type operators involving Charlier polynomials. Kajla and Agrawal [24] introduced the Szász-Durrmeyer-type operators involving Charlier polynomials. Varma et al. [41] proposed Szász-type operators based on Brenke-type polynomials. Sucu et al. [37] defined Szász-type operators involving Boas-Buck polynomials that include Brenke-type polynomials, Sheffer polynomials, and Appell polynomials. In the literature, several sequences of linear positive operators have been defined, involving orthogonal polynomials, and their Durrmeyer and Kantorovich-type variants have been investigated. Our aim in this chapter is to make a survey of the research work available on approximation by the linear positive operators defined by using orthogonal polynomials.

Throughout this chapter, we shall use the following notations and definitions.

2.1.1 NOTATIONS

$\mathbb{R}^+$	$(0,\infty)$,
$C[0,\infty)$	the space of all continuous functions on $[0,\infty)$,
$C_B[0,\infty)$	the space of all continuous and bounded functions on $[0,\infty)$,
E_1	$\left\{f:[0,\infty)\to\mathbb{R}: x\in[0,\infty), \lim_{x\to\infty}\frac{f(x)}{1+x^2} \text{ exists}\right\}$,
$\tilde{C}[0,\infty)$	the space of all uniformly continuous functions on $[0,\infty)$,
$\tilde{C}^r[0,\infty)$	the space of r-times differentiable functions such that $f^{(r)}$ is uniformly continuous on $[0,\infty)$,
$\tilde{C}_B[0,\infty)$	the space of all real-valued bounded and uniformly continuous functions f on $[0,\infty)$, endowed with the norm $\parallel f\parallel=\sup_{x\in[0,\infty)}\lvert f(x)\rvert$,
$C_\gamma[0,\infty)$	$\{f\in C[0,\infty): \lvert f(x)\rvert\le M_f(1+x^\gamma), \gamma>0\}$,
$B_2[0,\infty)$	the space of all real-valued functions on $[0,\infty)$ satisfying the condition $\lvert f(x)\rvert\le M_f(1+x^2)$, where M_f is a positive constant depending only on f and $1+x^2$ is a weight function,
$C_2[0,\infty)$	the space of all continuous functions in $B_2[0,\infty)$ with the norm $\lVert f\rVert_2:=\sup_{x\in[0,\infty)}\frac{\lvert f(x)\rvert}{1+x^2}$
$C_2^*[0,\infty)$	$\left\{f\in C_2[0,\infty): \lim_{x\to\infty}\frac{\lvert f(x)\rvert}{1+x^2} \text{ is finite}\right\}$,
E_2	$\{f:[0,\infty)\to\mathbb{R}: \lvert f(x)\rvert\le Me^{Ax}, \text{ for some } A,M\in\mathbb{R}^+\}$,
E_3	$\{f:[0,\infty)\to\mathbb{R}: \lvert f(x)\rvert=\lvert\int_0^x f(s)ds\rvert\le Ke^{Bx}, \text{ for some } B,K\in\mathbb{R}^+\}$.

2.1.2 DEFINITIONS

Let $f\in\tilde{C}[0,\infty)$ and $\delta>0$. The modulus of continuity $\omega(f;\delta)$ of the function f is defined by

$$\omega(f;\delta)=\sup_{0<|h|<\delta}\ \sup_{x,x+h\in[0,\infty)}|f(x+h)-f(x)|. \tag{2.1}$$

Lipschitz-type maximal function of order r introduced is defined by Lenze [29] as

$$\widetilde{\omega}_r(f,x)=\sup_{t\neq x,\ t\in[0,\infty)}\frac{|f(t)-f(x)|}{|t-x|^r},\quad x\in[0,\infty) \text{ and } r\in(0,1]. \tag{2.2}$$

In what follows, let $\tilde{C}_B[0,\infty)$ be the space of all real-valued bounded and uniformly continuous functions f on $[0,\infty)$, endowed with the norm $||f||_{\tilde{C}_B[0,\infty)}=\sup_{x\in[0,\infty)}|f(x)|$.

Further, let us define the following Peetre's K-functional:

$$K_2(f,\delta) = \inf_{g\in W^2}\{||f-g||+\delta||g''||\},\ \delta>0,$$

where $W^2 = \{g \in \tilde{C_B}[0,\infty) : g', g'' \in \tilde{C_B}[0,\infty)\}$ and the norm

$$\| f \|_{W^2}=\| f \|_{\tilde{C_B}[0,\infty)} + \| f' \|_{\tilde{C_B}[0,\infty)} + \| f'' \|_{\tilde{C_B}[0,\infty)} .$$

By ([9], p. 177, Theorem 2.4), there exists an absolute constant $M > 0$ such that

$$K_2(f,\delta) \le M\left\{\omega_2(f,\sqrt{\delta}) + \min(1,\delta)\ \| f \|_{\tilde{C_B}[0,\infty)}\right\}, \tag{2.3}$$

where the second-order modulus of smoothness is defined as

$$\omega_2(f,\sqrt{\delta}) = \sup_{0<|h|\le\sqrt{\delta}}\ \sup_{x\in[0,\infty)} |f(x+2h)-2f(x+h)+f(x)|.$$

The modulus $\omega_{\phi^\tau}(f,t), 0 \le \tau \le 1$, is given by

$$\omega_{\phi^\tau}(f,t) = \sup_{0\le h\le t}\ \sup_{x\pm\frac{h\phi^\tau(x)}{2}\in[0,\infty)} \left|f\left(x+\frac{h\phi^\tau(x)}{2}\right) - f\left(x-\frac{h\phi^\tau(x)}{2}\right)\right|. \tag{2.4}$$

Now for $0 < r \le 1$, the Lipschitz-type space [26] is defined as:

$$Lip_M^*(r) := \left\{f \in C[0,\infty) : |f(t)-f(x)| \le M_f\frac{|t-x|^r}{(t+x)^{\frac{r}{2}}}; t\in(0,\infty), x>0\right\}, \tag{2.5}$$

for some $M_f > 0$.

For $f \in C_2[0,\infty)$, Ispir and Atakut [20] introduced the weighted modulus of continuity as follows:

$$\Omega(f;\delta) = \sup_{0\le|h|<\delta, x\in[0,\infty)} \frac{f(x+h)-f(x)|}{(1+x^2)(1+h^2)}. \tag{2.6}$$

If $f \in C_2^*[0,\infty)$, then $\lim_{\delta\to 0}\Omega(f;\delta) = 0$.

2.1.3 APPELL POLYNOMIALS

Appell [4] introduced a sequence of polynomials $P_n(x)$ of degree n which satisfies the differential equation

$$DP_n(x) = nP_{n-1}(x),\ D \equiv \frac{d}{dx},$$

known as Appell polynomials. In [35], Sheffer extended the class of Appell polynomials and called these polynomials as zero-type polynomials. The very well-known Szász operator is defined in the following.

For $f \in C[0,\infty)$ and $n \in \mathbb{N}$, we have

$$S_n(f;x) = e^{-nx} \sum_{k=0}^{\infty} \frac{(nx)^k}{k!} f\left(\frac{k}{n}\right) \quad x \geq 0, \tag{2.7}$$

whenever the series on the right side converges. Using Appell polynomials, Jakimovski and Leviatan [21] introduced a generalization of the Szász operators as

$$P_n(f;x) = \frac{e^{-nx}}{g(1)} \sum_{k=0}^{\infty} p_k(nx) f\left(\frac{k}{n}\right), \tag{2.8}$$

where

$$g(u)e^{ux} = \sum_{k=0}^{\infty} p_k(x) u^k \tag{2.9}$$

is the generating function for the Appell polynomials $p_k(x) \geq 0$, with $g(u) = \sum_{n=0}^{\infty} a_n u^n$, $|u| < r$, $r > 1$, and $g(1) \neq 0$, and established several approximation properties of these operators.
Later, Ismail [19] gave another generalization of Szász operators by means of Sheffer polynomials. Let $A(z) = \sum_{k=0}^{\infty} a_k z^k$, $(a_0 \neq 0)$, and $H(z) = \sum_{k=1}^{\infty} h_k z^k$ $(h_k \neq 0$ $\forall k$ be the analytic functions in the disk $|z| \leq R$ $(R > 1)$, where a_k and h_k are real. The Sheffer polynomials $p_k(x)$ have the generating functions of the form $A(t)e^{xH(t)} = \sum_{k=0}^{\infty} p_k(x) t^k$, $|t| < R$. By assuming $p_k(x) \geq 0$, $\forall x \in [0,\infty)$, $A(1) \neq 0$, and $H'(1) = 1$, Ismail introduced and studied the convergence properties of the positive linear operators defined by

$$T_n(f;x) = \frac{e^{-nx}}{A(1)} \sum_{k=0}^{\infty} p_k(nx) f\left(\frac{k}{n}\right).$$

In [8], Ciupa considered the following operators of general form

$$D_n(f;x) = \frac{e^{-nx}}{g(1)} \sum_{k=0}^{\infty} p_k(nx) A_{k,n}(f),\ x \geq 0,$$

where $p_k(x)$ are the Appell polynomials defined by the relation

$$g(u)e^{ux} = \sum_{k=0}^{\infty} p_k(u) u^k,\ g(1) \neq 0.$$

The main result of this paper is that the rth derivative of $D_n(f;x)$ is given by

$$(D_n(f))^{(r)}(x) = \frac{n^r}{g(1)} \sum_{k=0}^{\infty} e^{-nx} p_k(nx) \sum_{i=0}^{r} (-1)^i \binom{r}{i} A_{k+r-i,n}(f).$$

Choosing different expressions for $A_{k,n}(f)$, the known formulas for the cases such as the Jakimovski-Leviatan operators, Favard-Szász operators, the Durrmeyer variant of

Jakimovski-Leviatan operators, and the Durrmeyer variant of Favard-Szász operators may be obtained.

Brenke-type polynomials [7] have the generating functions of the type

$$A(t)B(xt) = \sum_{k=0}^{\infty} p_k(x)t^k, \tag{2.10}$$

where $A(t)$ and $B(t)$ are the analytic functions given by

$$A(t) = \sum_{i=0}^{\infty} a_i t^i, \; a_0 \neq 0 \tag{2.11}$$

$$B(t) = \sum_{j=0}^{\infty} b_j t^j, \; b_j \neq 0, \; \forall\, j \tag{2.12}$$

and $p_k(x) = \sum_{i=0}^{k} a_{k-i} b_i x^i$, $k = 0,1,2...$, which includes Appell polynomials for $B(t) = e^t$, as a particular case. Assuming that $A(1) \neq 0$, $\dfrac{a_{k-i}b_i}{A(1)} \geq 0$, $0 \leq i \leq k$, $k = 0,1,2...$, $B : [0,\infty) \to (0,\infty)$ and (2.10)–(2.12) converge for $|t| < R$ $(R > 1)$, Verma et al. [40] proposed the following operators:

$$L_n(f;x) = \frac{1}{A(1)B(nx)} \sum_{k=0}^{\infty} p_k(nx) f\left(\frac{k}{n}\right), \; x \in [0,\infty) \tag{2.13}$$

and studied some results concerning the rate of convergence of these operators. In particular, if $B(t) = e^t$ and $A(t) = g(t)$, then the operators given by (2.13) reduce to the operators defined by (2.8).

2.1.4 BOAS-BUCK-TYPE POLYNOMIALS

For $f \in C[0,\infty)$, Sucu et. al [37] considered the Szász operators based on Boas-Buck-type polynomials as follows:

$$B_n(f;x) = \frac{1}{A(1)G(nxH(1))} \sum_{k=0}^{\infty} p_k(nx) f\left(\frac{k}{n}\right), \; x \geq 0, \; n \in \mathbb{N},$$

where the generating function of the Boas-Buck-type polynomials is given by

$$A(u)G(xH(u)) = \sum_{k=0}^{\infty} p_k(x)u^k,$$

and $A(u)$, $G(u)$, and $H(u)$ are the analytic functions described by $A(u) = \sum_{k=0}^{\infty} a_k u^k$, $(a_0 \neq 0$, $G(u) = \sum_{k=0}^{\infty} g_k u^k$, $(g_k \neq 0, \forall k)$, and $H(u) = \sum_{k=1}^{\infty} h_k u^k$, $(h_1 \neq 0$, and studied some approximation properties of these operators. As a particular case, Boas-Buck polynomials include Sheffer polynomials and Appell polynomials.

Yilik et al. [44] estimated the approximation degree of the operators B_n for bounded variation functions using some results of the probability theory. The authors also investigated the Voronovskaja and Grüss-Voronovskaja-type theorems in the quantitative form.

2.1.5 CHARLIER POLYNOMIALS

In 1950, Szász [38] introduced the following linear positive operators:

$$S_n(f;x) = e^{-nx} \sum_{k=0}^{\infty} \frac{(nx)^k}{k!} f\left(\frac{k}{n}\right), \tag{2.14}$$

where $x \in [0,\infty)$ and $f(x)$ is a continuous function on $[0,\infty)$ whenever the above sum converges uniformly. These polynomials [18] have the generating functions of the form

$$e^t\left(1-\frac{t}{a}\right)^u = \sum_{k=0}^{\infty} C_k^{(a)}(u)\frac{t^k}{k!}, \ |t|<a, \tag{2.15}$$

where $C_k^{(a)}(u) = \sum_{r=0}^{k} \binom{k}{r}(-u)_r\left(\frac{1}{a}\right)^r$ and $(m)_0 = 1, (m)_j = m(m+1)\cdots(m+j-1)$ for $j \geq 1$. For $C_A[0,\infty)$, Varma and Taşdelen [42] defined the following Szász-type operators involving Charlier polynomials

$$\mathcal{L}_n(f;x,a) = e^{-1}\left(1-\frac{1}{a}\right)^{(a-1)nx} \sum_{k=0}^{\infty} \frac{C_k^{(a)}(-(a-1)nx)}{k!} f\left(\frac{k}{n}\right), \tag{2.16}$$

where $a > 1$ and $x \in [0,\infty)$, and studied the degree of approximation of these operators.

2.1.6 APPROXIMATION BY APPELL POLYNOMIALS

For $f \in C_B[0,\infty)$, Karaisa [26] introduced a Durrmeyer-type modification based on Appell polynomials as follows:

$$\begin{aligned} L_n(f;x) &= \frac{e^{-nx}}{g(1)} \sum_{k=1}^{\infty} \frac{p_k(nx)}{B(n+1,k)} \int_0^{\infty} b_{n,k}(t) f(t)dt \\ &+ \frac{e^{-nx}}{g(1)} a_0 f(0), \ x \geq 0, \end{aligned} \tag{2.17}$$

where $B(k+1,n)$ is the beta function, $b_{n,k}(t) = \frac{t^{k-1}}{(1+t)^{n+k+1}}$, and studied the local approximation properties and the convergence in a weighted space of functions. A Voronovskaja-type theorem for these operators was also proved. Gupta and Agrawal [14] established the rate of convergence for a Lipschitz-type space and obtained the degree of approximation in terms of Lipschitz-type maximal function for the Durrmeyer-type modification of Jakimovski-Leviatan operators based on Appell polynomials.

2.1.7 APPROXIMATION BY OPERATORS INCLUDING GENERALIZED APPELL POLYNOMIALS

For $f \in C[0,\infty)$, Icoz et al. [17] considered the linear operators

$$M_n(f;x) := \frac{1}{A(g(1))B(nxg(1))} \sum_{k=0}^{\infty} p_k(nx) f\left(\frac{k}{n}\right), \tag{2.18}$$

where $p_k(x)$ are the generalized Appell polynomials [4] having the generating functions of the following form:

$$A(g(t))B(xg(t)) = \sum_{k=0}^{\infty} p_k(x) t^k \tag{2.19}$$

and A, B, and g are the analytic functions such that

$$A(t) = \sum_{k=0}^{\infty} a_k t^k \ (a_0 \neq 0), \quad B(t) = \sum_{k=0}^{\infty} b_k t^k \ (b_k \neq 0), \quad g(t) = \sum_{k=1}^{\infty} g_k t^k \ (g_1 \neq 0). \tag{2.20}$$

Assuming that the generalized Appell polynomials (2.19) satisfy

1. $A(g(1)) \neq 0,\ g'(1) = 1,\ p_k(x) \geq 0\ k = 0,1,2,\ldots,$
2. $B : \mathbb{R} \to (0,\infty),$
3. (2.19) and the power series (2.20) converges for $|t| < R\ (R > 1),$

it turns out that the linear operators M_n defined by (2.18) are positive.

Remark 2.1. *Let $g(t) = t$, the operators* (2.18) *reduce to the operators L_n involving Brenke-type polynomials.*

Remark 2.2. *Let $g(t) = t$ and $B(t) = e^t$, one can get the operators* (2.8). *In addition, if we choose $A(t) = 1$, we meet the well-known Szász operators* (2.7).

Icoz et al. [17] proved the qualitative and quantitative approximation theorems. Icoz et al. [17] obtained the qualitative convergence result by means of M_n operators with the help of universal Korovkin-type property with respect to positive linear operators. Next, they stated the quantitative results for estimating the error of approximation using the classical approach, the second modulus of continuity, and Peetre's K-functional in the continuous functions space and the Lipschitz class.

Neer et al. [32] defined Baskakov-Durrmeyer-type operators based on the generalized Appell polynomials and established some results in weighted approximation besides the quantitative Voronovskaja and Grüss-Voronovskaja-type theorems. The approximation for functions having derivatives of bounded variation was also studied.

2.1.8 SZÁSZ-TYPE OPERATORS INVOLVING MULTIPLE APPELL POLYNOMIALS

First, let us recall the definition of multiple Appell polynomials [28]. A set of polynomials $\{p_{k_1,k_2}(x)\}_{k_1,k_2=0}^{\infty}$ with degree (k_1+k_2) for $k_1,k_2 \geq 0$, is called multiple polynomial system (multiple PS), and a multiple PS is called multiple Appell if it is generated by the relation

$$A(t_1,t_2)e^{x(t_1+t_2)} = \sum_{k_1=0}^{\infty}\sum_{k_2=0}^{\infty}\frac{p_{k_1,k_2}(x)}{k_1!k_2!}t_1^{k_1}t_2^{k_2}, \tag{2.21}$$

where $A(t_1,t_2)$ is given by

$$A(t_1,t_2) = \sum_{k_1=0}^{\infty}\sum_{k_2=0}^{\infty}\frac{a_{k_1,k_2}}{k_1!k_2!}t_1^{k_1}t_2^{k_2}, \tag{2.22}$$

with $A(0,0) = a_{0,0} \neq 0$.

Theorem 2.1. *For multiple PS, $\{p_{k_1,k_2}(x)\}_{k_1,k_2=0}^{\infty}$, the following statements are equivalent:*

a. $\{p_{k_1,k_2}(x)\}_{k_1,k_2=0}^{\infty}$ *is a set of multiple Appell polynomials.*
b. *There exists a sequence* $\{a_{k_1,k_2}\}_{k_1,k_2=0}^{\infty}$ *with* $a_{0,0} \neq 0$ *such that*

$$p_{k_1,k_2}(x) = \sum_{r_1=0}^{\infty}\sum_{r_2=0}^{\infty}\binom{k_1}{r_1}\binom{k_2}{r_2}a_{k_1-r_1,k_2-r_2}x^{r_1+r_2}.$$

c. *For every* $k_1+k_2 \geq 1$, *we have*

$$p'_{k_1,k_2}(x) = k_1 p_{k_1-1,k_2}(x) + k_2 p_{k_1,k_2-1}(x).$$

Utilizing these polynomials, for any $f \in C[0,\infty)$, Varma [40] defined a sequence of linear positive operators as

$$K_n(f;x) = \frac{e^{-nx}}{A(1,1)}\sum_{k_1=0}^{\infty}\sum_{k_2=0}^{\infty}\frac{p_{k_1,k_2}\left(\frac{nx}{2}\right)}{k_1!k_2!}f\left(\frac{k_1+k_2}{n}\right), \tag{2.23}$$

provided $A(1,1) \neq 0$, $\frac{a_{k_1,k_2}}{A(1,1)} \geq 0$ for $k_1,k_2 \in \mathbb{N}$, and series (2.21) and (2.22) converge for $|t_1| < R_1$, $|t_2| < R_2$ $(R_1,R_2 > 1)$, respectively.

Remark 2.3. *Let us note the following special cases:*

1. *For* $t_2 = 0$, *the generating functions given by* (2.21) *reduce to the generating functions for the Appell polynomials given by* (2.9).
2. *For* $t_2 = 0$ *and* $A(t_1,0) = 1$, *from the generating function given by* (2.21), *we easily find* $p_k(x) = x^k$. *From this fact, one can obtain Szász operators* (2.7).

We use the following notation for the partial derivatives:

$$\frac{\partial A}{\partial t_i} = A_{t_i} \text{ and } A_{t_it_j} = \frac{\partial^2 A}{\partial t_i \partial t_j} \quad i,j = 1,2,3,4.$$

From definition (2.23) and relation (2.21), we obtain the following auxiliary result

Lemma 2.1. *[40] The operator (2.23) satisfies the following inequalities:*

1. $K_n(1;x) = 1$;
2. $K_n(s;x) = x + \frac{A_{t_1}(1,1)+A_{t_2}(1,1)}{nA(1,1)}$;
3. $K_n(s^2;x) = x^2 + \frac{x}{n}\left(1 + \frac{2(A_{t_1}(1,1)+A_{t_2}(1,1))}{A(1,1)}\right)$
$+ \frac{1}{n^2A(1,1)}\{A_{t_1}(1,1) + A_{t_2}(1,1) + A_{t_1t_1}(1,1) + 2A_{t_1t_2}(1,1) + A_{t_2t_2}(1,1)\}$,

where $x \geq 0$. The following theorem shows that the operator defined by (2.23) is an approximation process for continuous functions contained in E.

Theorem 2.2. *[40] Let $f \in C[0,\infty) \cap E_1$. Then,*

$$\lim_{n\to\infty} K_n(f;x) = f(x),$$

the convergence being uniform in each compact subset of $[0,\infty)$.

Theorem 2.3. *[40] Let $f \in \tilde{C}[0,\infty) \cap E_1$, we have the following inequality for the operator* (2.23)

$$|K_n(f;x) - f(x)|$$

$$\leq \left\{1 + \sqrt{x + \frac{1}{nA(1,1)}\{A_{t_1}(1,1) + A_{t_2}(1,1) + A_{t_1t_1}(1,1) + 2A_{t_1t_2}(1,1) + A_{t_2t_2}(1,1)\}}\right\}$$
$$\times\, \omega\left(f; \frac{1}{\sqrt{n}}\right).$$

The following theorem shows that the operators (2.23) possess the simultaneous approximation.

Theorem 2.4. *[40] For $f \in \tilde{C}^r[0,\infty) \cap E_1$, the following inequality holds:*

$$|K_n^{(r)}(f;x) - f^{(r)}(x)|$$

$$\leq \left\{1 + \sqrt{x + \frac{1}{nA(1,1)}\{A_{t_1}(1,1) + A_{t_2}(1,1) + A_{t_1t_1}(1,1) + 2A_{t_1t_2}(1,1) + A_{t_2t_2}(1,1)\}}\right\}$$
$$\times\, \omega\left(f^{(r)}; \frac{1}{\sqrt{n}} + \frac{r}{n}\right) + \omega\left(f^{(r)}; \frac{r}{n}\right).$$

2.1.9 KANTOROVICH-TYPE GENERALIZATION OF K_n OPERATORS

Also, Varma [40] gave a Kantorovich-type generalization of the operators K_n (2.23) with the same restrictions

$$K_n^*(f;x) = \frac{ne^{-nx}}{A(1,1)} \sum_{k_1=0}^{\infty}\sum_{k_2=0}^{\infty} \frac{p_{k_1,k_2}\left(\frac{nx}{2}\right)}{k_1!k_2!} \int_{\frac{k_1+k_2}{n}}^{\frac{k_1+k_2+1}{n}} f(s)\mathrm{ds}. \tag{2.24}$$

In order to study the approximation properties of the operators K_n^*, the first author gave a lemma for the moments of these operators:

Lemma 2.2. *[40] The operators (2.24) satisfy the following equalities:*

1. $K_n^*(1;x) = 1;$
2. $K_n^*(s;x) = x + \frac{A_{t_1}(1,1)+A_{t_2}(1,1)}{nA(1,1)} + \frac{1}{2n};$
3. $K_n^*(s^2;x) = x^2 + \frac{2x}{n}\left(1 + \frac{(A_{t_1}(1,1)+A_{t_2}(1,1))}{A(1,1)}\right)$ $+\frac{1}{n^2A(1,1)}\{2(A_{t_1}(1,1)+A_{t_2}(1,1)+A_{t_1t_2}(1,1))A_{t_1t_1}(1,1)+A_{t_2t_2}(1,1)\}+\frac{1}{3n^2};$
4. $K_n^*(s^3;x) = \frac{1}{4n^3A(1,1)}\left[\left(36A_{t_1}(1,1)+36A_{t_2}(1,1)+14A(1,1)+12A_{t_2t_2}(1,1)+ 12A_{t_1t_1}(1,1)+24A_{t_1t_2}(1,1)\right)nx+\left(48A_{t_1}(1,1)+48A_{t_2}(1,1)+72A(1,1)\right)\frac{n^2x^2}{4} +32A(1,1)\frac{n^3x^3}{8}+\left(14A_{t_1}(1,1)+14A_{t_2}(1,1)+18A_{t_2t_2}(1,1)+18A_{t_1t_1}(1,1)+ 36A_{t_1t_2}(1,1)+4A_{t_2t_2}(1,1)+4A_{t_1t_1t_1}(1,1)+12A_{t_1t_2t_2}(1,1)+12A_{t_1t_1t_2}(1,1)\right)\right];$
5. $K_n^*(s^4;x) = \frac{1}{n^4A(1,1)}\left[\frac{nx}{2}\left(12A(1,1)+60A_{t_1}(1,1)+60A_{t_2}(1,1)+48A_{t_1t_1}(1,1)+ 48A_{t_2t_2}(1,1)+8A_{t_1t_1t_1}(1,1)+8A_{t_2t_2t_2}(1,1)+96A_{t_1t_2}(1,1)+12A_{t_1t_1t_2}(1,1)+ 12A_{t_1t_2t_2}(1,1)\right)+\frac{n^2x^2}{4}\left(54A(1,1)+78A_{t_2}(1,1)+78A_{t_2}(1,1)+18A_{t_1t_1}(1,1)+ 18A_{t_2t_2}(1,1)+24A_{t_1t_2}(1,1)\right)+\frac{n^3x^3}{8}\left(64A(1,1)+32A_{t_1}(1,1)+32A_{t_2}(1,1)\right)+ n^4x^4A(1,1)+4A_{t_1}(1,1)+4A_{t_2}(1,1)+13A_{t_1t_1}(1,1)+13A_{t_2t_2}(1,1)+ 9A_{t_1t_1t_1}(1,1)+9A_{t_2t_2t_2}(1,1)+6A_{t_1t_1t_2t_2}(1,1)+6A_{t_1t_1t_2}(1,1)+6A_{t_1t_2t_2}+16A_{t_1t_1}+ 6A_{t_1t_2t_2}+\frac{1}{5}\right].$

In the following, the authors show that the uniform convergence of $K_n^*(f)$ to f is established.

Theorem 2.5. *[40] Let $f \in C[0,\infty) \cap E_1$. Then,*

$$\lim_{n\to\infty} K_n^*(f;x) = f(x),$$

the convergence being uniform in each compact subset of $[0,\infty)$.

In the next theorem, the authors obtained the order of approximation $K_n^*(f)$ to f.

Theorem 2.6. *[40] Let $f \in \tilde{C}[0,\infty) \cap E_1$, we have the following inequality for the operator* (2.24)

$|K_n^*(f;x)-f(x)|$

$$\le\left\{1+\sqrt{x+\frac{1}{nA(1,1)}\{2(A_{t_1}(1,1)+A_{t_2}(1,1)+A_{t_1t_2}(1,1))+A_{t_1t_1}(1,1)+A_{t_2t_2}(1,1)\}+\frac{1}{3n}}\right\}$$
$$\times\,\omega\left(f;\frac{1}{\sqrt{n}}\right).$$

Lemma 2.3. *[15] For all $x\ge 0$ and $n>2$, we have*

$$K_n^*(|s-x|;x)\le\sqrt{\gamma_n(x)},$$

where $\gamma_n(x)=K_n^*((s-x)^2;x)$.

Theorem 2.7. *[15] Let $0<r\le 1$ and $f\in Lip_M^*(r)$. Then for all $x>0$ and $n>2$, we have*

$$|K_n^*(f;x)-f(x)|\le M\left(\frac{\gamma_n(x)}{x}\right)^{\frac{r}{2}},$$

where $\gamma_n(x)$ is defined as in Lemma 2.3.

In order to study the estimate of the error in terms of the second-order modulus of continuity via Peetre's K-functional, we define an auxiliary operator

$$K_n^{**}(f;x)=K_n^*(f;x)+f(x)-f\left(x+\frac{A_{t_1}(1,1)+A_{t_2}(1,1)}{nA(1,1)}+\frac{1}{2n}\right),\tag{2.25}$$

Applying Lemma 2.2, we have $K_n^{**}(1;x)=1$ and $K_n^{**}(t;x)=x$.

Theorem 2.8. *[15] Let $f\in\tilde{C}_B[0,\infty)$. Then for all $x\ge 0$, the following inequality holds:*

$$|K_n^*(f;x)-f(x)|\le C\omega_2(f,\sqrt{\psi_n(x)})+\omega\left(f;\frac{2(A_{t_1}(1,1)+A_{t_2}(1,1))+A(1,1)}{2A(1,1)n}\right),$$

where $\psi_n(x)=\left(\gamma_n(x)+\left(\frac{2(A_{t_1}(1,1)+A_{t_2}(1,1))+A(1,1)}{2A(1,1)n}\right)^2\right).$

In order to obtain the next result, the authors adopt an approach of Steklov mean. For $f\in\tilde{C}_B[0,\infty)$, the Steklov mean is defined as

$$f_h(x)=\frac{4}{h^2}\int_0^{\frac{h}{2}}\int_0^{\frac{h}{2}}[2f(x+\xi+\eta)-f(x+2(\xi+\eta))]d\xi d\eta.\tag{2.26}$$

The Steklov mean verifies the following inequalities:

1. $\|f_h-f\|\le\omega_2(f,h)$;
2. $f_h',f_h''\in\tilde{C}_B[0,\infty)$ and $\|f_h'\|\le\frac{5}{n}\omega(f,h),\|f_h''\|\le\frac{9}{h^2}\omega_2(f,h)$.

Theorem 2.9. *[15] Let $f \in \tilde{C}_B[0,\infty)$. Then, for every $x \geq 0$, the following inequality holds:*

$$|K_n^*(f;x) - f(x)| \leq 5\omega(f, \sqrt{\gamma_n(x)}) + \frac{13}{2}\omega_2(f, \sqrt{\gamma_n(x)}).$$

Theorem 2.10. *[15] For any $f \in C_B^1[0,\infty)$ and $x \in [0,\infty)$, we have*

$$|K_n^*(f,x) - f(x)| \leq |f'(x)| \left| \frac{2(A_{t_1}(1,1) + A_{t_2}(1,1)) + A(1,1)}{2A(1,1)n} \right| + 2\omega(f',\delta)\sqrt{\gamma_n(x)}.$$

Theorem 2.11. *[15] Let $f \in \tilde{C}_B[0,\infty)$, then for every $x \in (0,\infty)$, we have*

$$|K_n^*(f;x) - f(x)| \leq C\omega_\tau\left(f; \sqrt{\frac{\mu}{n}}\right).$$

where $\mu \geq \max\left\{1, \left(\frac{2A_{t_1}(1,1)+2A_{t_2}(1,1)+2A_{t_1t_2}(1,1)A_{t_1t_1}(1,1)+A_{t_2t_2}(1,1)}{A(1,1)} + \frac{1}{3}\right)\right\}$ is any real constant.

In the next result, the authors establish a quantitative Voronovskaja-type theorem by means of the weighted modulus of continuity defined in (2.6).

Theorem 2.12. *Let $f \in C_2^*[0,\infty)$ such that $f'' \in C_2^*[0,\infty)$. Then, the inequality*

$$\left| n(K_n^*(f;x) - f(x)) - f'(x)\left(\frac{A_{t_2}(1,1) + A_{t_1}(1,1)}{A(1,1)} + \frac{1}{2}\right) - \frac{xf''(x)}{2} \right.$$
$$\left. - \frac{f''(x)}{2n}\left(\frac{2A_{t_1}(1,1) + 2A_{t_2}(1,1) + 2A_{t_1t_2t_2}(1,1) + A_{t_1t_2}(1,1) + A_{t_1t_1t_2t_2}(1,1)}{A(1,1)} + \frac{1}{3}\right) \right|$$
$$= O(1)\Omega(f''; 1/\sqrt{n}), \quad as \ n \to \infty$$

holds true.

Now the Grüss-Voronovskaja-type theorem for the operators K_n^*, we state.

Theorem 2.13. *Let $f', g', f'', g'' \in C_2^*[0,\infty)$, then for every $x \in [0,\infty)$*

$$\lim_{n\to\infty} n(K_n^*(fg)(x) - K_n^*(f;x)K_n^*(g;x)) = xf'(x)g'(x).$$

Ansari et al. [3] introduced Jakimovski-Leviatan-Durrmeyer-type operators involving multiple Appell polynomials, and investigated Korovkin-type theorem and the order of convergence by means of the modulus of continuity. Some results in the weighted approximation were also discussed in this paper.

2.1.10 KANTOROVICH VARIANT OF SZÁSZ OPERATORS BASED ON BRENKE-TYPE POLYNOMIALS

For $f \in C[0,\infty)$, Varma et al. [41] defined another generalization of the Szász operators by using Brenke-type polynomials, which are as follows:

$$L_n(f;x) = \frac{1}{A(1)B(nx)} \sum_{k=0}^{\infty} p_k(nx) f\left(\frac{k}{n}\right), \tag{2.27}$$

where $x \geq 0$ and $n \in \mathbb{N}$. Taşdelen et al. [39] introduced the Kantorovich modification of the operators (2.27), which is as follows:

$$K_n(f;x) = \frac{n}{A(1)B(nx)} \sum_{k=0}^{\infty} p_k(nx) \int_{k/n}^{(k+1)/n} f(t)\,dt, \tag{2.28}$$

where $n \in \mathbb{N}, \quad x \geq 0$, and $f \in C[0,\infty)$. If we take $B(t) = e^t$ and $A(t) = 1$, then it reduces to the Szász Mirakyan-Kantorovich operators defined by

$$K_n(f;x) = ne^{-nx} \sum_{k=0}^{\infty} \frac{(nx)^k}{k!} \int_{k/n}^{(k+1)/n} f(t)\,dt. \tag{2.29}$$

Mursaleen and Ansari [30] introduced a Chlodowsky-type generalization of Szász operators defined by means of the Brenke-type polynomials, proved basic convergence theorem, and also established the order of approximation with the aid of the second-order modulus of continuity and the Peetre's K-functional.

Öksüzer et al. [33] estimated the rate of convergence of the operators (2.28) for the functions of bounded variations using the techniques of probability theory. The authors rewrite the operator (2.28) as:

$$K_n(f;x) = \int_0^{\infty} f(t) M_n(x,t)dt, \tag{2.30}$$

where

$$M_n(x,t) = \frac{n}{A(1)B(nx)} \sum_{k=0}^{\infty} p_k(nx)\chi_{n,k}(t)$$

$\chi_{n,k}(t)$ is the characteristic function of the interval $[k/n,(k+1)/n]$ with respect to $I = [0,\infty)$.

Lemma 2.4. *[39] For the operator K_n, there hold the equalities*

1. $K_n(1;x) = 1$;
2. $K_n(t;x) = \dfrac{B'(nx)}{B(nx)} + \dfrac{aA'(1)+A(1)}{2nA(1)}$;
3. $K_n(t^2;x) = \dfrac{B''(nx)}{B(nx)}x^2 + \dfrac{2B'(nx)[A'(1)+A(1)]}{nA(1)B(nx)}x + \dfrac{1}{n^2A(1)}\left\{A''(1)+2A'(1)+\dfrac{A(1)}{3}\right\}$;
4. $K_n((t-x)^2;x) = \left\{\dfrac{B''(nx)-2B'(nx)+B(nx)}{B(nx)}\right\}x^2 + \left\{\dfrac{2A'(1)[B'(nx)-B(nx)]+A(1)[2B'(nx)-B(nx)]}{}\right\}x + \dfrac{1}{n^2A(1)}\left\{A''(1)+2A'(1)+\dfrac{A(1)}{3}\right\}$.

Theorem 2.14. *[39] Let*

$$\lim_{y\to\infty}\frac{B'(y)}{B(y)}=1,\quad \lim_{y\to\infty}\frac{B''(y)}{B(y)}=1.$$

If $f\in C[0,\infty)\cap E_1$, *then*

$$\lim_{n\to\infty}K_n(f;x)=f(x),$$

and the operators K_n *converge uniformly in each compact subset of* $[0,\infty)$.

Next theorem gives the rate of convergence for the operator K_n.

Theorem 2.15. *[39] Let* $f\in\tilde{C}[0,\infty)\cap E$, *the operators* K_n *satisfy the following inequality:*

$$|K_n(f;x)-f(x)|\le 2\omega\left(f;\sqrt{\delta_n(x)}\right),$$

where $\delta_n(x)=K_n\left((t-x)^2;x\right)$.

Theorem 2.16. *[39] Let* $f\in C_B^2[0,\infty)$. *There holds the inequality*

$$|K_n(f;x)-f(x)|\le\zeta||f||_{C_B^2},$$

where

$$\begin{aligned}\zeta=\zeta_n(x)=&\left\{\frac{B''(nx)-2B'(nx)+B(nx)}{2B(nx)}\right\}+\left\{\frac{2A'(1)\left(B'(nx)-B(nx)\right)}{2nA(1)B(nx)}\right.\\&\left.+\frac{A(1)\left(2(n+1)B'(nx)-(2n+1)B(nx)\right)}{2nA(1)B(nx)}\right\}x+\frac{2A'(1)+A(1)}{2nA(1)}\\&+\frac{1}{2n^2A(1)}\left\{A''(1)+2A(1)+\frac{A(1)}{3}\right\}.\end{aligned}$$

Theorem 2.17. *[39] Let* $f\in\tilde{C}_B[0,\infty)$. *There holds*

$$|K_n(f;x)-f(x)|\le 2\,M\left\{\omega_2(f;\sqrt{\delta})+\min(1,\delta)||f||_{C_B}\right\},$$

where

$$\delta=\delta_n(x)=\frac{1}{2}\zeta_n(x),$$

and $M>0$ *is a constant, which is independent of* f *and* δ.

Atakut and Buyukyazici [6] considered a generalization of the Kantorovich-Szász-type operators involving Brenke-type polynomials as follows:

$$L_n^{\alpha_n,\beta_n}(f;x)=\frac{\beta_n}{A(1)B(\alpha_nx)}\sum_{k=0}^{\infty}p_k(\alpha_nx)\int_{\frac{k}{\beta_n}}^{\frac{k+1}{\beta_n}}f(t)dt, \tag{2.31}$$

where α_n and β_n are strictly increasing sequence of positive numbers such that $\lim_{n\to\infty}\frac{1}{\beta_n}=0, \frac{\alpha_n}{\beta_n}=1+O\left(\frac{1}{\beta_n}\right)$, as $n\to\infty$ and $p_k(x), k=1,1,2...$ are Brenke-type polynomials.

Assuming that $\lim_{x\to\infty}\frac{B^{(k)}(x)}{B(x)}=1$ as $k=1,2,3,4$, Atakut and Buyukyazici obtained the order of convergence of the operators given by (2.31). Subsequently, Garg et al. [12] investigated the degree of approximation of these operators by means of Peetre's K-functional and the Ditizian-Totik modulus of smoothness. The rate of approximation of functions having derivatives equivalent with a function of bounded variation was also obtained.

2.1.11 OPERATORS DEFINED BY MEANS OF BOAS-BUCK-TYPE POLYNOMIALS

Sidharth et al. [36] introduced the Szász-Durrmeyer-type polynomials based on Boas-Buck-type polynomials and established a direct approximation theorem with the aid of unified Ditzian-Totik modulus of smoothness. The approximation of functions whose derivative are locally of bounded variation are also discussed. For a function $f\in C_\gamma[0,\infty)$, Sidharth et al. [36] defined the Szǻsz-Durrmeyer-type operators based on Boas-Buck-type polynomials as follows:

$$M_n(f;x)=\frac{1}{A(1)G(nxH(1))}\sum_{k=1}^{\infty}\frac{p_k(nx)}{B(k,n+1)}\int_0^\infty\frac{t^{k-1}}{(1+t)^{n+k+1}}f(t)dt$$
$$+\frac{a_0b_0}{A(1)G(nxH(1))}f(0), \qquad (2.32)$$

where $B(k,n+1)$ is the beta function and $x\geq 0,\ n\in\mathbb{N}$.

Lemma 2.5. *[36] For the operator M_n, the following equalities hold:*

1. $M_n(1;x)=1$;
2. $M_n(t;x)=\frac{1}{n}\left(\frac{G'(nxH(1))}{G(nxH(1))}nx+\frac{A'(1)}{A(1)}\right)$;
3. $M_n(t^2;x)=\frac{1}{n(n-1)}\left[\frac{G''(nxH(1))}{G(nxH(1))}n^2x^2+\left(2\frac{A'(1)}{A(1)}+H''(1)+2\right)\right.$ $\left.\times\frac{G'(nxH(1))}{G(nxH(1))}nx+2\frac{A'(1)}{A(1)}+\frac{A''(1)}{A(1)}\right]$;
4. $M_n(t^3;x)=\frac{1}{n(n-1)(n-2)}\left[\frac{G'''(nxH(1))}{G(nxH(1))}n^3x^3+\left(3\frac{A'(1)}{A(1)}+6+3H''(1)\right)\right.$ $\times\frac{G''(nxH(1))}{G(nxH(1))}n^2x^2+\left(12\frac{A'(1)}{A(1)}+H''(1)+3\frac{A'(1)}{A(1)}H''(1)+H'''(1)+4\right)$ $\left.\times\frac{G'(nxH(1))}{G(nxH(1))}nx+7\frac{A''(1)}{A(1)}+6\frac{A'(1)}{A(1)}\right]$;

5. $M_n(t^4;x) = \frac{1}{n(n-1)(n-2)(n-3)}\Bigg[\frac{G^{iv}(nxH(1))}{G(nxH(1))}n^4x^4 + \left(4\frac{A'(1)}{A(1)} + 6H''(1) + 12\right)\frac{G'''(nxH(1))}{G(nxH(1))}n^3x^3 + \left(6\frac{A''(1)}{A(1)} + 12\frac{A'(1)}{A(1)}H''(1) + 21\frac{A'(1)}{A(1)} + 3\frac{H''(1)}{A(1)} + 4H'''(1) + 18H''(1) + 3(H''(1))^2 + 21\right)\frac{G''(nxH(1))}{G(nxH(1))}n^2x^2 + \left(4\frac{A''(1)}{A(1)} + 6\frac{A''(1)}{A(1)}H''(1) + 36\frac{A''(1)}{A(1)}H''(1) + 42\frac{A'(1)}{A(1)} + 4\frac{A'(1)}{A(1)}H'''(1) + 36\frac{A''(1)}{A(1)} + H^{iv}(1) + 12H'''(1) + 36H''(1) + 24\right)\frac{G'(nxH(1))}{G(nxH(1))}nx + \frac{A^{iv}(1)}{A(1)} + 48\frac{A''(1)}{A(1)} + 13\frac{A'(1)}{A(1)} + 11\Bigg].$

Lemma 2.6. *[36] For the operator (2.32), the following results hold:*

1. $M_n((t-x);x) = \left(\frac{G'(nxH(1))}{G(nxH(1))} - 1\right)x + \frac{A'(1)}{nA(1)};$

2. $M_n((t-x)^2;x) = \left(\frac{n}{n-1}\frac{G''(nxH(1))}{G(nxH(1))} - 2\frac{G'(nxH(1))}{G(nxH(1))} + 1\right)x^2 + \left(\frac{1}{n-1}\left(2\frac{A'(1)}{A(1)} + H''(1) + 2\right)\frac{G'(nxH(1))}{G(nxH(1))} - \frac{2}{n}\frac{A'(1)}{A(1)}\right)x + \frac{1}{n(n-1)}\left(2\frac{A'(1)}{A(1)} + \frac{A''(1)}{A(1)}\right);$

3. $M_n((t-x)^4;x) = \Bigg\{\frac{n^3}{(n-1)(n-2)(n-3)}\frac{G^{iv}(nxH(1))}{G(nxH(1))} - \frac{4n^2}{(n-1)(n-2)}\frac{G'''(nxH(1))}{G(nxH(1))} + \frac{6n}{(n-1)}\frac{G''(nxH(1))}{G(nxH(1))} - 4\frac{G'(nxH(1))}{G(nxH(1))} + 1\Bigg\}x^4 + \Bigg\{\frac{n^2}{(n-1)(n-2)(n-3)}\frac{G'''(nxH(1))}{G(nxH(1))}\left(4\frac{A'(1)}{A(1)} + 6H''(1) + 12\right) - \frac{4n}{(n-1)(n-2)}\frac{G''(nxH(1))}{G(nxH(1))}\left(3\frac{A'(1)}{A(1)} + 3H''(1) + 6\right) + \frac{6}{(n-1)}\frac{G'(nxH(1))}{G(nxH(1))} \times \left(2\frac{A'(1)}{A(1)} + H''(1) + 2\right) - \frac{4}{n}\frac{A'(1)}{A(1)}\Bigg\}x^3 + \Bigg\{\frac{n}{(n-1)(n-2)(n-3)}\frac{G''(nxH(1))}{G(nxH(1))} \times \left(6\frac{A''(1)}{A(1)} + 12\frac{A'(1)}{A(1)}H''(1) + 21\frac{A'(1)}{A(1)} + 3\frac{H''(1)}{A(1)} + 4H'''(1) + 18H''(1) + 3(H''(1))^2 + 21\right) - \frac{4}{(n-1)(n-2)}\frac{G'(nxH(1))}{G(nxH(1))}\left(12\frac{A'(1)}{A(1)} + H''(1) + 3\frac{A'(1)}{A(1)}H''(1) + H''(1) + 4\right)\Bigg\}x^2 + \Bigg\{\frac{1}{(n-1)(n-2)(n-3)}\frac{G'(nxH(1))}{G(nxH(1))}\Bigg(4\frac{A''(1)}{A(1)} + 6\frac{A''(1)}{A(1)}H''(1) + 36\frac{A''(1)}{A(1)}H''(1) + 42\frac{A'(1)}{A(1)} + 4\frac{A'(1)}{A(1)}H'''(1) + 36\frac{A''(1)}{A(1)}$

$$+H^{iv}(1)+12H'''(1)+36H''(1)+24\Big)-\frac{4}{n(n-1)(n-2)}\left(7\frac{A''(1)}{A(1)}+6\frac{A'(1)}{A(1)}\right)\Big\}x$$
$$+\frac{1}{(n-1)(n-2)(n-3)}\left(\frac{A^{iv}(1)}{A(1)}+48\frac{A''(1)}{A(1)}+13\frac{A'(1)}{A(1)}+11\right).$$

Lemma 2.7. *[36] For the operator (2.32), the equalities hold:*

1. $\lim_{n\to\infty} nM_n((t-x);x)=l_1(x)x+\frac{A'(1)}{A(1)}$;
2. $\lim\limits_{n\to\infty} nM_n((t-x)^2;x)=l_2(x)x^2+x(H''(1)+2)=\eta(x),(say)$;
3. $\lim\limits_{n\to\infty} n^2M_n((t-x)^4;x)=l_4(x)x^4+l_3(x)x^3+\Big(6\frac{A''(1)}{A(1)}-27\frac{A'(1)}{A(1)}+\frac{H''(1)}{A(1)}+$
$$14H''(1)+3(H''(1))^2+5\Big)=\nu(x),(say).$$

The following theorem shows that the operators defined by (2.32) are an approximation process for $f\in C_\gamma[0,\infty)$, using the Bohman-Korovkin theorem.

Theorem 2.18. *[36] Let $f\in C_\gamma[0,\infty)$. Then,*

$$\lim_{n\to\infty} M_n(f;x)=f(x),$$

holds uniformly in $x\in[0,a]$, $a>0$.

Proof. From Lemma 2.5, it follows that

$$\lim_{n\to\infty} M_n(t^i;x)=x^i,\ i=0,1,2$$

uniformly in $x\in[0,a]$. Hence by Bohman-Korovkin theorem, the required result is immediate. □

Theorem 2.19. *[36] Let $f\in C_\gamma[0,\infty)$, admitting a derivative of second order at a point $x\in[0,\infty)$, then there holds*

$$\lim_{n\to\infty} n(M_n(f;x)-f(x))=\left\{l_1(x)x+\frac{A'(1)}{A(1)}\right\}f'(x)+\Big\{l_2(x)x^2 \tag{2.33}$$
$$+x(H''(1)+2)\Big\}\frac{f''(x)}{2}. \tag{2.34}$$

If f'' is continuous on $[0,\infty)$, then the limit in (2.33) holds uniformly in $x\in[0,a]\subset[0,\infty), a>0$.

In 2018, Mursaleen et al. [31] considered a Chlodowsky variant of generalized Szász-type operators involving Boas-Buck-type polynomials and studied some approximation properties of these operators in a weighted space of continuous functions on $[0,\infty)$.

2.1.12 OPERATORS DEFINED BY MEANS OF CHARLIER POLYNOMIALS

In 1950, Szász [38] introduced the following linear positive operators

$$S_n(f;x) = e^{-nx} \sum_{k=0}^{\infty} \frac{(nx)^k}{k!} f\left(\frac{k}{n}\right), \tag{2.35}$$

where $x \in [0,\infty)$ and $f(x)$ is a continuous function on $[0,\infty)$ whenever the above sum converges uniformly. These polynomials [18] have the generating functions of the form

$$e^t \left(1 - \frac{t}{a}\right)^u = \sum_{k=0}^{\infty} C_k^{(a)}(u) \frac{t^k}{k!}, \ |t| < a, \tag{2.36}$$

where $C_k^{(a)}(u) = \sum_{r=0}^{k} \binom{k}{r} (-u)_r \left(\frac{1}{a}\right)^r$ and $(m)_0 = 1, (m)_j = m(m+1)\cdots(m+j-1)$ for $j \geq 1$. For $f \in C[0,\infty) \cap E_2$, Varma and Taşdelen [42] defined the following Szász-type operators involving Charlier polynomials

$$\mathcal{L}_n(f;x,a) = e^{-1}\left(1 - \frac{1}{a}\right)^{(a-1)nx} \sum_{k=0}^{\infty} \frac{C_k^{(a)}(-(a-1)nx)}{k!} f\left(\frac{k}{n}\right), \tag{2.37}$$

where $a > 1$ and $x \in [0,\infty)$. For the special case, $a \to \infty$ and $x - \frac{1}{n}$ instead of x, these operators reduce to the operators (2.35).

Lemma 2.8. *[42] The operators (2.37) satisfy the following equalities:*

1. $L_n(1;x,a) = 1$;
2. $L_n(t;x,a) = x + \frac{1}{n}$;
3. $L_n(t^2;x,a) = x^2 + \frac{x}{n}\left(3 + \frac{1}{a-1}\right) + \frac{2}{n^2}$

where $x \geq 0$.

Theorem 2.20. *[42] Let* $f \in C[0,\infty) \cap E_2$, *then*

$$\lim_{n\to\infty} L_n(f;x,a) = f(x),$$

the sequence of operators given by (2.37) *converges uniformly in each compact subset of* $[0,\infty)$.

Theorem 2.21. *[42] Let* $f \in \tilde{C}[0,\infty) \cap E_2$. *For the operators* L_n *given by* (2.37), *there holds the inequality:*

$$|L_n(f;x,a) - f(x)| \leq \left\{1 + \sqrt{x\left(1 + \frac{1}{a-1}\right) + \frac{2}{n}}\right\} \omega\left(f; \frac{1}{\sqrt{n}}\right).$$

Also, Varma and Taşdelen [42] considered Kantorovich-type generalization of the operators $L_n(f;x,a)$ for a function

$$f \in \bar{C}[0,\infty) := \left\{ f \in C[0,\infty) : |F(x)| = \left| \int_0^x f(s)ds \right| \le Ke^{Bx}, B \in \mathbb{R} \text{ and } K \in \mathbb{R}^+ \right\}$$

as follows :

$$L_{n,a}^*(f;x) = ne^{-1}\left(1 - \frac{1}{a}\right)^{(a-1)nx} \sum_{k=0}^{\infty} \frac{C_k^{(a)}(-(a-1)nx)}{k!} \int_{\frac{k}{n}}^{\frac{k+1}{n}} f(t)dt, \quad (2.38)$$

where $a > 1$ and $x \ge 0$, and studied the uniform convergence of $L_{n,a}^*(f;x)$ to f on each compact subset of $[0,\infty)$ and the degree of approximation in terms of the classical modulus of continuity.

Lemma 2.9. *[42] The operators given by* (2.38) *satisfy the following equalities:*

1. $L_n^*(1;x,a) = 1;$
2. $L_n^*(t;x,a) = x + \frac{3}{2n};$
3. $L_n^*(t^2;x,a) = x^2 + \frac{x}{n}\left(4 + \frac{1}{a-1}\right) + \frac{10}{3n^2}.$

Theorem 2.22. *[42] Let $f \in C[0,\infty) \cap E_3$, then*

$$\lim_{n\to\infty} L_n^*(f;x,a) = f(x),$$

holds uniformly in each compact subset of $[0,\infty)$.

Theorem 2.23. *[42] Let $f \in \tilde{C}[0,\infty) \cap E_3$. For the operators L_n*, there holds the following inequality:*

$$|L_n^*(f;x,a) - f(x)| \le \left\{ 1 + \sqrt{x + \left(1 + \frac{1}{a-1}\right) + \frac{10}{3n}} \right\} \omega\left(f; \frac{1}{\sqrt{n}}\right).$$

Further, for the operator L_n (2.37), Kajla and Agrawal [23] proved some approximation theorems. Let $e_i(x) = x^i, i = 0,1,2,\cdots$

Lemma 2.10. *[23] For the operators $\mathcal{L}_n(e_s(t);x,a)$, $s = 3,4$, we have*

1. $\mathcal{L}_n(e_3(t);x,a) = x^3 + \frac{x^2}{n}\left(6 + \frac{3}{a-1}\right) + \frac{2x}{n^2}\left(\frac{1}{(a-1)^2} + \frac{3}{a-1} + 5\right) + \frac{5}{n^3};$
2. $\mathcal{L}_n(e_4(t);x,a) = x^4 + \frac{x^3}{n}\left(10 + \frac{6}{a-1}\right) + \frac{x^2}{n^2}\left(31 + \frac{30}{a-1} + \frac{11}{(a-1)^2}\right)$
$+ \frac{x}{n^3}\left(67 + \frac{31}{a-1} + \frac{20}{(a-1)^2} + \frac{6}{(a-1)^3}\right) + \frac{15}{n^4}.$

Let $e_i^x(t) = (t-x)^i, i = 0,1,2\cdots$

Lemma 2.11. *[23] For the operators $\mathcal{L}_n(f;x,a)$, we have*

1. $\mathcal{L}_n(e_1^x(t);x,a) = \frac{1}{n}$;
2. $\mathcal{L}_n(e_2^x(t);x,a) = \frac{ax}{n(a-1)} + \frac{2}{n^2}$;
3. $\mathcal{L}_n(e_4^x(t);x,a) = \frac{x}{n^3}\left(17 + \frac{49}{(a-1)} - \frac{20}{(a-1)^2} + \frac{6}{(a-1)^3}\right)$
$+ \frac{x^2}{n^2}\left(19 - \frac{46}{(a-1)} + \frac{3}{(a-1)^2}\right) + \frac{15}{n^4}$.

Proof. Using ([42], p. 119, Lemma 1) and Lemma 2.10, the proof of this lemma easily follows. Hence, the details are omitted. □

To study the rate of convergence of functions having a derivative of bounded variation, let us rewrite the operators (2.37) as

$$\mathcal{L}_n(f;x,a) = \int_0^\infty f(w)\frac{\partial}{\partial w}\{\mathcal{K}_n(x,w,a)\}\,dw, \tag{2.39}$$

where

$$\mathcal{K}_n(x,w,a) = \begin{cases} \sum\limits_{k\le nw} e^{-1}\left(1-\frac{1}{a}\right)^{(a-1)nx} \frac{C_k^{(a)}(-(a-1)nx)}{k!}, & \text{if } 0 < w < \infty, \\ 0, & \text{if } w = 0. \end{cases}$$

From Lemma 2.11, for $x \in (0,\infty)$ and sufficiently large n, we have

$$\mathcal{L}_n(|e_1^x(t)|;x,a) \le \left(\mathcal{L}_n((t-x)^2;x,a)\right)^{1/2} \le \sqrt{\frac{\lambda(a)x}{n}}, \tag{2.40}$$

where $\lambda(a)$ is some positive constant depending on a.
Also the authors get, for $r \ge 2$ and fixed $x \in [0,\infty)$,

$$\mathcal{L}_n(e_{2r}^x(t);x,a) = O(n^{-r});\ n \to \infty. \tag{2.41}$$

Lemma 2.12. *[23] For all $x \in (0,\infty)$ and sufficiently large n, we have*

1. $\vartheta_{n,a}(x,t) = \int_0^t \frac{\partial}{\partial w}\{\mathcal{K}_n(x,w,a)\}\,dw \le \frac{1}{(x-t)^2}\frac{\lambda(a)x}{n},\ 0 \le t < x$;
2. $1 - \vartheta_{n,a}(x,z) = \int_z^\infty \frac{\partial}{\partial w}\{\mathcal{K}_n(x,w,a)\}\,dw \le \frac{1}{(z-x)^2}\frac{\lambda(a)x}{n},\ x < z < \infty$,

where $\lambda(a)$ is a constant as described in (2.40).

Proof. First, we prove (i).

$$\vartheta_{n,a}(x,t) = \int_0^t \frac{\partial}{\partial w}\{\mathcal{K}_n(x,w,a)\}\,dw \le \int_0^t \left(\frac{x-w}{x-t}\right)^2 \frac{\partial}{\partial w}\{\mathcal{K}_n(x,w,a)\}\,dw$$

$$\leq \frac{1}{(x-t)^2}\mathcal{L}_n((w-x)^2;x,a)$$
$$\leq \frac{1}{(x-t)^2}\frac{\lambda(a)x}{n}.$$

The proof of (ii) is similar, and hence, it is omitted. □

Theorem 2.24. *[23] Let $f \in Lip_M^*(r)$ and $r \in (0,1]$. Then, for all $x \in (0,\infty)$, we have*

$$|\mathcal{L}_n(f;x,a) - f(x)| \leq M\left(\frac{\zeta_{n,a}(x)}{a_1x^2+a_2x}\right)^{\frac{r}{2}},$$

where $\zeta_{n,a}(x) = \mathcal{L}_n((t-x)^2;x,a)$.

Proof. Applying the Hölder's inequality with $p = \frac{2}{r}$ and $q = \frac{2}{2-r}$, we find that

$$|\mathcal{L}_n(f;x,a) - f(x)| \leq \left\{e^{-1}\left(1-\frac{1}{a}\right)^{(a-1)nx}\sum_{k=0}^{\infty}\frac{C_k^{(a)}(-(a-1)nx)}{k!}\left|f\left(\frac{k}{n}\right)-f(x)\right|^{\frac{2}{r}}\right\}^{r/2}$$
$$\leq M\left\{e^{-1}\left(1-\frac{1}{a}\right)^{(a-1)nx}\sum_{k=0}^{\infty}\frac{C_k^{(a)}(-(a-1)nx)}{k!}\frac{\left(\frac{k}{n}-x\right)^2}{\left(\frac{k}{n}+a_1x^2+a_2x\right)}\right\}^{r/2}.$$

Since $f \in Lip_M^*(r)$ and $\frac{1}{\sqrt{\frac{k}{n}+a_1x^2+a_2x}} < \frac{1}{\sqrt{a_1x^2+a_2x}}, \forall x \in (0,\infty)$, we have

$$|\mathcal{L}_n(f;x,a) - f(x)| \leq \frac{M}{(a_1x^2+a_2x)^{\frac{r}{2}}}\left\{e^{-1}\left(1-\frac{1}{a}\right)^{(a-1)nx}\right.$$
$$\left.\times\sum_{k=0}^{\infty}\frac{C_k^{(a)}(-(a-1)nx)}{k!}\left(\frac{k}{n}-x\right)^2\right\}^{r/2}$$
$$= \frac{M}{(a_1x^2+a_2x)^{\frac{r}{2}}}(\mathcal{L}_n|t-x|;x,a)^r$$
$$\leq M\left(\frac{\zeta_{n,a}(x)}{a_1x^2+a_2x}\right)^{\frac{r}{2}}.$$

This completes the proof of the theorem. □

Theorem 2.25. *[23] Let $f \in \tilde{C}_B[0,\infty)$ and $0 < r \leq 1$. Then, for all $x \in [0,\infty)$, we have*

$$|\mathcal{L}_n(f;x,a) - f(x)| \leq \widetilde{\omega}_r(f,x)(\zeta_{n,a}(x))^{\frac{r}{2}}.$$

Proof. From equation (2.2), we have

$$|\mathcal{L}_n(f;x,a) - f(x)| \leq \widetilde{\omega}_r(f,x)\mathcal{L}_n(|t-x|^r;x,a).$$

Applying the Hölder's inequality with $p = \frac{2}{r}$ and $q = \frac{2}{2-r}$, we get

$$|\mathcal{L}_n(f;x,a) - f(x)| \leq \widetilde{\omega}_r(f,x)(\mathcal{L}_n((t-x)^2;x,a))^{\frac{r}{2}} \leq \widetilde{\omega}_r(f,x)(\zeta_{n,a}(x))^{\frac{r}{2}}.$$

Thus, the proof is completed. □

Theorem 2.26. *[23] Let $f \in C_2^*[0,\infty)$. Then, we have*

$$\lim_{n\to\infty} \|\mathcal{L}_n(f;\cdot,a) - f\|_2 = 0. \tag{2.42}$$

Proof. From [10], we know that it is sufficient to verify the following three equations

$$\lim_{n\to\infty} \|\mathcal{L}_n(e_m;\cdot,a) - e_m\|_2 = 0,\ m = 0,1,2. \tag{2.43}$$

Since $\mathcal{L}_n(1;x,a) = 1$, the condition in (2.43) holds true for $m = 0$.

By ([42], Lemma 1), we have

$$\|\mathcal{L}_n(e_1;\cdot,a) - e_1\|_2 = \sup_{x\geq 0} \frac{1}{1+x^2}\left|x + \frac{1}{n} - x\right| \leq \frac{1}{n}.$$

Thus, $\lim_{n\to\infty} \|\mathcal{L}_n(t;\cdot,a) - e_1\|_2 = 0$. Similarly, we get

$$\begin{aligned}\|\mathcal{L}_n(e_2;\cdot,a) - e_2\|_2 &= \sup_{x\geq 0} \frac{1}{1+x^2}\left|x^2 + \frac{x}{n}\left(3 + \frac{1}{a-1}\right) + \frac{2}{n^2} - x^2\right| \\ &\leq \frac{1}{n}\left(3 + \frac{1}{(a-1)}\right) + \frac{2}{n^2},\end{aligned}$$

which implies that $\lim_{n\to\infty} \|\mathcal{L}_n(e_2;\cdot,a) - e_2\|_2 = 0$. This completes the proof. □

Next, the authors [23] gave a theorem to approximate all functions in $C_2[0,\infty)$. This type of result is discussed in [11] for locally integrable functions.

Theorem 2.27. *[23] For each $f \in C_2[0,\infty)$ and $\beta > 0$, we have*

$$\lim_{n\to\infty} \sup_{x\in[0,\infty)} \frac{|\mathcal{L}_n(f;x,a) - f(x)|}{(1+x^2)^{1+\beta}} = 0.$$

Proof. For any fixed $x_0 > 0$,

$$\begin{aligned}\sup_{x\in[0,\infty)} \frac{|\mathcal{L}_n(f;x,a) - f(x)|}{(1+x^2)^{1+\beta}} &\leq \sup_{x\leq x_0} \frac{|\mathcal{L}_n(f;x,a) - f(x)|}{(1+x^2)^{1+\beta}} + \sup_{x\geq x_0} \frac{|\mathcal{L}_n(f;x,a) - f(x)|}{(1+x^2)^{1+\beta}} \\ &\leq \| \mathcal{L}_n(f) - f \|_{C[0,x_0]} + \| f \|_2 \sup_{x\geq x_0} \frac{|\mathcal{L}_n(1+t^2;x,a)|}{(1+x^2)^{1+\beta}}\end{aligned}$$

$$+\sup_{x\geq x_0}\frac{|f(x)|}{(1+x^2)^{1+\beta}},$$
$$=: I_1+I_2+I_3, \text{ say.} \tag{2.44}$$

Since $|f(x)|\leq ||f||_2(1+x^2)$, we have

$$I_3=\sup_{x\geq x_0}\frac{|f(x)|}{(1+x^2)^{1+\beta}}\leq \sup_{x\geq x_0}\frac{||f||_2}{(1+x^2)^{\beta}}\leq\frac{||f||_2}{(1+x_0^2)^{\beta}}.$$

Let $\varepsilon>0$ be arbitrary.

$$\frac{M_f}{(1+x_0^2)^{\gamma}}<\frac{\varepsilon}{3}, \tag{2.45}$$

In view of ([42], Theorem 1), there exists $n_1\in\mathbb{N}$ such that

$$\| f \|_2\frac{|\mathcal{L}_n(1+t^2;x,a)|}{(1+x^2)^{1+\beta}}<\frac{1}{(1+x^2)^{1+\beta}}\| f \|_2\left((1+x^2)+\frac{\varepsilon}{3||f||_2}\right), \forall n\geq n_1$$
$$<\frac{||f||_2}{(1+x^2)^{\beta}}+\frac{\varepsilon}{3}, \forall n\geq n_1. \tag{2.46}$$

Hence, $||f||_2\sup\limits_{x\geq x_0}\dfrac{|\mathcal{L}_n(1+t^2;x,a)|}{(1+x^2)^{1+\beta}}<\dfrac{||f||_2}{(1+x_0^2)^{\beta}}+\dfrac{\varepsilon}{3},\ \forall n\geq n_1.$

Thus, $I_2+I_3<\dfrac{2||f||_\varphi}{(1+x_0^2)^{\beta}}+\dfrac{\varepsilon}{3},\ \forall n\geq n_1.$

Now, let us choose x_0 to be so large that $\dfrac{||f||_\varphi}{(1+x^2)^{\beta}}<\dfrac{\varepsilon}{6}.$

Then,

$$I_2+I_3<\frac{2\varepsilon}{3},\ \forall n\geq n_1. \tag{2.47}$$

By Theorem (2.28), there exists $n_2\in\mathbb{N}$ such that

$$I_1=\| \mathcal{L}_n(f)-f \|_{C[0,x_0]}<\frac{\varepsilon}{3},\ \forall n\geq n_2. \tag{2.48}$$

Let $n_0=\max(n_1,n_2)$. Then, combining (2.44)–(2.48)

$$\sup_{x\in[0,\infty)}\frac{|\mathcal{L}_n(f;x,a)-f(x)|}{(1+x^2)^{1+\beta}}<\varepsilon,\ \forall n\geq n_0.$$

This completes the proof. □

Theorem 2.28. *[23] Let $f\in C_2[0,\infty)$. Then, we have*

$$||\mathcal{L}_n(f;\cdot,a)-f||_{C[0,b]}\leq 4M_f(1+b^2)\zeta_{n,a}(b)+2\omega_{b+1}(f,\sqrt{\zeta_{n,a}(b)}),$$

where $\zeta_{n,a}(b)=\dfrac{ab}{n(a-1)}+\dfrac{2}{n^2}.$

Proof. From [16], for $x \in [0,b]$ and $t \geq 0$, we have

$$|f(t)-f(x)| \leq 4M_f(1+x^2)(t-x)^2 + \left(1+\frac{|t-x|}{\delta}\right)\omega_{b+1}(f,\delta), \delta > 0.$$

Hence applying Cauchy-Schwarz inequality, we get

$$\begin{aligned}|\mathcal{L}_n(f;x,a)-f(x)| &\leq 4M_f(1+x^2)\mathcal{L}_n((t-x)^2;x,a) \\ &\quad + \omega_{b+1}(f,\delta)\left(1+\frac{1}{\delta}\mathcal{L}_n(|t-x|;x,a)\right) \\ &\leq 4M_f(1+x^2)\zeta_{n,a}(x) + \omega_{b+1}(f,\delta)\left(1+\frac{1}{\delta}\sqrt{\zeta_{n,a}(x)}\right) \\ &\leq 4M_f(1+b^2)\zeta_{n,a}(b) + \omega_{b+1}(f,\delta)\left(1+\frac{1}{\delta}\sqrt{\zeta_{n,a}(b)}\right).\end{aligned}$$

Choosing $\delta = \sqrt{\zeta_{n,a}(b)}$, we get the desired result. □

Further, Kajla and Agrawal [25] considered Kantorovich-type generalization of the operators $L_n(f;x,a)$ for a function $f \in \tilde{C}[0,\infty) := \{f \in C[0,\infty) : |F(x)| = |\int_0^x f(s)ds| \leq Ke^{Bx}, B \in \mathbb{R} \text{ and } K \in \mathbb{R}^+\}$, and established some more approximation properties of the operators $L^*_{n,a}$ such as weighted approximation, A-statistical convergence, and approximation of functions with a derivative of bounded variation. They also presented some moment estimates and a result needed to study approximation of functions with derivatives of bounded variation. They discussed the main results of this paper wherein we establish approximation in a Lipschitz-type space, weighted approximation theorems, and A-statistical convergence properties for the operators $L^*_{n,a}$. Lastly, they obtained the rate of convergence for functions having a derivative of bounded variation on every finite subinterval of $[0,\infty)$, for these operators.

In 2016, Agrawal and Ispir [2] introduced a new bivariate operator associated with a combination of Chlodowsky and generalized Szász-Charlier-type operators as follows:

$$T_{n,m}(f;x,y,a) = \sum_{k=0}^{n}\sum_{j=0}^{\infty} p_{n,k}\left(\frac{x}{\alpha_n}\right) s_{m,j}(\beta_m y,a) f\left(\frac{k\alpha_n}{n},\frac{j}{\gamma_m}\right).$$

for all $n,m \in \mathbb{N}$, $f \in C(A_{\alpha_n})$ with $A_{\alpha_n} = \{(x,y) : 0 \leq x \leq \alpha_n,\ 0 \leq y \leq \infty\}$ and $C(A_{\alpha_n}) = \{f : A_{\alpha_n} \to \mathbb{R} \text{ is continuous}\}$. Here, α_n is an unbounded sequence of positive numbers such that $\lim_{n\to\infty}\left(\frac{\alpha_n}{n}\right) = 0$, and also γ_m and β_m denote the unbounded sequences of positive numbers such that $\frac{\beta_m}{\gamma_m} = 1 + O\left(\frac{1}{\gamma_m}\right)$. Also, the basic elements are $p_{n,k}\left(\frac{x}{\alpha_n}\right) = \binom{n}{k}\left(\frac{x}{\alpha_n}\right)\left(1-\frac{x}{\alpha_n}\right)^{n-k}$ and $s_{m,j}(\beta_m y,a) = e^{-1}\left(1-\frac{1}{a}\right)^{(a-1)\beta_m y}\frac{C_j^{(a)}(-(a-1)\beta_m y)}{j!}$, where $C_j^{(a)}(u)$ is Charlier polynomial and $a > 1$. For the special case, $a \to \infty$ and $y - \left(\frac{x}{m}\right)$ instead of y, the operators $T_{n,m}$ reduce to the operators studied by Gazanfer and Buyukyazici [13]. They give the degree of approximation for these bivariate operators by means of the complete and partial modulus of continuity, and also by using weighted modulus of continuity. Furthermore, they construct a GBS (generalized Boolean sum) operator of bivariate Chlodowsky-Szász-Charlier type and estimate the order of approximation in terms of mixed modulus of continuity.

Motivated by the above development, Wafi et al. [43] defined a new sequence of Kantorovich-Szász-type operators which preserves constant and quadratic test functions, i.e., $e_0(x)$ and $e_2(x)$, where $e_i(x) = x^i$, i = 0,2

$$K_{n,a}^*(f; r_{n,a}^*(x)) = ne^{-1}\left(1-\frac{1}{a}\right)^{(a-1)nr_{n,a}^*(x)} \sum_{k=0}^{\infty} \frac{C_k^{(a)}(-(a-1)nr_{n,a}^*(x))}{k!} \int_{\frac{k}{n}}^{\frac{k+1}{n}} f(t)dt \tag{2.49}$$

where $r_{n,a}^*(x) = \frac{(-4+\frac{1}{a-1})+\sqrt{(4+\frac{1}{a-1})^2+4(n^2x^2-\frac{10}{3})}}{2n}$ and $r_{n,a}^*(x) \geq 0$ for $x \in \left[\frac{\sqrt{\frac{10}{3}}}{n}, \infty\right)$.

For a fixed $x \in \left[\frac{\sqrt{\frac{10}{3}}}{n}, \infty\right)$, $r_{n,a}^*(x) \to x$, as $n \to \infty$, and the operators (2.49) reduce to operators (2.38). Wafi et al. [43] discussed the rate of convergence by determining better error estimates. Further, the authors investigated the order of approximation by means of local approximation results with the help of Ditzian-Totik modulus of smoothness, second-order modulus of continuity, Peetre's K-functional, and Lipschitz class.

2.1.13 OPERATORS DEFINED BY USING Q-CALCULUS

First, we shall give some definitions.

Following [5,22], for any fixed real number $q > 0$, satisfying the condition $0 < q < 1$, the q-integer $[k]_q$, for $k \in \mathbb{N}$, and q-factorial $[k]_q!$ are defined as

$$[k]_q = \begin{cases} \frac{(1-q^k)}{(1-q)}, & \text{if } q \neq 1 \\ k, & \text{if } q = 1, \end{cases}$$

and

$$[k]_q! = \begin{cases} [k]_q[k-1]_q \ldots 1, & \text{if } k \geq 1 \\ 1, & \text{if } k = 0, \end{cases}$$

respectively. For any integers n,k satisfying $0 \leq k \leq n$, the q-binomial coefficient is given by

$$\binom{n}{k}_q = \frac{[n]_q!}{[n-k]_q![k]_q!}.$$

The q-analogue of $(1-x)^n$ is given by

$$(1-x)_q^n = \begin{cases} \prod_{j=0}^{n-1} (1-q^j x), & n = 1,2,... \\ 1, & n = 0. \end{cases}$$

The q-integration in the interval [0,a] is defined by

$$\int_0^a f(t) d_q t = a(1-q) \sum_{n=0}^{\infty} f(aq^n) q^n \quad 0 < q < 1,$$

provided the series converges.

Karaisa [27] introduced a Stancu-type generalization of the q-Favard-Szász operator as follows:

$$T_{n,t}^{\alpha,\beta}(f;q,x) = \frac{E_q^{-[n]_q t}}{A(1)} \sum_{k=0}^{\infty} \frac{P_k(q;[n]_q t)}{[k]_q!} f\left(x + \frac{[k]_q + \alpha}{[n]_q + \beta}\right),$$

where $P_k(q;.) \geq 0$ for each k is a q-Appell generated by

$$B(u) e^{[n]_q t u} = \sum_{k=0}^{\infty} \frac{P_k(q;[n]_q t) u^k}{[k]_q!}$$

and $B(u)$ is defined by

$$B(u) = \sum_{k=0}^{\infty} a_k u^k,$$

and studied Korovkin-type statistical approximation properties and rate of convergence using modulus of continuity. He also obtained some local approximation results for these operators.

For $f \in C_\gamma[0,\infty)$, Agrawal and Gupta [1] defined the q-analogue of the operators (2.17) defined in [26] as follows:

$$\mathcal{L}_n(f;q,x) = \frac{E^{-[n]_q x}}{g(1)} \sum_{k=0}^{\infty} \frac{P_k(q;[n]_q x)}{[k]_q! B_q(n+1,k)} q^{\frac{k(k-1)}{2}}$$
$$\int_0^{\infty} \frac{t^{k-1}}{(1+t)_q^{n+k+1}} f(q^k t) d_q t,$$

and obtained the rate of convergence in terms of the weighted modulus of continuity and a Lipschitz-type maximal function for these operators.

ACKNOWLEDGMENT

Many thanks to our TEX-pert for developing this class file.

REFERENCES

[1] P. N. Agrawal and P. Gupta, Durrmeyer variant of q-Favard-Szász operators based on Appell polynomials. *Creat. Math. Inform.* **26**(1) (2017) 9–7.

[2] P. N. Agrawal, and N. Ispir, Degree of approximation for bivariate Chlodowsky-Szász-Charlier type operators. *Results Math.,* **69**(3–4) (2016) 369–385.

[3] K. J. Ansari, M. Mursaleen and S. Rahman, Approximation by Jakimovski-Leviatan operators of Durrmeyer type involving multiple Appell polynomials. *RACSAM* **113** (2019) 1007–1024.

[4] P. E. Appell, Sur une classe de polynomes. *Annales scientifique de l'E.N.S.* **2**(9) (1880) 119–144.

[5] A. Aral, V. Gupta and R. P. Agarwal, *Application of q-Calculus in Operator Theory.* Springer, New York (2013).

[6] Ç. Atakut and I. Büyükyazici, Approximation by Kantorovich-Szász type operators based on Brenke type polynomials. *Numer. Funct. Anal. Optim.* **37(12)** (2016), 1488–1502.

[7] T. S. Chihara, *An Introduction to Orthogonal Polynomials*, Gordon and Breach, NewYork, 1978.

[8] A. Ciupa, Positive linear operators obtained by means of Appell polynomials. in *Approximation and optimization,* Vol. II (Cluj-Napoca, 1996), 63–68, Transilvania, Cluj.

[9] R. A. Devore and G. G. Lorentz, *Constructive Approximation.* Springer, Berlin (1993).

[10] A. D. Gadjiev., On P. P. Korovkin type theorems. *Math. Zametki* **20**(5) (1976) 781–786.

[11] A. D. Gadjiev, R. O. Efendiyev and E. Ibikli, On Korovkin type theorems in space of locally integrable functions. *Czechoslovak Math. J.* **53(128)**(1) (2003) 45–53.

[12] T. Garg, P. N. Agrawal and S. Araci, Rate of convergence by Kantorovich-Szász type operators based on Brenke type polynomials. *J. Inequal. Appl.* 2017 **(156)**.

[13] A. F. Gazanfer and I. Büyükyazici, Approximation by certain linear positive operators of two variables. *Abstr. Appl. Anal.* 2014, Art. ID 782080 (2014).

[14] P. Gupta and P. N. Agrawal, Jakimovski-Leviatan operators of Durrmeyer type involving Appell polynomials. *Turk. J. Math.* **42** (2018) 1457–1470.

[15] P. Gupta and P. N. Agrawal, Quantitative Voronovskaja and Grüss Voronovskaja-type theorems for operators of Kantorovich type involving multiple Appell polynomials. *Iran. J. Sci. Technol. Trans. Sci.* **43** (2019) 1679–1687.

[16] E. Ibikli and E. A. Gadjiev, The order of approximation of some unbounded function by the sequences of positive linear operators. *Turk. J. Math.* **19**(3) (1995) 331–337.

[17] G. İçöz, S. Varma and S. Sucu, Approximation by operators including generalized Appell polynomials, *Filomat* **30** (2) (2016) 429–440.

[18] M. E. H. Ismail, *Classical and Quantum Orthogonal Polynomials in One Variable.* Cambridge University Press, Cambridge (2005).

[19] M. E. H. Ismail, On a generalization of Szász operators. *Mathematica (Cluj),* **39** (1974) 259–267.

[20] N. Ispir and Ç. Atakut, Approximation by modified Szász-Mirakjan operators on weighted space. *Proc. Indian Acad. Sci. Math. Sci.* **112**(4) (2002) 571–578.

[21] A. Jakimovski and D. Leviatan, Generalized Szász operators for the approximation in the infinite interval. *Mathematica (Cluj),* **34** (1969) 97–103.

[22] V. Kac and P. Cheung, *Quantum Calculus.* Universitext, Springer, New York (2002).

[23] A. Kajla and P. N. Agrawal, Approximation properties of Szász type operators based on Charlier polynomials. *Turk. J. Math.* **39** (2015) 990–1003.

[24] A. Kajla and P. N. Agrawal, Szász-Durrmeyer type operators based on Charlier polynomials. *Appl. Math. Comput.* **268** (2015) 1001–1014.

[25] A. Kajla and P. N. Agrawal, Szász-Kantorovich type operators based on Charlier polynomials. *Kyungpook Math. J.* **56**(3) (2016) 877–897.

[26] A. Karaisa, Approximation by Durrmeyer type Jakimovski-Leviatan operators. *Math. Meth. Appl. Sci.* **39**(9) (2016) 2401–2410.

[27] A. Karaisa, D. Turgut and Y. Asar, Stancu type generalization of q-Favard-Szász operators. *Appl. Math. Comput.* **264** (2015) 249–257.

[28] D. Lee, On multiple Appell polynomials. *Proc. Amer. Math. Soc.* **139**(6) (2011), 2133–2141.

[29] B. Lenze, On Lipschitz type maximal functions and their smoothness spaces. *Nederl. Akad. Indag. Math.* **50** (1) (1988) 53–63.

[30] M. Mursaleen and K. J. Ansari, On Chlodowsky variant of Szász operators by Brenke type polynomials. *Appl. Math. Comput.* **271** (2015) 991–1003.

[31] M. Mursaleen, A. H. Al-Abied and A. M. Acu, Approximation by Chlodowsky type of Szász operators based on Boas-Buck type polynomials. *Turk. J. Math.* **42** (2018) 2243–2259.

[32] T. Neer, A. M. Acu and P. N. Agrawal, Baskakov-Durrmeyer type operators involving generalized Appell polynomials. *Math. Meth. Appl. Sci.* 2019: 1–13, http://doi.org/10.1002/mma.6089.

[33] Ö. Öksüzer, H. Karsli and F. Taşdelen, Approximation by Kantorovich variant of Szász operators based on Brenke type polynomials. Mediterr. *J. Math.* **13** (2016) 3327–3340.

[34] M. A. Özarslan and O. Duman, Local approximation behaviour of modified SMK operators. *Miskolc Math. Notes.* **11**(1) (2010) 87–99.

[35] I. M. Sheffer, Some properties of polynomial sets of type zero. *Duke Math. J.* **5** (1939) 590–622.

[36] M. Sidharth, P. N. Agrawal and S. Araci, Szász-Durrmeyer operators involving Boas-Buck polynomials of blending type. *J. Inequal. Appl.* 2017, **127**.

[37] S. Sucu, G. İçöz and S. Varma, *On some extension of Szász operators including Boas-Buck type polynomials*, Absr. Appl. Anal. volume 2012. Hindawi Publishing Corporation, (2012).

[38] O. Szász, Generalization of S. Bernstein's polynomials to the infinite interval. *J. Res. Nat. Bur. Standards* **45** (1950) 239–245.

[39] F. Taşdelen, R. Aktaş, and A. Altin, A Kantorovich type of Szász operators including Brenke type polynomials , *Abs. Appl. Anal.* Artical Number 867203 (2012) DOI: 10.1155/2012/867203.

[40] S. Varma, On a generalization of Szász operators by multiple Appell polynomials. Studia Universitatis Babes-Bolyai, *Mathematica* **58**(3) (2013) 361–369.

[41] S. Varma, S. Sucu and G. Icoz, Generalization of Szász operators involving Brenke type polynomials, *Comput. Math. Appl.* **64**(2) (2012) 121–127.

[42] S. Varma and F. Taşdelen, Szász type operators involving Charlier polynomials. *Math. Comput. Modelling,* **56** (2012) 118–122.

[43] A. Wafi, N. Rao and Deepmala, On Kantorovich form of generalized Szász type operators using Charlier polynomials. *Korean J. Math.* **25**(1) (2017) 99–116.

[44] Ö Ö Yilik, T. Garg and P. N. Agrawal, Convergence rate of Szász operators involving Boas-Buck type polynomials. *Proc. Natl. Acad. Sci., India, Sect. A Phys. Sci.,* DOI 10.1007/s40010-020-00663-3.

3 Approximation by Kantorovich Variant of $\lambda-$Schurer Operators and Related Numerical Results

Faruk Özger
İzmir Katip Çelebi University

Kamil Demirci and Sevda Yıldız
Sinop University

CONTENTS

3.1 INTRODUCTION

Functions have been widely approximated by positive linear operators over the past decades. Sergei Natanovich Bernstein first used the known polynomials in the approximation theory to prove Weierstrass' well-known theorem [6]. Bernstein polynomials of order n are given by

$$B_n(h;z) = \sum_{i=0}^{n} \binom{n}{i} z^i \, (1-z)^{n-i} h\left(\frac{i}{n}\right) \quad (z \in [0,1]), \tag{3.1}$$

for any continuous function $h(z)$ defined on $C[0,1]$, the space of all real-valued continuous functions on $[0,1]$ endowed with the norm $\|h\|_{C[0,1]} = \sup_{z\in[0,1]} |h(z)|$. The Bernstein basis polynomials play an important role in approximations of functions, computer-aided geometric design, differential equations, numerical analysis, and constructing Bézier curves [12,13,15,21,22].

A shape parameter $\lambda \in [-1,1]$ was used to construct new Bernstein bases by Ye et al. in [30] since shape parameter λ provides more modeling flexibility. Shape parameter λ was also used by Cai et al. [7] to construct λ-Bernstein operators:

$$B_{n,\lambda}(h;z) = \sum_{i=0}^{n} \tilde{b}_{n,i}(\lambda;z)\, h\left(\frac{i}{n}\right). \tag{3.2}$$

Considering two nonnegative parameters α and β, which satisfy $0 \leq \alpha \leq \beta$, the λ-Stancu operators $S_n^{\alpha,\beta}(h;z;\lambda) : C[0,1] \longrightarrow C[0,1]$ were defined by Srivastava et al. [23] as

$$S_n^{\alpha,\beta}(h;z;\lambda) = \sum_{i=0}^{n} h\left(\frac{i+\alpha}{n+\beta}\right) \tilde{b}_{n,i}(\lambda;z),$$

for any $n \in \mathbb{N}$. Many approximation properties of the bivariate variant of these operators were also investigated. The direct approximation theorem with the help of second-order modulus of smoothness was established, and the rate of convergence was calculated via Lipschitz-type function. Srivastava et al. also constructed the bivariate case of Stancu-type λ-Bernstein operators and studied their approximation behaviors.

Kantorovich modified the well-known Bernstein polynomials (3.1) to approximate the Lebesgue integrable functions on the interval $[0,1]$ in [14]. Acu et al. [3] introduced the following integral modification of Bernstein operators:

$$K_{n,\lambda}(h;z) = (n+1)\sum_{i=0}^{n} \tilde{b}_{n,i}(\lambda;z) \int_{\frac{i}{n+1}}^{\frac{i+1}{n+1}} h(t)dt. \tag{3.3}$$

They generalized both Bernstein (3.1) and λ-Bernstein (3.2) operators and obtained some asymptotic type results. Also, Cai proposed λ-Kantorovich operators (3.3) and the following Bézier variant of Kantorovich-type λ-Bernstein operators [8]:

$$L_{n,\lambda,\beta}(h;z) = (n+1)\sum_{i=0}^{n} Q_{n,i}^{(\beta)}(\lambda;z) \int_{\frac{i}{n+1}}^{\frac{i+1}{n+1}} h(t)dt,$$

where

$$Q_{n,i}^{(\beta)}(\lambda;z) = \left[J_{n,i}^{(\beta)}(\lambda;z)\right]^{\beta} - \left[J_{n,i+1}^{(\beta)}(\lambda;z)\right]^{\beta} \quad \text{and} \quad J_{n,i}^{(\beta)}(\lambda;z) = \sum_{j=i}^{n} \tilde{b}_{n,j}(\lambda;z).$$

Note that Bézier variants of Kantorovich-type λ-Bernstein operators reduce to the λ-Kantorovich operators (3.3) when $\beta = 0$.

Approximation properties of λ-Kantorovich operators with shifted knots were introduced by Rahman et al. [24] as

$$K_{n,\lambda}^{(\alpha,\beta)}(h;z) = \left(\frac{n+\beta}{n}\right)^{n} (n+\beta+1)\sum_{i=0}^{n} \hat{b}_{n,i}^{(\alpha,\beta)}(\lambda;z) \int_{\frac{i+\alpha}{n+\beta+1}}^{\frac{i+\alpha+1}{n+\beta+1}} h(t)dt.$$

A family of GBS operators of bivariate tensor product of λ-Kantorovich type was constructed in [9]. An estimate for the rate of convergence of such operators was given for B-continuous and B-differentiable functions. Also, a Voronovskaja-type asymptotic formula was established for the bivariate case. Many researchers established some Kantorovich-type operators by modifying Bernstein-type operators to have better error estimation [16,24,26].

More recently, statistical approximation properties of univariate and bivariate λ-Kantorovich operators were considered by Özger in [28]. The rate of weighted A-statistical convergence was estimated. A Voronovskaja-type approximation theorem by a family of linear operators was proved using the notion of weighted A-statistical convergence. Some estimates for differences of λ-Bernstein and λ-Durrmeyer, and λ-Bernstein and λ-Kantorovich operators were given. Finally, a Voronovskaja-type approximation theorem by weighted A-statistical convergence was established for the bivariate case of operators.

Acu et al. extended λ-Bernstein operators to introduce a new type genuine Bernstein-Durrmeyer operators in [4]:

$$U^{\rho}_{n,\lambda}(h;z) = \sum_{i=1}^{n-1} \int_0^1 \left[\frac{t^{i\rho-1}(1-t)^{(n-i)\rho-1}}{B(i\rho,(n-i)\rho)} h(t)dt \right] \tilde{b}_{n,i}(\lambda;z)$$
$$+ h(0)\tilde{b}_{n,0}(\lambda;z) + h(1)\tilde{b}_{n,n}(\lambda;z).$$

Here, $B(a,b)$ is Euler's beta function and $\tilde{b}_{n,i}(\lambda;z)$ is the new Bernstein base, which was introduced by Ye et al. in [30].

In [29], Qi et al. constructed λ-Szász-Mirakian operators

$$M_{n,\lambda}(h;z) = \sum_{i=0}^{n} \tilde{m}_{n,i}(\lambda;z)\, h\left(\frac{i}{n}\right), \quad z \in [0,\infty),$$

where

$$\tilde{m}_{n,0}(\lambda;z) = m_{n,0}(z) - \frac{\lambda}{n+\alpha+1} m_{n+1,1}(z),$$
$$\tilde{m}_{n,i}(\lambda;z) = m_{n,i}(z) + \frac{\lambda}{n^2-1}\left[(n-2i+1)m_{n+1,i}(z) - (n-2i-1)m_{n+1,i+1}(z)\right],$$
$$\tilde{m}_{n,n}(\lambda;z) = m_{n,n}(z) - \frac{\lambda}{n+1} m_{n+1,n}(z),$$

and

$$m_{n,,i}(z) = e^{-nz}\frac{(nz)^i}{i!}.$$

The Schurer polynomials $s_{n,i}(z)$ were introduced by Frans Schurer in [20] as

$$s_{n,i}(z) = \binom{n+\alpha}{i} z^i\,(1-z)^{n+\alpha-i} \qquad (i = 0,1,\ldots,n+\alpha), \tag{3.4}$$

where α is a nonnegative integer. The operators generated by these polynomials are called Schurer operators, which were introduced to extend the domain of function from $C[0,1]$ to $C[0,1+\alpha]$. Some relevant works about the Schurer polynomials and Schurer operators may be found in [5,17–19].

The bases in [30] were modified by adding parameter α to introduce the following new bases in [27]:

$$\begin{aligned}\tilde{s}_{n,0}(\lambda;z) &= s_{n,0}(z) - \frac{\lambda}{n+\alpha+1}s_{n+1,1}(z),\\ \tilde{s}_{n,i}(\lambda;z) &= s_{n,i}(z) + \frac{\lambda}{(n+\alpha)^2-1}\left[(n+\alpha-2i+1)s_{n+1,i}(z)\right.\\ &\quad \left.-(n+\alpha-2i-1)s_{n+1,i+1}(z)\right]\ (i=1,2\ldots,n+\alpha-1),\\ \tilde{s}_{n,n+\alpha}(\lambda;z) &= s_{n,n+\alpha}(z) - \frac{\lambda}{n+\alpha+1}s_{n+1,n+\alpha}(z), \end{aligned} \tag{3.5}$$

where shape parameter $\lambda \in [-1,1]$. In the same work, the λ-Schurer operators were introduced, and some approximation and statistical approximation properties were studied. Also, an estimate for the rate of weighted A-statistical convergence was obtained. Moreover, two Voronovskaja-type theorems, including a Voronovskaja-type approximation theorem using weighted A-statistical convergence, were proved.

The rest of this chapter is organized as follows. In Section 3.2, we construct the λ-Schurer-Kantorovich operators, and find moments and central moments. Section 3.3 investigates the approximation behavior of defined λ-Schurer-Kantorovich operators (3.6) and obtains a global and a local direct estimate for the rate of convergence. Section 3.4 establishes three Voronovskaja-type theorems, including a quantitative type. Section 3.5 analyzes the approximation behavior of the defined operators by supporting theoretical results with numerical experiments and graphs. Finally, Section 3.6 summarizes our work and gives some suggestions for readers to carry out further after reading the work.

3.2 AUXILIARY RESULTS

Considering a given nonnegative integer α, we now introduce λ-Schurer-Kantorovich operators $K^{\lambda}_{n,\alpha} : C[0,1+\alpha] \longrightarrow C[0,1]$ for any $n \in \mathbb{N}$ as

$$K^{\lambda}_{n,\alpha}(h;z) = (n+\alpha+1)\sum_{i=0}^{n+\alpha}\tilde{s}_{n,i}(\lambda;z)\int_{\frac{i}{n+\alpha+1}}^{\frac{i+1}{n+\alpha+1}} h(t)dt, \tag{3.6}$$

where new Schurer polynomials $\tilde{s}_{n,i}(\lambda;z)$ are given in (3.5).

Based on the above definition, we reveal the special cases of our new Schurer operators.

Remark 3.1. *We have the following results:*

a. *The λ-Schurer-Kantorovich operators (3.6) are reduced to the classical Schurer-Kantorovich operators if $\lambda = 0$.*

b. *The λ-Schurer-Kantorovich operators (3.6) are reduced to the classical Bernstein-Kantorovich operators if $\alpha=\lambda=0$.*

c. *The λ-Schurer-Kantorovich operators (3.6) are reduced to the λ-Bernstein-Kantorovich operators if $\alpha=0$.*

Now, we find moments and central moments of λ-Schurer-Kantorovich operators (3.6). First, we recall the following lemma.

Lemma 3.1. *[27] We have the following results for λ-Schurer operators:*

$$
\begin{aligned}
S_{n,\alpha}^{\lambda}(1;z) &= 1,\\
S_{n,\alpha}^{\lambda}(t;z) &= \frac{n+\alpha}{n}z+\frac{1-2z+z^{n+\alpha+1}-(1-z)^{n+\alpha+1}}{n(n+\alpha-1)}\lambda,\\
S_{n,\alpha}^{\lambda}(t^2;z) &= \frac{(n+\alpha)^2}{n^2}z^2+\frac{n+\alpha}{n^2}(z-z^2)\\
&\quad+\frac{2(n+\alpha)z-1-4(n+\alpha)z^2+(2(n+\alpha)+1)z^{n+\alpha+1}+(1-z)^{n+\alpha+1}}{n^2(n+\alpha-1)}\lambda.
\end{aligned}
$$

Lemma 3.2. *Let $\lambda\in[-1,1]$ and α be a nonnegative integer, then the moments of λ-Schurer-Kantorovich operators are as follows:*

$$
\begin{aligned}
K_{n,\alpha}^{\lambda}(1;z) &= 1,\\
K_{n,\alpha}^{\lambda}(t;z) &= \frac{1+2(n+\alpha)z}{2(n+\alpha+1)}+\frac{1-2z+z^{n+\alpha+1}-(1-z)^{n+\alpha+1}}{(n+\alpha)^2-1}\lambda,\\
K_{n,\alpha}^{\lambda}(t^2;z) &= \frac{(n+\alpha)^2}{(n+\alpha+1)^2}z^2+\frac{n+\alpha}{(n+\alpha+1)^2}z(2-z)+\frac{1+6z\lambda}{3(n+\alpha+1)^2}\\
&\quad+\frac{2z^{n+\alpha+1}}{(n+\alpha)^2-1}\lambda-\frac{4(n+\alpha)z^2}{(n+\alpha+1)((n+\alpha)^2-1)}\lambda.
\end{aligned}
$$

Proof. Proof of the first part of the lemma is provided by the partition of unity property of Schurer and Kantorovich operators. Bearing in mind the definition of operators (3.6) and Lemma 3.1, we have

$$
\begin{aligned}
K_{n,\alpha}^{\lambda}(t;z) &= (n+\alpha+1)\sum_{i=0}^{n+\alpha}\tilde{s}_{n,i}(\lambda;z)\int_{\frac{i}{n+\alpha+1}}^{\frac{i+1}{n+\alpha+1}}t\,dt\\
&= \sum_{i=0}^{n+\alpha}\tilde{s}_{n,i}(\lambda;z)\frac{2i+1}{2(n+\alpha+1)}\\
&= \frac{n}{n+\alpha+1}S_{n,\alpha}^{\lambda}(t;z)+\frac{1}{2(n+\alpha+1)}S_{n,\alpha}^{\lambda}(1;z)\\
&= \frac{1+2(n+\alpha)z}{2(n+\alpha+1)}+\frac{1-2z+z^{n+\alpha+1}-(1-z)^{n+\alpha+1}}{(n+\alpha)^2-1}\lambda,
\end{aligned}
$$

which completes the proof of the second part. Now, we prove the third part:

$$
\begin{aligned}
K_{n,\alpha}^{\lambda}(t^2;z) &= (n+\alpha+1)\sum_{i=0}^{n+\alpha}\tilde{s}_{n,i}(\lambda;z)\int_{\frac{i}{n+\alpha+1}}^{\frac{i+1}{n+\alpha+1}} t^2\,dt \\
&= \sum_{i=0}^{n+\alpha}\tilde{s}_{n,i}(\lambda;z)\,\frac{3i^2+3i+1}{3(n+\alpha+1)^2} \\
&= \frac{n^2}{(n+\alpha+1)^2}S_{n,\alpha}^{\lambda}(t^2;z)+\frac{n}{(n+\alpha+1)^2}S_{n,\alpha}^{\lambda}(t;z)+\frac{1}{3(n+\alpha+1)^2}S_{n,\alpha}^{\lambda}(1;z) \\
&= \frac{(n+\alpha)^2}{(n+\alpha+1)^2}z^2+\frac{n+\alpha}{(n+\alpha+1)^2}z(2-z)+\frac{1+6z\lambda}{3(n+\alpha+1)^2} \\
&\quad+\frac{2z^{n+\alpha+1}}{(n+\alpha)^2-1}\lambda-\frac{4(n+\alpha)z^2}{(n+\alpha+1)((n+\alpha)^2-1)}\lambda.
\end{aligned}
$$

Thus, Lemma 3.2 is proved. □

Corollary 3.1. *Let $z\in[0,1]$, α be a nonnegative integer, $\lambda\in[-1,1]$, and $\psi_z=t-z$, then we have the following central moments:*

$$
\begin{aligned}
K_{n,\alpha}^{\lambda}(\psi_z;z) &= \frac{1-2z}{2(n+\alpha+1)}+\frac{1-2z+z^{n+\alpha+1}-(1-z)^{n+\alpha+1}}{(n+\alpha)^2-1}\lambda, \\
K_{n,\alpha}^{\lambda}(\psi_z^2;z) &= \frac{z-z^2}{n+\alpha+1}+\frac{1+6z\lambda}{3(n+\alpha+1)^2} \\
&\quad-\frac{z(1-2z)+z^{n+\alpha+1}(z-1)-z(1-z)^{n+\alpha+1}}{(n+\alpha)^2-1}2\lambda \\
&\quad-\frac{4(n+\alpha)z^2}{(n+\alpha+1)((n+\alpha)^2-1)}\lambda.
\end{aligned}
$$

Corollary 3.2. *The following relations hold for operators $K_{n,\alpha}^{\lambda}(\psi_z;z)$ and $K_{n,\alpha}^{\lambda}(\psi_z^2;z)$:*

$$
\begin{aligned}
\lim_{n\to\infty} n\,K_{n,\alpha}^{\lambda}(\psi_z;z) &= \frac{1-2z}{2}, \\
\lim_{n\to\infty} n\,K_{n,\alpha}^{\lambda}(\psi_z^2;z) &= z-z^2.
\end{aligned}
$$

3.3 APPROXIMATION BEHAVIOR OF λ-SCHURER-KANTOROVICH OPERATORS

The following theorem gives the uniform convergence property of λ-Schurer-Kantorovich operators (3.6) by the well-known Bohman-Korovkin-Popoviciu theorem:

Theorem 3.1. *Let $h\in C[0,1+\alpha]$, then*

$$\lim_{n\to\infty} K_{n,\alpha}^{\lambda}(h;z)=h(z)$$

uniformly on $[0,1]$.

Proof. As stated in the Bohman-Korovkin-Popoviciu theorem, it is sufficient to show that

$$\lim_{n\to\infty} K_{n,\alpha}^{\lambda}(t^i;z) = z^i, \quad i = 0,1,2$$

uniformly on $[0,1]$. Using Lemma 3.2, we can easily get these three conditions. Hence, we have the result. □

We need the following notions and notation to achieve a local direct estimate for the rate of convergence of λ-Schurer-Kantorovich operators (3.6) and obtain a global approximation formula.

Definition 3.1. *Global approximation formulas via Ditzian-Totik uniform modulus of smoothness of first and second orders are defined by*

$$\omega_{\xi}(h,\delta) := \sup_{0<|\rho|\leq\delta} \sup_{z,z+\rho\xi(z)\in[0,1]} \{|h(z+\rho\xi(z)) - h(z)|\};$$

$$\omega_2^{\beta}(h,\delta) := \sup_{0<|\rho|\leq\delta} \sup_{z,z\pm\rho\beta(z)\in[0,1]} \{|h(z+\rho\beta(z)) - 2h(z) + h(z-\rho\beta(z))|\},$$

respectively, where β is an admissible step-weight function on $[a,b]$, i.e., $\beta(z) = [(z-a)(b-z)]^{1/2}$ if $z\in[a,b]$, [11]. We denote absolutely continuous functions by AC, and so K-functional is

$$K_{2,\beta(z)}(h,\delta) = \inf_{g\in W^2(\beta)} \{||h-g||_{C[0,1+\alpha]} + \delta||\beta^2 g''||_{C[0,1+\alpha]} : g\in C^2[0,1+\alpha]\},$$

where $\delta>0$, $W^2(\beta) = \{g\in C[0,1+\alpha] : g'\in AC[0,1+\alpha],\ \beta^2 g''\in C[0,1+\alpha]\}$ and $C^2[0,1+\alpha] = \{g\in C[0,1+\alpha] : g',g''\in C[0,1+\alpha]\}$.

Remark 3.2. *[10] There is an absolute constant $\Omega>0$, such that*

$$\Omega^{-1}\omega_2^{\beta}(h,\sqrt{\delta}) \leq K_{2,\beta(z)}(h,\delta) \leq \Omega\omega_2^{\beta}(h,\sqrt{\delta}). \tag{3.7}$$

First, we obtain the global approximation formula in terms of Ditzian-Totik uniform modulus of smoothness of first and second orders.

Theorem 3.2. *Let $\beta(z)$ $(\beta\neq 0)$ be an admissible step-weight function of modulus of smoothness in Definition 3.1 such that β^2 is concave and let $\psi_z = t-z$, $h\in C[0,1+\alpha]$, $z\in[0,1]$, and $\lambda\in[-1,1]$. Then for $\Omega>0$, λ-Schurer-Kantorovich operators (3.6) verify*

$$|K_{n,\alpha}^{\lambda}(h;z) - h(z)| \leq \Omega\ \omega_2^{\beta}\left(h, \frac{[\chi_{n,\alpha,\lambda}(z) + \bar{\chi}_{n,\alpha,\lambda}^2(z)]^{1/2}}{2[(z-a)(b-z)]^{1/2}}\right) + \omega_{\xi}\left(h, \frac{\bar{\chi}_{n,\alpha,\lambda}(z)}{\xi(z)}\right),$$

where $\bar{\chi}_{n,\alpha,\lambda}(z) = K_{n,\alpha}^{\lambda}(\psi_z;z)$ and $\chi_{n,\alpha,\lambda}(z) = K_{n,\alpha}^{\lambda}(\psi_z^2;z)$ are given in Corollary 3.1.

Proof. We define the following operators:

$$\breve{K}_{n,\alpha}^{\lambda}(h;z) = K_{n,\alpha}^{\lambda}(h;z) + h(z) - h\left(\frac{1+2(n+\alpha)z}{2(n+\alpha+1)} + \frac{1-2z+z^{n+\alpha+1}-(1-z)^{n+\alpha+1}}{(n+\alpha)^2-1}\lambda\right) \tag{3.8}$$

where $h \in C[0,1+\alpha]$, $z \in [0,1]$, and $\lambda \in [-1,1]$. We have the following relations

$$\breve{K}_{n,\alpha}^{\lambda}(1;z) = 1 \text{ and } \breve{K}_{n,\alpha}^{\lambda}(t;z) = x$$

by Lemma 3.2. These relations imply $\breve{K}_{n,\alpha}^{\lambda}(\psi_z;z) = 0$.

Let $u = \rho z + (1-\rho)t$, $\rho \in [0,1]$. Since β^2 is concave on $[0,1]$, it follows that $\beta^2(u) \geq \rho\beta^2(z) + (1-\rho)\beta^2(t)$ and

$$\frac{|t-u|}{\beta^2(u)} \leq \frac{\rho|z-t|}{\rho\beta^2(z)+(1-\rho)\beta^2(t)} \leq \frac{|\psi_z|}{\beta^2(z)}. \tag{3.9}$$

Hence, the following inequalities hold:

$$\begin{aligned} |\breve{K}_{n,\alpha}^{\lambda}(h;z) - h(z)| &\leq |\breve{K}_{n,\alpha}^{\lambda}(h-g;z)| + |\breve{K}_{n,\alpha}^{\lambda}(g;z) - g(z)| + |h(z)-g(z)| \\ &\leq 4\|h-g\|_{C[0,1+\alpha]} + |\breve{K}_{n,\alpha}^{\lambda}(g;z) - g(z)|. \end{aligned} \tag{3.10}$$

Applying Taylor's formula, we obtain

$$\begin{aligned} &|\breve{K}_{n,\alpha}^{\lambda}(g;z) - g(z)| \\ &\leq K_{n,\alpha}^{\lambda}\left(\left|\int_z^t |t-u|\,|g''(u)|du\right|;z\right) + \left|\int_z^{z+\bar{\chi}_{n,\alpha,\lambda}(z)} |z+\bar{\chi}_{n,\alpha,\lambda}(z)-u|\,|g''(u)|\,du\right| \\ &\leq \|\beta^2 g''\|_{C[0,1+\alpha]} K_{n,\alpha}^{\lambda}\left(\left|\int_z^t \frac{|t-u|}{\beta^2(u)}du\right|;z\right) + \|\beta^2 g''\|_{C[0,1+\alpha]} \\ &\times \left|\int_z^{z+\bar{\chi}_{n,\alpha,\lambda}(z)} \frac{|z+\bar{\chi}_{n,\alpha,\lambda}(z)-u|}{\beta^2(u)}\,du\right| \\ &\leq \beta^{-2}(z)\|\beta^2 g''\|_{C[0,1+\alpha]} K_{n,\alpha}^{\lambda}(\psi_z^2;z) + \beta^{-2}(z)\|\beta^2 g''\|_{C[0,1+\alpha]} \bar{\chi}_{n,\alpha,\lambda}^2(z). \end{aligned} \tag{3.11}$$

By definition of K-functional with relation (3.7) and inequalities (3.10)–(3.11), we have

$$\begin{aligned} |\breve{K}_{n,\alpha}^{\lambda}(h;z) - h(z)| &\leq 4\|h-g\|_{C[0,1+\alpha]} + \beta^{-2}(z)\|\beta^2 g''\|_{C[0,1+\alpha]}(\chi_{n,\alpha,\lambda}(z) + \bar{\chi}_{n,\alpha,\lambda}^2(z)) \\ &\leq \Omega\,\omega_2^{\beta}\left(h, \frac{[\chi_{n,\alpha,\lambda}(z) + \bar{\chi}_{n,\alpha,\lambda}^2(z)]^{1/2}}{2\beta(z)}\right). \end{aligned}$$

Also, by the help of uniform modulus of smoothness of first order in Definition 3.1, we have

$$|h(z+\bar{\chi}_{n,\alpha,\lambda}(z)) - h(z)| = \left|h\left(z + \xi(z)\frac{\bar{\chi}_{n,\alpha,\lambda}(z)}{\xi(z)}\right) - h(z)\right| \leq \omega_{\xi}\left(h, \frac{\bar{\chi}_{n,\alpha,\lambda}(z)}{\xi(z)}\right).$$

Therefore, the following inequalities, which complete the proof, hold:

$$|K^{\lambda}_{n,\alpha}(h;z)-h(z)| \leq |\breve{K}^{\lambda}_{n,\alpha}(h;z)-h(z)|+\left|h(z+\bar{\chi}_{n,\alpha,\lambda}(z))-h(z)\right|$$
$$\leq \Omega\, \omega_2^{\beta}\left(h, \frac{[\chi_{n,\alpha,\lambda}(z)+\bar{\chi}^2_{n,\alpha,\lambda}(z)]^{1/2}}{2[(z-a)(b-z)]^{1/2}}\right)+\omega_{\xi}\left(h, \frac{\bar{\chi}_{n,\alpha,\lambda}(z)}{\xi(z)}\right).$$

□

Theorem 3.3. *Let $h, h' \in C[0,1+\alpha]$ and $z \in [0,1]$, then the following inequality is satisfied:*

$$|K^{\lambda}_{n,\alpha}(h;z)-h(z)| \leq |\bar{\chi}_{n,\alpha,\lambda}(z)|\ |h'(z)|+2\sqrt{\chi_{n,\alpha,\lambda}(z)}w(h', \sqrt{\chi_{n,\alpha,\lambda}(z)}\ ),$$

where $\chi_{n,\alpha,\lambda}(z)$ and $\bar{\chi}_{n,\alpha,\lambda}(z)$ are given in Theorem 3.2.

Proof. We have the following relation

$$h(t)-h(z)=\psi_z h'(z)+\int_z^t (h'(u)-h'(z))du \tag{3.12}$$

for any $t, z \in [0,1]$. Applying operators $K^{\lambda}_{n,\alpha}(h;z)$ to both sides of (3.12), we have

$$K^{\lambda}_{n,\alpha}(h(t)-h(z);z)=h'(z)K^{\lambda}_{n,\alpha}(\psi_z;z)+K^{\lambda}_{n,\alpha}\left(\int_z^t (h'(u)-h'(z))du;z\right).$$

The following inequality holds for any $\delta>0$, $u \in [0,1]$, and $h \in C[0,1+\alpha]$:

$$|h(u)-h(z)| \leq w(h,\delta)\left(\frac{|u-z|}{\delta}+1\right),$$

With above inequality, we get

$$\left|\int_z^t (h'(u)-h'(z))du\right| \leq w(h',\delta)\left(\frac{\psi_z^2}{\delta}+|\psi_z|\right).$$

Hence, we have

$$|K^{\lambda}_{n,\alpha}(h;z)-h(z)| \leq |h'(z)|\ |K^{\lambda}_{n,\alpha}(\psi_z;z)|+w(h',\delta)\left\{\frac{1}{\delta}K^{\lambda}_{n,\alpha}(\psi_z^2;z)+K^{\lambda}_{n,\alpha}(\psi_z;z)\right\}.$$

Applying Cauchy-Schwarz inequality on the right-hand side of above inequality, we have

$$|K^{\lambda}_{n,\alpha}(h;z)-h(z)| \leq h'(z)|\bar{\chi}_{n,\alpha,\lambda}(z)|+w(h',\delta)\left\{\frac{1}{\delta}\sqrt{K^{\lambda}_{n,\alpha}(\psi_z^2;z)}+1\right\}\sqrt{K^{\lambda}_{n,\alpha}(|\psi_z|;z)}.$$

□

In the following theorem, we obtain a local direct estimate of the rate of convergence via Lipschitz-type function involving two parameters for the operators $K_{n,\alpha}^{\lambda}$. Before proceeding further, let us recall that

$$Lip_M^{(k_1,k_2)}(\eta) := \left\{h \in C[0,1] : |h(t)-h(z)| \leq M\frac{|\psi_z|^{\eta}}{(k_1z^2+k_2z+t)^{\frac{\eta}{2}}};\ z \in (0,1], t \in [0,1]\right\}$$

for $k_1 \geq 0, k_2 > 0$, where $\eta \in (0,1]$ and M is a positive constant (see [25]).

Theorem 3.4. *Let $\lambda \in [-1,1]$, $z \in (0,1]$, and $\eta \in (0,1]$ $h \in Lip_M^{(k_1,k_2)}(\eta)$, then*

$$|K_{n,\alpha}^{\lambda}(h;z)-h(z)| \leq M\sqrt{\frac{\chi_{n,\alpha,\lambda}^{\eta}(z)}{(k_1z^2+k_2z)^{\eta}}},$$

where $\chi_{n,\alpha,\lambda}(z)$ is defined in Theorem 3.2.

Proof. Let $h \in Lip_M^{(k_1,k_2)}(\eta)$ and $\eta \in (0,1]$. First, we show that the statement holds for $\eta = 1$. Since we can express λ-Schurer-Kantorovich operators as

$$K_{n,\alpha}^{\lambda}(h;z) = (n+\alpha+1)\sum_{i=0}^{n+\alpha} \tilde{s}_{n,i}(\lambda;z) \int_0^1 h\left(\frac{i+t}{n+\alpha+1}\right)dt,$$

we have

$$\begin{aligned}|K_{n,\alpha}^{\lambda}(h;z)-h(z)| &\leq |K_{n,\alpha}^{\lambda}(|h(t)-h(z)|;z)| + h(z)\,|K_{n,\alpha}^{\lambda}(1;z)-1| \\ &\leq \sum_{i=0}^{n+\alpha}\left|h\left(\frac{i+t}{n+\alpha+1}\right)-h(z)\right|\tilde{s}_{n,i}(\lambda;z) \\ &\leq M\sum_{i=0}^{n+\alpha}\frac{|\frac{i+t}{n+\alpha+1}-z|}{(k_1z^2+k_2z+t)^{\frac{1}{2}}}\tilde{s}_{n,i}(\lambda;z)\end{aligned}$$

for $h \in Lip_M^{(k_1,k_2)}(1)$. By using

$$(k_1z^2+k_2z+t)^{-1/2} \leq (k_1z^2+k_2z)^{-1/2} \quad (k_1 \geq 0, k_2 > 0)$$

and applying Cauchy-Schwarz inequality, we obtain

$$\begin{aligned}|K_{n,\alpha}^{\lambda}(h;z)-h(z)| &\leq M(k_1z^2+k_2z)^{-1/2}\sum_{i=0}^{n+\alpha}\left|\frac{i+t}{n+\alpha+1}-z\right|\tilde{s}_{n,i}(\lambda;z) \\ &= M(k_1z^2+k_2z)^{-1/2}|K_{n,\alpha}^{\lambda}(\psi_z;z)| \\ &\leq M|\chi_{n,\alpha,\lambda}(z)|^{1/2}(k_1z^2+k_2z)^{-1/2}.\end{aligned}$$

Hence, the statement is true for $\eta = 1$. By the monotonicity of $K^{\lambda}_{n,\alpha}(h;z)$ and applying Hölder's inequality two times with $a = 2/\eta$ and $b = 2/(2-\eta)$, we can see that the statement is true for $\eta \in (0,1]$, as follows:

$$\begin{aligned}
\left|K^{\lambda}_{n,\alpha}(h;z) - h(z)\right| &\leq \sum_{i=0}^{n+\alpha}\left|h\left(\frac{i+t}{n+\alpha+1}\right) - h(z)\right| \tilde{s}_{n,i}(\lambda;z) \\
&\leq \left(\sum_{i=0}^{n+\alpha}\left|h\left(\frac{i+t}{n+\alpha+1}\right) - h(z)\right|^{\frac{2}{\eta}} \tilde{s}_{n,i}(\lambda;z)\right)^{\frac{\eta}{2}} \left(\sum_{i=0}^{n+\alpha}\tilde{s}_{n,i}(\lambda;z)\right)^{\frac{2-\eta}{2}} \\
&\leq M\left(\sum_{i=0}^{n+\alpha}\frac{\left(\frac{i+t}{n+\alpha+1} - z\right)^2 \tilde{s}_{n,i}(\lambda;z)}{\frac{i+t}{n+\alpha+1} + k_1z^2 + k_2z}\right)^{\frac{\eta}{2}} \\
&\leq M(k_1z^2 + k_2z)^{-\eta/2}\left\{\sum_{i=0}^{n+\alpha}\left(\frac{i+t}{n+\alpha+1} - z\right)^2 \tilde{s}_{n,i}(\lambda;z)\right\}^{\frac{\eta}{2}} \\
&\leq M(k_1z^2 + k_2z + t)^{-\eta/2}\left[K^{\lambda}_{n,\alpha}(\psi_z^2;z)\right]^{\frac{\eta}{2}} \\
&= M\sqrt{\frac{\chi^{\eta}_{n,\alpha,\lambda}(z)}{(k_1z^2 + k_2z)^{\eta}}}.
\end{aligned}$$

□

3.4 VORONOVSKAJA-TYPE APPROXIMATION THEOREMS

In this section, we establish a quantitative Voronovskaja-type and a Grüss-Voronovskaja type theorem for $K^{\lambda}_{n,\alpha}(h;z)$ using modulus of smoothness, which is given in Definition 3.1. This modulus is defined as

$$\omega_\beta(h,\delta) := \sup_{0<|\rho|\leq\delta}\left\{\left|h\left(z + \frac{\rho\beta(z)}{2}\right) - h\left(z - \frac{\rho\beta(z)}{2}\right)\right|, z \pm \frac{\rho\beta(z)}{2} \in [0,1]\right\}.$$

Here, $\beta(z) = (z - z^2)^{1/2}$ and $h \in C[0,1+\alpha]$, and the related Peetre's K-functional is known as

$$K_\beta(h,\delta) = \inf_{g\in W_\beta[0,1+\alpha]}\{||h-g|| + \delta||\beta g'|| : g \in C^1[0,1+\alpha], \delta > 0\},$$

where $W_\beta[0,1+\alpha] = \{g : \ g \in AC_{loc}[0,1+\alpha], \ ||\beta g'|| < \infty\}$ and $AC_{loc}[0,1+\alpha]$ is the class of absolutely continuous functions defined on $[a,b] \subset [0,1+\alpha]$. There is a constant $\Omega > 0$ so that

$$K_\beta(h,\delta) \leq \Omega\ \omega_\beta(h,\delta).$$

Theorem 3.5. *Let $h \in C^2[0,1+\alpha]$, then we have*

$$\left|K^{\lambda}_{n,\alpha}(h;z) - h(z) - \bar{\chi}_{n,\alpha,\lambda}(z)h'(z) - \frac{\chi_{n,\alpha,\lambda}(z)}{2}h''(z)\right| \leq \frac{\Omega}{n}\beta^2(z)\omega_\beta\left(h'', \frac{1}{\sqrt{n}}\right)$$

for every $z \in [0,1]$ and sufficiently large n, where Ω is a positive constant, and $\chi_{n,\alpha,\lambda}(z)$ and $\bar{\chi}_{n,\alpha,\lambda}(z)$ are defined in Theorem 3.2.

Proof. Consider the following equality:

$$h(t) - h(z) - \psi_z h'(z) = \int_z^t (t-u)h''(u)du$$

for $h \in C[0,1+\alpha]$. It means we have

$$h(t) - h(z) - \psi_z h'(z) - \frac{\psi_z^2}{2}h''(z) = \int_z^t (t-u)[h''(u) - h''(z)]du. \tag{3.13}$$

Applying $K_{n,\alpha}^{\lambda}(h;z)$ to both sides of (3.13), we obtain

$$\begin{aligned}&\left|K_{n,\alpha}^{\lambda}(h;z) - h(z) - K_{n,\alpha}^{\lambda}(\psi_z;z)h'(z) - \frac{K_{n,\alpha}^{\lambda}(\psi_z^2;z)}{2}h''(z)\right| \\ &\leq K_{n,\alpha}^{\lambda}\left(\left|\int_z^t |t-u|\,|h''(u) - h''(z)|\,du\right|;z\right).\end{aligned} \tag{3.14}$$

The expression in the right-hand side of (3.14) can be estimated as

$$\left|\int_z^t |t-u|\,|h''(u) - h''(z)|\,du\right| \leq 2\|h'' - g\|\psi_z^2 + 2\|\beta g'\|\beta^{-1}(z)|\psi_z|^3, \tag{3.15}$$

where $g \in W_\beta[0,1+\alpha]$. There exists $\Omega > 0$ such that

$$K_{n,\alpha}^{\lambda}(\psi_z^2;z) \leq \frac{\Omega}{2n}\beta^2(z) \quad \text{and} \quad K_{n,\alpha}^{\lambda}(\psi_z^4;z) \leq \frac{\Omega}{2n^2}\beta^4(z) \tag{3.16}$$

for sufficiently large n. Using Cauchy-Schwarz inequality, we have

$$\begin{aligned}&\left|K_{n,\alpha}^{\lambda}(h;z) - h(z) - K_{n,\alpha}^{\lambda}(\psi_z;z)h'(z) - \frac{K_{n,\alpha}^{\lambda}(\psi_z^2;z)}{2}h''(z)\right| \\ &\leq 2\|h'' - g\|K_{n,\alpha}^{\lambda}(\psi_z^2;z) + 2\|\beta g'\|\beta^{-1}(z)K_{n,\alpha}^{\lambda}(|\psi_z|^3;z) \\ &\leq \frac{\Omega}{n}(z - z^2)\|h'' - g\| + 2\|\beta g'\|\beta^{-1}(z)\{K_{n,\alpha}^{\lambda}(\psi_z^2;z)\}^{1/2}\{K_{n,\alpha}^{\lambda}(\psi_z^4;z)\}^{1/2} \\ &\leq \frac{\Omega}{n}\beta^2(z)\left\{\|h'' - g\| + n^{-1/2}\|\beta g'\|\right\}\end{aligned}$$

by (3.14)–(3.16). Taking infimum on the right-hand side over all $g \in W_\beta[0,1+\alpha]$, we deduce

$$\left|K_{n,\alpha}^{\lambda}(h;z) - h(z) - \bar{\chi}_{n,\alpha,\lambda}(z)h'(z) - \frac{\chi_{n,\alpha,\lambda}(z)}{2}h''(z)\right| \leq \frac{\Omega}{n}\beta^2(z)\omega_\beta\left(h'', \frac{1}{\sqrt{n}}\right).$$

The result follows immediately by applying Corollaries 3.1 and 3.2. □

As an immediate consequence of Theorem 3.5, we have the following result.

Corollary 3.3. *Let $h \in C^2[0,1+\alpha]$, then*

$$\lim_{n\to\infty} n\left[K_{n,\alpha}^{\lambda}(h;z) - h(z) - \bar{\chi}_{n,\alpha,\lambda}(z)h'(z) - \frac{\chi_{n,\alpha,\lambda}(z)}{2}h''(z)\right] = 0,$$

where $\chi_{n,\alpha,\lambda}(z)$ and $\bar{\chi}_{n,\alpha,\lambda}(z)$ are defined in Theorem 3.2.

A Grüss inequality for the positive linear operators was first given by Acu et al. in [2] by using the least concave majorant of the modulus of continuity. Then, Acar et al. [1] obtained a Grüss-type approximation theorem and a Grüss-Voronovskaja-type theorem for a class of sequences of linear positive operators. Now, we will obtain a Grüss-Voronovskaja-type theorem for λ-Schurer-Kantorovich operators $K_{n,\alpha}^{\lambda}$:

Theorem 3.6. *Let $h,k \in C^2[0,1+\alpha]$. Then, for each $z \in [0,1]$,*

$$\lim_{n\to\infty} n\left\{K_{n,\alpha}^{\lambda}((hk);z) - K_{n,\alpha}^{\lambda}(h;z)K_{n,\alpha}^{\lambda}(k;z)\right\} = h'(z)k'(z)\left(z-z^2\right).$$

Proof. It can be easily seen that the following equality holds:

$$\begin{aligned}
&K_{n,\alpha}^{\lambda}((hk);z) - K_{n,\alpha}^{\lambda}(h;z)K_{n,\alpha}^{\lambda}(k;z)\\
&= K_{n,\alpha}^{\lambda}((hk);z) - h(z)k(z) - (hk)'(z)\overline{\chi}_{n,\alpha,\lambda}(z) - (hk)''(z)\frac{\chi_{n,\alpha,\lambda}(z)}{2}\\
&\quad - k(z)\left\{K_{n,\alpha}^{\lambda}(h;z) - h(z) - h'(z)\overline{\chi}_{n,\alpha,\lambda}(z) - h''(z)\frac{\chi_{n,\alpha,\lambda}(z)}{2}\right\}\\
&\quad - K_{n,\alpha}^{\lambda}(h;z)\left\{K_{n,\alpha}^{\lambda}(k;z) - k(z) - k'(z)\overline{\chi}_{n,\alpha,\lambda}(z) - k''(z)\frac{\chi_{n,\alpha,\lambda}(z)}{2}\right\}\\
&\quad + \frac{\chi_{n,\alpha,\lambda}(z)}{2}\left\{h(z)k''(z) + 2h'(z)k'(z) - k''(z)K_{n,\alpha}^{\lambda}(h;z)\right\}\\
&\quad + \overline{\chi}_{n,\alpha,\lambda}(z)\left\{h(z)k'(z) - k'(z)K_{n,\alpha}^{\lambda}(h;z)\right\}.
\end{aligned}$$

Hence, by using Theorem 3.1 and Corollary 3.3, we have

$$\begin{aligned}
&\lim_{n\to\infty} n\left\{K_{n,\alpha}^{\lambda}((hk);z) - K_{n,\alpha}^{\lambda}(h;z)K_{n,\alpha}^{\lambda}(k;z)\right\}\\
&= \lim_{n\to\infty} nk''(z)\left\{h(z) - K_{n,\alpha}^{\lambda}(h;z)\right\}\frac{\chi_{n,\alpha,\lambda}(z)}{2} + \lim_{n\to\infty} 2nh'(z)k'(z)\frac{\chi_{n,\alpha,\lambda}(z)}{2}\\
&\quad + \lim_{n\to\infty} nk'(z)\left\{h(z) - K_{n,\alpha}^{\lambda}(h;z)\right\}\overline{\chi}_{n,\alpha,\lambda}(z)\\
&= h'(z)k'(z)\left(z-z^2\right).
\end{aligned}$$

□

Finally, we obtain the following theorem by Taylor's expansion theorem and Lemma 3.2, and Corollaries 3.1 and 3.2:

Theorem 3.7. *Let $h \in C^2[0,1+\alpha]$, then for each $z \in [0,1]$*

$$\lim_{n\to\infty} n\{K_{n,\alpha}^{\lambda}(h;z)-h(z)\} = (1-2z)\,\frac{h'(z)}{2} + (z-z^2)\,\frac{h''(z)}{2}$$

uniformly on $[0,1]$.

Proof. We first write the following equality by Taylor's expansion theorem of function $h(z)$ in $C[0,1]$:

$$h(t) = h(z) + \psi_z h'(z) + \frac{1}{2}\psi_z^2 h''(z) + \psi_z^2\, r_z(t), \tag{3.17}$$

where $r_z(t)$ is Peano form of the remainder, $r_z \in C[0,1]$ and $r_z(t) \to 0$ as $t \to z$. Applying the operators $K_{n,\alpha}^{\lambda}(\cdot;z)$ to identity (3.17), we have

$$K_{n,\alpha}^{\lambda}(h;z)-h(z) = h'(z)K_{n,\alpha}^{\lambda}(\psi_z;z) + \frac{h''(z)}{2}K_{n,\alpha}^{\lambda}(\psi_z^2;z) + K_{n,\alpha}^{\lambda}(\psi_z^2 r_z(t);z).$$

Using Cauchy-Schwarz inequality, we have

$$K_{n,\alpha}^{\lambda}(\psi_z^2 r_z(t);z) \le \sqrt{K_{n,\alpha}^{\lambda}(\psi_z^4;z)}\sqrt{K_{n,\alpha}^{\lambda}(r_z^2(t);z)}. \tag{3.18}$$

We observe that $\lim_n K_{n,\alpha}^{\lambda}(r_z^2(t);z) = 0$ and hence

$$\lim_{n\to\infty} n\{K_{n,\alpha}^{\lambda}(\psi_z^2 r_z(t);z)\} = 0.$$

Thus,

$$\begin{aligned}\lim_{n\to\infty} n\{K_{n,\alpha}^{\lambda}(h;z)-h(z)\} = \lim_{n\to\infty} n\{&K_{n,\alpha}^{\lambda}(\psi_z;z)h'(z) + \frac{h''(z)}{2}K_{n,\alpha}^{\lambda}(\psi_z^2;z)\\ &+ K_{n,\alpha}^{\lambda}(\psi_z^2 r_z(t);z)\}.\end{aligned}$$

□

3.5 GRAPHICAL AND NUMERICAL RESULTS

The aim of this section is to support the results given in the previous sections by graphs and numerical examples. We first study on the behavior of polynomials $\tilde{s}_{n,i}(\lambda;z)$ for different values of shape parameter λ. In Figure 3.1, we demonstrate the classical Bernstein polynomials and λ-Schurer polynomials $\tilde{s}_{n,i}(\lambda;z)$ to see the difference.

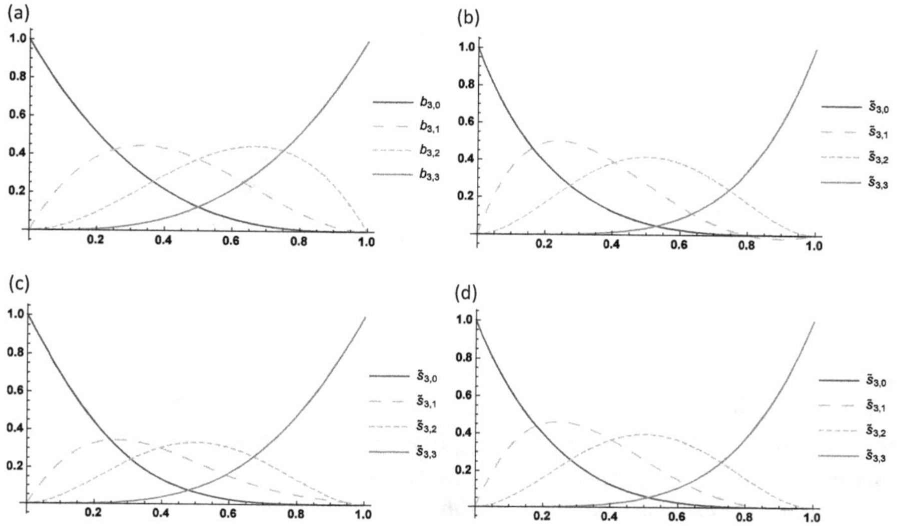

Figure 3.1 Behavior of polynomials $b_{n,i}(z)$ and $\tilde{s}_{n,i}(\lambda;z)$ for $n=3$ and $i=0,1,\ldots,3$. (a) $b_{3,i}(z)$ for $i=0,1,\ldots,3$, (b) $\tilde{s}_{3,i}(\lambda;z)$ with $\lambda=1$ and $\alpha=1$, (c) $\tilde{s}_{3,i}(\lambda;z)$ with $\lambda=-1$ and $\alpha=1$, and (d) $\tilde{s}_{3,i}(\lambda;z)$ with $\lambda=0.5$ and $\alpha=1$.

Example 3.1. *Let $\alpha=2$, $\lambda=0.5$, $h(z)=\cos(\pi z)$, and $E^{\lambda}_{n,\alpha}(h;z)=\left|h(z)-K^{\lambda}_{n,\alpha}(h;z)\right|$ be the error function of λ-Schurer-Kantorovich operators. The graphs of operators $K^{\lambda}_{n,\alpha}$ for $n=20$, $n=50$, $n=100$ and the graph of the function h are illustrated in Figure 3.2a. The error functions $E^{\lambda}_{n,\alpha}$ are given in Figure 3.2b. Hence, we show the convergence and error of approximation of λ-Schurer-Kantorovich operators to the function h. Also, we see the error estimation from Table 3.1.*

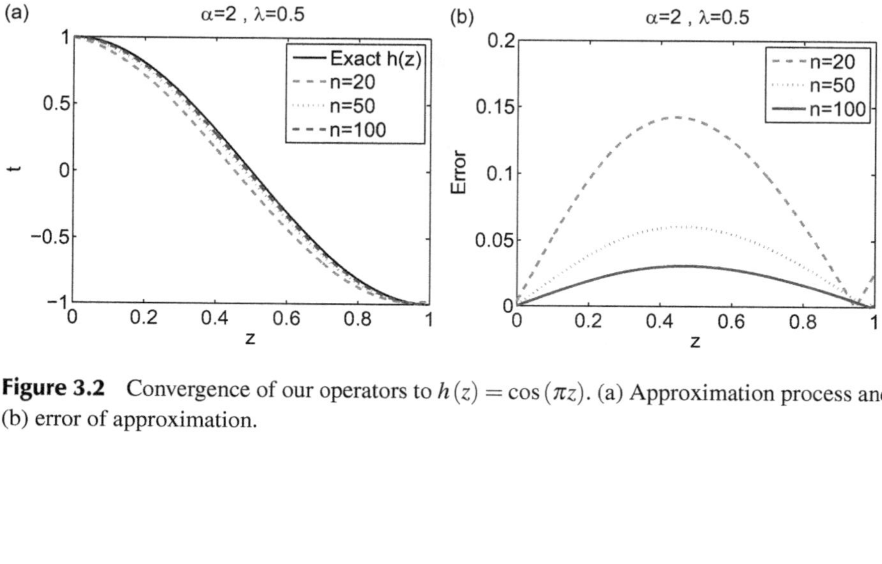

Figure 3.2 Convergence of our operators to $h(z)=\cos(\pi z)$. (a) Approximation process and (b) error of approximation.

Table 3.1
The Errors of the Approximation

n	α	λ	Error
20	2	0.5	1.42e−01
50	2	0.5	6.05e−02
100	2	0.5	3.09e−02

Example 3.2. *Let $n = 10$, $h(z) = \left(z - \frac{1}{2}\right)\left(z - \frac{3}{8}\right)\left(z - \frac{1}{4}\right)$. In Figure 3.3a, we see the graphs of λ-Schurer-Kantorovich operators $K^1_{n,2}$, λ-Kantorovich operators $K^1_{n,0}$ that are given by Acu et al.[3], and $K^0_{10,0}$ that are the classical Kantorovich operators and the graph of function h. The error functions $E^{\lambda}_{n,\alpha}$ are given in Figure 3.3b. Also, we see the error estimation from Table 3.2. Hence, we show that the error for λ-Schurer-Kantorovich operators $K^1_{n,2}$ is smaller than the ones for $K^1_{n,0}$ and $K^0_{10,0}$.*

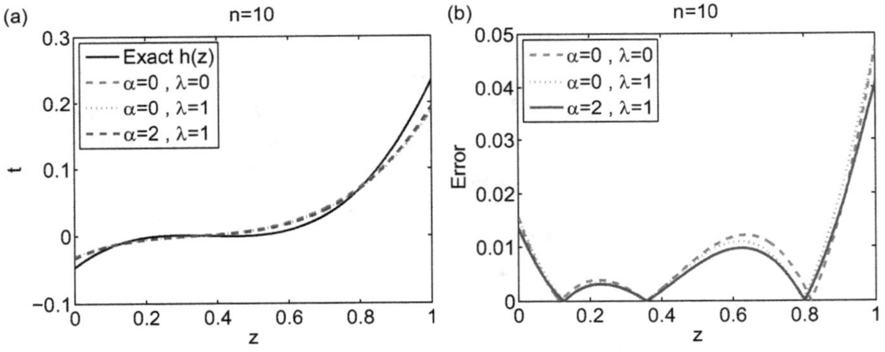

Figure 3.3 Convergence of our operators to $h(z) = \left(z - \frac{1}{2}\right)\left(z - \frac{3}{8}\right)\left(z - \frac{1}{4}\right)$. (a) Approximation process and (b) error of approximation.

Table 3.2
Comparison of Operators via Errors

n	α	λ	Error
10	0	0	4.76e−02
10	0	1	4.76e−02
10	2	1	4.09e−02

When we compare the numerical results of this section and the results in [3,7,8,27], we see that our operators have less error of approximation.

3.6 CONCLUSION

One of the main contributions of our work is to approximate the Lebesgue integrable functions on the interval $[0,1]$ more precisely. This is achieved by considering new Schurer polynomials defined in [27] instead of the polynomials defined in [7].

We provide a comprehensive literature review about λ-Bernstein-type operators, Bernstein operators, Schurer operators, and Schurer and Bernstein polynomials. We obtain some Voronovskaja-type theorems, including a Grüss-Voronovskaja and a quantitative Voronovskaja-type theorem for λ-Schurer-Kantorovich operators. We explore the convergence of our operators, and we see that our new operators have several advantages. As it is theoretically and numerically shown, the new operators defined in this chapter generalize the results in [3,5–9,17,20,27]. We investigate some approximation properties and show the relevance of the results by examples, and hence, we get better error estimation for our newly defined operators.

This study may be extended by considering q and (p,q) analogues of λ-Schurer, λ-Schurer-Kantorovich, λ-Schurer-Stancu, and λ-Schurer-Durrmeyer operators after introducing λ-Schurer-Durrmeyer and λ-Schurer-Stancu operators. All these operators may be considered for the functions of two variables, too.

REFERENCES

[1] Acar T, Aral A, Rasa I. The new forms of Voronovskaja's theorem in weighted spaces. *Positivity* 2016; 20(1):25–40.

[2] Acu A M, Gonska H, Rasa I. Grüss-type and Ostrowski-type inequalities in approximation theory. *Ukr. Math. J.* 2011; 63(6):843–864.

[3] Acu AM, Manav N, Sofonea S. Approximation properties of λ-Kantorovich operators. *J Inequal Appl.* 2018; 2018:202.

[4] Acu A M, Acar T, Radu V A. Approximation by modified $U^{\rho}_{n,\lambda}$ operators. *RACSAM* 2019; 113:2715–2729.

[5] Bărbosu D, Bărbosu M. Some properties of the fundamental polynomials of Bernstein-Schurer. Bul Ştiinţ Univ Baia Mare Ser B, *Mat-Inf.* 2002; 18(2):133–136.

[6] Bernstein SN. Démonstration du théorème de Weierstrass fondée sur le calcul des probabilités. *Communications of the Kharkov Mathematical Society* 1912; 13(2):1–2.

[7] Cai QB, Lian B-Y, Zhou G. Approximation properties of λ-Bernstein operators. *J Ineq and App* 2018; 2018:61.

[8] Cai QB. The Bézier variant of Kantorovich type λ-Bernstein operators. *J Inequal Appl.* 2018; 2018:90.

[9] Cai QB, Zhou G. Blending type approximation by GBS operators of bivariate tensor product of λ-Bernstein-Kantorovich type. *J Inequal Appl.* 2018; 2018:268.

[10] DeVore RA, Lorentz GG. *Constructive Approximation*. Berlin: Springer; 1993.

[11] Ditzian Z, Totik V. *Moduli of Smoothness*. New York: Springer; 1987.

[12] Farouki RT. The Bernstein polynomials basis: a centennial retrospective. *Comput Aided Geom Des.* 2012; 29:379–419.

[13] Goldman R. *Pyramid algorithms, a dynamic programming approach to curves and surfaces for geometric modeling*. The Morgan Kaufmann Series in Computer Graphics and Geometric Modeling San Francisco: Elsevier Science; 2002.

[14] Kantorovich LV. Sur certains developements suivant les polynmes de la forme de S. Bernstein I, II. *Dokl Akad Nauk SSSR* 1930; (563)568:595–600.

[15] Lorentz GG. *Bernstein polynomials*. Chelsea Pub. Comp: New York, 1986.

[16] Mohiuddine SA, F. Özger. Approximation of functions by Stancu variant of Bernstein-Kantorovich operators based on shape parameter α, *Rev. R. Acad. Cienc. Exactas Fís. Nat. Ser. A Math. RACSAM* 2020; 114:70.

[17] Muraru C. On the monotonicity of Schurer type polynomials. *Carpathian J Math.* 2005; 21(1–2):89–94.

[18] Mursaleen M, Ahasan M, Ansari KJ. Bivariate BernsteinSchurerStancu type GBS operators in (p,q)-analogue. *Adv Differ Equ.* 2020; 76(2020).

[19] Mursaleen M, Al-Abeid AH, Ansari KJ. On approximation properties of Baskakov-Schurer-Szász-Stancu operators based on q-integers, *Filomat* 2018; 32(41):359-1378.

[20] Schurer F. Linear positive operators in approximation theory. Math Inst Techn Univ Delft Report 1962.

[21] Simsek Y. Construction a new generating function of Bernstein type polynomials. *Appl Math Comput.* 2011; 218:1072–1076.

[22] Simsek Y. Analysis of the Bernstein basis functions: an approach to combinatorial sums involving binomial coefficients and Catalan numbers. *Math Method Appl Sci.* 2015; 38:3007–3021.

[23] Srivastava HM, Özger F, Mohiuddine SA. Construction of Stancu-type Bernstein operators based on Bézier bases with shape parameter λ. *Symmetry* 2019; 11(3):316. DOI:10.3390/symxx010005.

[24] Rahman S, Mursaleen M, Acu AM. Approximation properties of λ-Bernstein Kantorovich operators with shifted knots. *Math Meth Appl Sci.* 2019; 42(11):042–4053.

[25] Özarslan MA, Aktuğlu H. Local approximation for certain King type operators. *Filomat* 2013; 27(1):173–181.

[26] Özarslan MA, Duman O. Smoothness properties of modified Bernstein-Kantorovich operators. *Numer Func Anal Opt.* 2016; 37(1):92–105.

[27] Özger F. On new Bézier bases with Schurer polynomials and corresponding results in approximation theory. *Commun Fac Sci Univ Ank Ser A1 Math Stat.* 2019; 69(1): 376–393.

[28] Özger F. Weighted statistical approximation properties of univariate and bivariate λ-Kantorovich operators. *Filomat* 2019; 33(11):3473–3486.

[29] Qi Q, Guo D, Yang G. Approximation Properties of λ-Szász-Mirakian Operators. *International Journal of Engineering Research and Technology.* 2019; 12(5):662–669.

[30] Ye Z, Long X, Zeng X-M. Adjustment algorithms for Bézier curve and surface. *International Conference on Computer Science and Education* 2010; 1712–1716. DOI: 10.1109/ICCSE.2010.5593563.

4 Characterizations of Rough Fractional-Type Integral Operators on Variable Exponent Vanishing Morrey-Type Spaces

Ferit Gürbüz
Hakkari University

Shenghu Ding, Huili Han, and Pinhong Long
Ningxia University

CONTENTS

4.1 INTRODUCTION

In this work, we mainly focus on some operators and commutators on the variable exponent-generalized Morrey-type space. Precisely, our aim is to characterize the boundedness for the maximal operator, fractional integral operator, and fractional maximal operator with rough kernel as well as the corresponding commutators on the variable exponent vanishing generalized Morrey spaces.

We first list a series of (somewhat standard) notation needed for later sections.

1. Let $x=(x_1,x_2,\dots,x_n)$, $\xi=(\xi_1,\xi_2,\dots,\xi_n)$ .... etc. be the points of the real n-dimensional space $\mathbb{R}^n$. Let $x.\xi=\sum\limits_{i=1}^{n}x_i\xi_i$ stand for the usual dot product in $\mathbb{R}^n$ and $|x|=\left(\sum\limits_{i=1}^{n}x_i^2\right)^{\frac{1}{2}}$ for the Euclidean norm of x.
2. By x', we always mean the unit vector corresponding to x, i.e., $x'=\frac{x}{|x|}$ for any $x\neq 0$.
3. $S^{n-1}=\{x\in\mathbb{R}^n:|x|=1\}$ represents the unit sphere in Euclidean n-dimensional space $\mathbb{R}^n$ $(n\geq 2)$, and dx' is its surface measure.
4. $B(x,r)=\{y\in\mathbb{R}^n:|x-y|<r\}$ denotes x-centered Euclidean ball with radius r, $B^C(x,r)$ denotes its complement, and $|B(x,r)|$ is the Lebesgue measure of the ball $B(x,r)$, $|B(x,r)|=v_nr^n$, where $v_n=|B(0,1)|=\frac{2\pi^{\frac{n}{2}}}{n\Gamma\left(\frac{n}{2}\right)}$ and $\tilde{B}(x,r)=B(x,r)\cap E$, where $E\subset\mathbb{R}^n$ is an open set. Finally, we use the notation

$$f_{B(x,r)}=\frac{1}{|B(x,r)|}\int\limits_{\tilde{B}(x,r)}f(y)dy.$$

5. C stands for a positive constant that can change its value in each statement without explicit mention.
6. The exponents $p'(\cdot)$ and $s'(\cdot)$ always denote the conjugate index of any exponent $1<p(x)<\infty$ and $1<s(x)<\infty$, i.e., $\frac{1}{p'(x)}:=1-\frac{1}{p(x)}$ and $\frac{1}{s'(x)}:=1-\frac{1}{s(x)}$.
7. In the sequel, for any exponent $1<p(x)<\infty$ and bounded sets $E\subset\mathbb{R}^n$, if we use

$$|p(x)-p(y)|\leq\frac{-C}{\log(|x-y|)}\qquad |x-y|\leq\frac{1}{2},\qquad x,y\in E,\tag{4.1}$$

where $C=C(p)>0$ does not depend on x, y, then we call that $p(\cdot)$ satisfies the local log-Hölder continuity condition or Dini-Lipschitz condition. The important role of the local log-Hölder continuity of $p(x)$ is well known in variable analysis. On the other hand, the condition

$$|p(x)-p(y)|\leq\frac{C}{\log(e+|x|)}\qquad |y|\geq|x|,\qquad x,y\in E,$$

introduced by Cruz-Uribe et al. in [4] is known as the log-Hölder decay condition used for unbounded sets E. It is equivalent to the condition that there exists a number $p_\infty\in[1,\infty)$ such that

$$\left|\frac{1}{p_\infty}-\frac{1}{p(x)}\right|\leq\frac{C_\infty}{\log(e+|x|)}\qquad\text{for all } x\in E,\tag{4.2}$$

where $p_\infty=\lim\limits_{|x|\to\infty}p(x)$.

If $p(\cdot)$ satisfies both (4.1) and (4.2), then we say that it is log-Hölder continuous.

8. Let $F, G \geq 0$. Here and henceforth, the symbol $F \approx G$ means that $F \lesssim G$ and $G \lesssim F$ happen simultaneously, while $F \lesssim G$ and $G \lesssim F$ mean that there exists a constant $C > 0$ such that $F \leq CG$.
9. Let $\Omega \in L_s(S^{n-1})$ with $1 < s \leq \infty$ be the homogeneous function of degree 0 on $\mathbb{R}^n$ and satisfy the integral zero property over the unit sphere S^{n-1}. Moreover, note that $\|\Omega\|_{L_s(S^{n-1})} := \left(\int\limits_{S^{n-1}} |\Omega(z')|^s \, d\sigma(z') \right)^{\frac{1}{s}}$ and

$$\begin{aligned}
\|\Omega(z-y)\|_{L_s(\tilde{B}(x,r))} &= \left(\int\limits_{\tilde{B}(x,r)} |\Omega((z-y))|^s \, dz \right)^{\frac{1}{s}} \\
&\lesssim \left(\int\limits_{\tilde{B}(x,r)} \Omega(\sigma)^s \int\limits_0^r \rho^{n-1} d\rho d\sigma \right)^{\frac{1}{s}} \\
&\lesssim \|\Omega\|_{L_s(S^{n-1})} r^{\frac{n}{s}},
\end{aligned} \tag{4.3}$$

for $z \in B(x,r)$.
10. Suppose that $0 < \alpha(x) < n, x \in E \subset \mathbb{R}^n$. Then, the rough Riesz-type potential operator with variable order $I_{\Omega,\alpha(\cdot)}$ and the corresponding rough fractional maximal operator with variable order $M_{\Omega,\alpha(\cdot)}$ are defined, respectively, by

$$I_{\Omega,\alpha(\cdot)} f(x) = \int\limits_E \frac{\Omega(x-y)}{|x-y|^{n-\alpha(x)}} f(y) dy$$

and

$$M_{\Omega,\alpha(\cdot)} f(x) = \sup_{r>0} |B(x,r)|^{\frac{\alpha(x)}{n}-1} \int\limits_{\tilde{B}(x,r)} |\Omega(x-y)| \, |f(y)| dy,$$

where $E \subset \mathbb{R}^n$ is an open set. On the other hand, if $\alpha(\cdot) = 0$, then the rough Calderón-Zygmund-type singular integral operator T_Ω in the sense of principal value Cauchy integral is defined by

$$T_\Omega f(x) = p.v. \int\limits_E \frac{\Omega(x-y)}{|x-y|^n} f(y) dy,$$

and especially in the limiting case $\alpha(\cdot) = 0$, the rough fractional maximal operator with variable order $M_{\Omega,\alpha}$ reduces to the rough Hardy-Littlewood maximal operator M_Ω, and M_Ω is also defined by

$$M_\Omega f(x) = \sup_{r>0} \frac{1}{|B(x,r)|} \int\limits_{\tilde{B}(x,r)} |\Omega(y)| \, |f(x-y)| \, dy,$$

where $E \subset \mathbb{R}^n$ is an open set. In fact, we can easily see that when $\Omega \equiv 1$, $M_{1,\alpha(\cdot)} \equiv M_{\alpha(\cdot)}$ and $I_{1,\alpha(\cdot)} \equiv I_{\alpha(\cdot)}$ are the fractional maximal operator with variable order and the Riesz-type potential operator with variable order, and similarly, M and T are the Hardy-Littlewood maximal operator and the standard Calderón-Zygmund-type singular integral operator, respectively.

11. The boundedness properties of commutator operators is also an important aspect of harmonic analysis as these are useful in the study of characterization of function spaces and regularity theory of partial differential equations. The commutators of the operators T_Ω, M_Ω with rough kernel Ω with locally integrable function b are given by

$$\begin{aligned}[b,T_\Omega] f(x) &= b(x) T_\Omega f(x) - T_\Omega (bf)(x) \\ &= p.v. \int\limits_E \frac{\Omega(x-y)}{|x-y|^n} (b(x) - b(y)) f(y) dy\end{aligned}$$

and

$$\begin{aligned}[b,M_\Omega] f(x) &= b(x) M_\Omega f(x) - M_\Omega (bf)(x) \\ &= \sup_{r>0} \frac{1}{|B(x,r)|} \int\limits_{\tilde{B}(x,r)} |\Omega(x-y)| (b(x) - |b(y)|) |f(y)| dy,\end{aligned}$$

similarly, define the rough commutators $[b, I_{\Omega,\alpha(\cdot)}]$, $[b, M_{\Omega,\alpha(\cdot)}]$ generated by the function b and the operators $I_{\Omega,\alpha(\cdot)}$ and $M_{\Omega,\alpha(\cdot)}$ with rough kernel Ω and variable order $\alpha(\cdot)(0 \le \alpha(\cdot) < n)$ by

$$\begin{aligned}[b, I_{\Omega,\alpha(\cdot)}] f(x) &= b(x) I_{\Omega,\alpha(\cdot)} f(x) - I_{\Omega,\alpha(\cdot)} (bf)(x) \\ &= \int\limits_E \frac{\Omega(x-y)}{|x-y|^{n-\alpha(x)}} (b(x) - b(y)) f(y) dy\end{aligned}$$

and

$$\begin{aligned}[b, M_{\Omega,\alpha(\cdot)}] f(x) &= b(x) M_{\Omega,\alpha(\cdot)} f(x) - M_{\Omega,\alpha(\cdot)} (bf)(x) \\ &= \sup_{r>0} |B(x,r)|^{\frac{\alpha(x)}{n}-1} \int\limits_{\tilde{B}(x,r)} |\Omega(x-y)| (b(x) - |b(y)|) |f(y)| dy.\end{aligned}$$

Later, Morrey spaces can complement the boundedness properties of operators that Lebesgue spaces can not handle. Morrey spaces that we have been handling are called classical Morrey spaces (see [16]). But classical Morrey spaces are not totally enough to describe the boundedness properties. To this end, we need to generalize parameters p and q, among others p, but this issue will exceed the scope of this chapter, so we pass this part. Though we do not consider the direct applications of

Morrey spaces to PDEs, Morrey spaces can be applied to PDEs. Applications to the second-order elliptic partial differential equations can be found in [9] and [21].

Recently, while we try out to resolve somewhat modern problems emerging inherently such that nonlinear elasticity theory, fluid mechanics, etc., it is well-known that classical function spaces are not anymore suitable spaces. It thus becomes essential to introduce and analyze the diverse function spaces from diverse viewpoints. One of such spaces is the variable exponent Lebesgue space $L^{p(\cdot)}$. This space is a generalization of the classical $L^p(\mathbb{R}^n)$ space, in which the constant exponent p is replaced by an exponent function $p(\cdot):\mathbb{R}^n \to (0,\infty)$, and it consists of all functions f such that $\int_{\mathbb{R}^n} |f(x)|^{p(x)}\,dx < \infty$. This theory got a boost in 1931 when Orlicz published his seminal paper [17]. The next major step in the investigation of variable exponent spaces was the comprehensive paper by Kováčik and Rákosník in the early 1990s [14]. Since then, the theory of variable exponent spaces was applied to many fields: Refer to [3,23] for the image processing, [2] for thermorheological fluids, [19] for electrorheological fluids, and [11] for the differential equations with nonstandard growth. For the nonweighted and weighted variable exponent settings, refer to [5–8]. On the other hand, Kováčik and Rákosník [14] established many of the basic properties of Lebesgue and Sobolev spaces. Moreover, since these authors clarified fundamental properties of the variable exponent Lebesgue and Sobolev spaces, there are many spaces studied, such as variable exponent Morrey, generalized Morrey, vanishing generalized Morrey, and Herz-Morrey spaces (see [1,10,12,13,15,20,22,24]). In the last decade, when the parameters that define the operator have changed from point to point, there has been a strong interest in fractional-type operators and the "variable setting" function spaces. The field called variable exponent analysis has become a fairly branched area with many interesting results obtained in the last decade, such as harmonic analysis, approximation theory, operator theory, and pseudo-differential operators. But the results in this paper lie in these spaces known as variable exponent Morrey-type spaces on the rough fractional-type operators with variable order of harmonic analysis, which has been extensively developed for the last ten years and continues to attract attention of researchers from various fields of mathematics. Many of problems about such spaces have been solved both in the classical setting and in the Euclidean setting, including fractional upper and lower dimensions. For example, in 2008, variable exponent Morrey spaces $L^{p(\cdot),\lambda(\cdot)}$ were introduced to study the boundedness of M and $I_{\alpha(\cdot)}$ in the Euclidean setting by Almeida et al. [1]. In 2010, variable exponent-generalized Morrey spaces $L^{p(\cdot),w(\cdot)}(E)$ were introduced to consider the boundedness of $M, I_{\alpha(\cdot)}, T$ for bounded sets $E \subset \mathbb{R}^n$ on $L^{p(\cdot),w(\cdot)}(E)$ in [10]. In 2016, variable exponent vanishing generalized Morrey spaces $VL_{\Pi}^{p(\cdot),w(\cdot)}(E)$ were introduced to characterize the boundedness of $M, I_{\alpha(\cdot)}, T$ for bounded or unbounded sets E on $VL_{\Pi}^{p(\cdot),w(\cdot)}(E)$ in [15]. These results inspire us to ask whether the above operators ($I_{\Omega,\alpha(\cdot)}$, $M_{\Omega,\alpha(\cdot)}$, T_{Ω}, and M_{Ω}) have the similar mapping properties on variable exponent $VL_{\Pi}^{p(\cdot),w(\cdot)}(E)$, which includes variable exponent $L^{p(\cdot),\lambda(\cdot)}$ and $L^{p(\cdot)}$ spaces. Our first results (see Theorem 4.3, Theorem 4.5, and Theorem 4.6 below) will give some affirmative answers to these questions. Another purpose of this chapter is to prove the boundedness of above operators ($\left[b, I_{\Omega,\alpha(\cdot)}\right]$, $\left[b, M_{\Omega,\alpha(\cdot)}\right]$, $[b, T_\Omega]$, and $[b, M_\Omega]$) on $VL_{\Pi}^{p(\cdot),w(\cdot)}(E)$ spaces (see Theorem 4.7 below).

4.2 PRELIMINARIES AND MAIN RESULTS

In this section, we first recall the definitions and some properties of basic spaces that we need and then give the main results.

4.2.1 VARIABLE EXPONENT LEBESGUE SPACES $L^{P(\cdot)}$

Unfortunately, the variable exponent Lebesgue spaces $L^{p(\cdot)}$ and the classical cases have some undesired properties. For example, the variable $L^{p(\cdot)}$ spaces are not translation invariant. As a consequence, the variable exponent Lebesgue spaces are not rearrangement-invariant Banach spaces, and so neither good-λ techniques nor rearrangement inequalities may be applied for a generalization of some standard results in classical Lebesgue spaces to the case of $L^{p(\cdot)}$.

Now, we begin with a brief and necessarily incomplete review of the variable exponent Lebesgue spaces $L^{p(\cdot)}$.

Definition 4.1. *[18] Given an open set $E \subset \mathbb{R}^n$ and a measurable function $p(\cdot) : E \to [1,\infty)$. We assume that $1 \le p_-(E) \le p_+(E) < \infty$, where $p_-(E) = \operatorname{essinf}_{x\in E} p(x)$ and $p_+(E) = \operatorname{esssup}_{x\in E} p(x)$. The variable exponent Lebesgue space $L^{p(\cdot)}(E)$ is the collection of all measurable functions f such that, for some $\lambda > 0$, $\rho(f/\lambda) < \infty$, where the modular is defined by*

$$\rho(f) = \rho_{p(\cdot)}(f) = \int_E |f(x)|^{p(x)}\, dx.$$

Then, the spaces $L^{p(\cdot)}(E)$ and $L_{loc}^{p(\cdot)}(E)$ are defined by

$$L^{p(\cdot)}(E) = \{f \text{ is measurable} : \rho_{p(\cdot)}(\lambda^{-1}f) < \infty \text{ for some } \lambda > 0\}$$

and

$$L_{loc}^{p(\cdot)}(E) = \left\{f \text{ is measurable} : f \in L^{p(\cdot)}(K) \text{ for all compact } K \subset E\right\}.$$

Moreover, we define $\|\cdot\|_{L^{p(\cdot)}(E)}$, which is called the variable Lebesgue norm or the Luxemburg norm as follows:

$$\|f\|_{L^{p(\cdot)}(E)} = \inf\left(\{\lambda \in (0,\infty) : \rho_{p(\cdot)}(\lambda^{-1}f) \le 1\} \cup \{\infty\}\right) \qquad f \in L^{p(\cdot)}(E). \quad (4.4)$$

Since $p_-(E) \ge 1$, $\|\cdot\|_{L^{p(\cdot)}(E)}$ is a norm and $\left(L^{p(\cdot)}(E), \|\cdot\|_{L^{p(\cdot)}(E)}\right)$ is a Banach space. However, if $p_-(E) < 1$, then $\|\cdot\|_{L^{p(\cdot)}(E)}$ is a quasinorm and $\left(L^{p(\cdot)}(E), \|\cdot\|_{L^{p(\cdot)}(E)}\right)$ is a quasi-Banach space. The variable exponent norm has the following property

$$\left\|f^\lambda\right\|_{L^{p(\cdot)}(E)} = \|f\|_{L^{\lambda p(\cdot)}(E)}^{\lambda},$$

for $\lambda \geq \frac{1}{p_-}$. Moreover, these spaces are referred to as the variable L^p spaces since they generalize the standard L^p spaces: If $p(x) = p$ is constant, then $L^{p(\cdot)}(E)$ is isometrically isomorphic to $L^p(E)$. As a result, using notations above ($p_-(E)$ and $p_+(E)$), we define a class of variable exponent as follows:

$$\Phi(E) = \{p(\cdot) : E \to [1,\infty),\ p_-(E) \geq 1,\ p_+(E) < \infty\}.$$

Now, we define two sets of exponents $p(x)$ with $1 \leq p_-(E) \leq p_+(E) < \infty$. These will be denoted as follows:

$$\mathcal{P}^{\log}(E) = \left\{ \begin{array}{c} p(\cdot) : p_-(E) \geq 1,\ p_+(E) < \infty \text{ and } p(\cdot) \text{ satisfy both conditions (4.1) and (4.2)} \\ \text{(the latter required if } E \text{ is unbounded)} \end{array} \right\}$$

and

$$\mathcal{B}(E) = \left\{p(\cdot) : p(\cdot) \in \mathcal{P}^{\log}(E), M \text{ is bounded on } L^{p(\cdot)}(E)\right\},$$

where M is the Hardy-Littlewood maximal operator. We recall that the generalized Hölder inequality on Lebesgue spaces with the variable exponent

$$\left|\int_E f(x)g(x)\,dx\right| \leq \int_E |f(x)g(x)|\,dx \leq C_p \|f\|_{L^{p(\cdot)}(E)} \|g\|_{L^{p'(\cdot)}(E)} \quad C_p = 1 + \frac{1}{p_-} - \frac{1}{p_+},$$

is known to hold for $p(\cdot) \in \Phi(E)$, $f \in L^{p(\cdot)}(E)$, and $g \in L^{p'(\cdot)}(E)$ (see Theorem 2.1 in [14]). Now, we recall some recent results for the rough Riesz-type potential operator with the variable order $I_{\Omega,\alpha(\cdot)}$ and the corresponding rough fractional maximal operator with variable order $M_{\Omega,\alpha(\cdot)}$ on the variable exponent Lebesgue space $L^{p(\cdot)}(E)$. The order $\alpha(x)$ of the potential is not assumed to be continuous. We assume that it is a measurable function on E satisfying the following assumptions:

$$\left. \begin{array}{c} \alpha_0 = \operatorname*{essinf}_{x\in E} \alpha(x) > 0 \\ \operatorname*{esssup}_{x\in E} \alpha(x)p(x) < n \end{array} \right\}. \tag{4.5}$$

First, the norm in the space $L^{p(\cdot)}(E)$ seems to be complicated in a sense, to be calculated or estimated. So the following basic estimation of the boundedness of an operator B:

$$\|Bf\|_{L^{p(\cdot)}(E)} \lesssim \|f\|_{L^{p(\cdot)}(E)} \tag{4.6}$$

is not easy. However, in the case of linear operators, the above inequality between the norm and the modular and the homogeneity property

$$\|B\|_{X\to X} = \sup_{f\in X} \frac{\|Bf\|_X}{\|f\|_X} = \sup_{\|f\|_X=1} \|Bf\|_X$$

allow us to replace checking of (4.6) by a work with a modular:

$$\int_E |Bf(x)|^{p(x)}\,dx, \text{ for all } f \text{ with } \|f\|_{L^{p(\cdot)}(E)} \leq 1,$$

which is certainly easier. In that respect, the boundedness of the rough Riesz-type potential operator from the space $L^{p(\cdot)}(\mathbb{R}^n)$ with the variable exponent $p(x)$ into the space $L^{q(\cdot)}(\mathbb{R}^n)$ with the limiting Sobolev exponent

$$\frac{1}{q(x)} = \frac{1}{p(x)} - \frac{\alpha(x)}{n} \tag{4.7}$$

was an open problem for a long time. It was solved in the case of bounded domains. First, in [18], in the case of bounded domains E, there has the following conditional result.

Theorem 4.1. *[18] Let E be a bounded open set, $\Omega \in L_s(S^{n-1})$ with $1 < s \leq \infty$, $p(x) \in \mathcal{P}^{\log}(E)$, $\alpha(x)$ satisfying assumptions (4.5) and $(p')_+ \leq s$. Define $q(x)$ by (4.7). Then, the rough Riesz-type potential operator $I_{\Omega,\alpha(\cdot)}$ is $\left(L^{p(\cdot)}(E) \to L^{q(\cdot)}(E)\right)$-bounded, i.e., the Sobolev-type theorem*

$$\left\|I_{\Omega,\alpha(\cdot)}f\right\|_{L^{q(\cdot)}(E)} \lesssim \|f\|_{L^{p(\cdot)}(E)} \tag{4.8}$$

is valid.

Corollary 4.1. *Let E be a bounded open set, $\Omega \in L_s(S^{n-1})$ with $1 < s \leq \infty$ being a homogeneous function of degree 0 on $\mathbb{R}^n$, $\frac{p}{s} \in \mathcal{B}(E)$, and $(p')_+ \leq s$. Under the conditions of Theorem 4.1 (taking $\alpha(\cdot) = 0$ there), the operator T_Ω is $\left(L^{p(\cdot)}(E) \to L^{p(\cdot)}(E)\right)$-bounded, i.e.,*

$$\|T_\Omega f\|_{L^{p(\cdot)}(E)} \lesssim \|f\|_{L^{p(\cdot)}(E)} \tag{4.9}$$

is valid.

On the other hand, the pointwise inequalities on variable exponent Lebesgue spaces are very useful. Indeed, we have

$$|f(x)| \leq |h(x)| \text{ implies that } \|f\|_{L^{p(\cdot)}(E)} \lesssim \|h\|_{L^{p(\cdot)}(E)}.$$

Thus, if one operator is pointwise dominated by another one:

$$|Bf(x)| \leq |Df(x)|,$$

and we know that the operator D is bounded, then the boundedness of the operator B immediately follows. For example, by Theorem 4.1, we get the following:

Theorem 4.2. *Under the conditions of Theorem 4.1, the operator $M_{\Omega,\alpha(\cdot)}$ is $\left(L^{p(\cdot)}(E) \to L^{q(\cdot)}(E)\right)$-bounded, i.e., the Sobolev-type theorem*

$$\left\|M_{\Omega,\alpha(\cdot)}f\right\|_{L^{q(\cdot)}(E)} \lesssim \|f\|_{L^{p(\cdot)}(E)} \tag{4.10}$$

is valid.

Corollary 4.2. *Let E be a bounded open set, $\Omega \in L_s(S^{n-1})$ with $1 < s \leq \infty$ being a homogeneous function of degree 0 on $\mathbb{R}^n$, $\frac{p}{s'} \in \mathcal{B}(E)$, and $(p')_+ \leq s$. Under the conditions of Theorem 4.2 (taking $\alpha(\cdot) = 0$ there), the operator M_Ω is $\left(L^{p(\cdot)}(E) \to L^{p(\cdot)}(E)\right)$-bounded, i.e.,*

$$\|M_\Omega f\|_{L^{p(\cdot)}(E)} \lesssim \|f\|_{L^{p(\cdot)}(E)} \tag{4.11}$$

is valid.

We are now in a place of proving (4.10) in Theorem 4.2.

Remark 4.1. *The conclusion of (4.10) is a direct consequence of the following Lemma 4.1 and (4.8). In order to do this, we need to define an operator by*

$$\widetilde{T}_{|\Omega|,\alpha(\cdot)}(|f|)(x) = \int\limits_E \frac{|\Omega(x-y)|}{|x-y|^{n-\alpha(x)}}|f(y)|\,dy \qquad 0 < \alpha(x) < n,$$

where $\Omega \in L_s(S^{n-1})\,(s > 1)$ is homogeneous of degree zero on $\mathbb{R}^n$.

Using the idea of proving Corollary 3.1. in [22], we can obtain the following pointwise relation.

Lemma 4.1. *Let $0 < \alpha(x) < n$ and $\Omega \in L_s(S^{n-1})\,(s > 1)$. Then, we have*

$$M_{\Omega,\alpha(\cdot)}(f)(x) \leq C\widetilde{T}_{|\Omega|,\alpha(\cdot)}(|f|)(x) \qquad \text{for } x \in \mathbb{R}^n, \tag{4.12}$$

where C does not depend on f and x.

Proof. To prove (4.12), we observe that for any $x \in \mathbb{R}^n$, there exists an $r = r_x$ such that

$$M_{\Omega,\alpha(\cdot)}(f)(x) \leq \frac{2}{|B(x,r_x)|^{n-\alpha(x)}} \int\limits_{B(x,r_x)} |\Omega(x-y)|\,|f(y)|dy,$$

and by the inequality above, we get

$$\begin{aligned}\widetilde{T}_{|\Omega|,\alpha(\cdot)}(|f|)(x) &\geq \int\limits_{B(x,r_x)} \frac{|\Omega(x-y)|}{|x-y|^{n-\alpha(x)}}|f(y)|\,dy \\ &\geq \frac{C}{|B(x,r_x)|^{n-\alpha(x)}} \int\limits_{B(x,r_x)} |\Omega(x-y)|\,|f(y)|\,dy.\end{aligned}$$

□

From the process proving (4.8) in [18], it is easy to see that the conclusions of (4.8) also hold for $\widetilde{T}_{|\Omega|,\alpha(\cdot)}$. Combining this with (4.12), we can immediately obtain (4.10), which completes the proof of Theorem 4.2.

Remark 4.2. *Taking $\alpha(\cdot)=0$ in Lemma 4.1 and the inequality*

$$M_{\Omega}(f)(x) \le C\widetilde{T}_{|\Omega|}(|f|)(x) \qquad \text{for } x \in \mathbb{R}^n,$$

which follows from the definitions of the operators.

The above theorems (Theorem 4.1 and Theorem 4.2) allow us to use the known results for the boundedness of the operators $M_{\Omega,\alpha(\cdot)}$ and $I_{\Omega,\alpha(\cdot)}$ that are transferred to the various function spaces. The following fact is known (see Lemma 3.1. in [15]).

Lemma 4.2. *[15] Let E be a bounded open set, $p(x) \in \mathcal{P}^{\log}(E)$ and $\alpha(x)$ satisfying assumptions (4.5). Then,*

$$\left\| |x-\cdot|^{\alpha(x)-n} \chi_{\tilde{B}(x,r)} \right\|_{L^{p(\cdot)}(E)} \lesssim r^{\alpha(x)-\frac{n}{p(x)}}.$$

We will also make use of the estimate provided by the following fact (see [15]).

$$\left\| \chi_{\tilde{B}(x,r)} \right\|_{L^{p(\cdot)}(E)} \lesssim r^{\psi_p(x,r)}, \qquad x \in E,\ p(x) \in \mathcal{P}^{\log}(E), \tag{4.13}$$

where

$$\psi_p(x,r) = \begin{cases} \frac{n}{p(x)}, & r \le 1 \\ \frac{n}{p(\infty)}, & r > 1. \end{cases}$$

4.2.2 VARIABLE EXPONENT MORREY SPACES $L^{p(\cdot),\lambda(\cdot)}$

Here, we recall the variable exponent Morrey spaces and the integral inequalities.

Definition 4.2. *[1] Let E be a bounded open set and $\lambda(x)$ be a measurable function on E with values in $[0,n]$. Then, the variable exponent Morrey space $L^{p(\cdot),\lambda(\cdot)} \equiv L^{p(\cdot),\lambda(\cdot)}(E)$ is defined by*

$$L^{p(\cdot),\lambda(\cdot)} \equiv L^{p(\cdot),\lambda(\cdot)}(E) = \left\{ \begin{array}{c} f \in L^{p(\cdot)}_{loc}(E): \\ \|f\|_{L^{p(\cdot),\lambda(\cdot)}} = \sup\limits_{x\in E, r>0} r^{-\frac{\lambda(x)}{p(x)}} \left\| f\chi_{\tilde{B}(x,r)} \right\|_{L^{p(\cdot)}(E)} < \infty \end{array} \right\}.$$

Note that $L^{p(\cdot),0}(E) = L^{p(\cdot)}(E)$ and $L^{p(\cdot),n}(E) = L^{\infty}(E)$. If $\lambda_- > n$, then $L^{p(\cdot),\lambda(\cdot)}(E) = \{0\}$.

Lemma 4.3. *Let E be a bounded open set, $\Omega \in L_s(S^{n-1})$ with $1 < s < \infty$, $p(x), q(x) \in \mathcal{P}^{\log}(E)$, $\alpha(x)$ satisfying the following assumptions:*

$$\left. \begin{array}{c} \alpha_0 = \operatorname*{essinf}\limits_{x\in E} \alpha(x) > 0 \\ \operatorname*{esssup}\limits_{x\in E} [\lambda(x) + \alpha(x)p(x)] < n \end{array} \right\} \tag{4.14}$$

and $(p')_+ \leq s$. *Define* $q(x)$ *by* $\frac{1}{q(\cdot)} = \frac{1}{p(\cdot)} - \frac{\alpha(\cdot)}{n-\lambda(\cdot)}$. *Then, the rough Riesz-type potential operator* $I_{\Omega,\alpha(\cdot)}$ *is* $\left(L^{p(\cdot),\lambda(\cdot)}(E) \to L^{q(\cdot),\lambda(\cdot)}(E)\right)$*-bounded. Moreover,*

$$\left\|I_{\Omega,\alpha(\cdot)}f\right\|_{L^{q(\cdot),\lambda(\cdot)}(E)} \lesssim \|f\|_{L^{p(\cdot),\lambda(\cdot)}(E)}.$$

Proof. By the embedding property in Lemma 7 in [1], we only need to prove that the operator $I_{\Omega,\alpha(\cdot)}$ is bounded in $L^{p(\cdot),\lambda(\cdot)}(E)$.

Hedberg's trick:

$$\begin{aligned} I_{\Omega,\alpha(\cdot)}f(x) &= \int_{B(x,2r)} \frac{\Omega(x-y)}{|x-y|^{n-\alpha(x)}} f(y)dy + \int_{B^C(x,2r)} \frac{\Omega(x-y)}{|x-y|^{n-\alpha(x)}} f(y)dy \\ &= \mathcal{F}(x,r) + \mathcal{G}(x,r). \end{aligned} \tag{4.15}$$

We may assume that $\|f\|_{L^{p(\cdot),\lambda(\cdot)}(E)} \leq 1$. For $\mathcal{F}(x,r)$, we first have to prove the following:

$$\mathcal{F}(x,r) := \left| \int_{|x-y|<r} \frac{\Omega(x-y)}{|x-y|^{n-\alpha(x)}} f(y)dy \right| \lesssim \frac{2^n r^{\alpha(x)}}{2^{\alpha(x)}-1} M_\Omega f(x). \tag{4.16}$$

Indeed, for $f(x) \geq 0$, we have

$$\begin{aligned} \mathcal{F}(x,r) &= \sum_{j=0}^{\infty} \int_{2^{-j-1}r \leq |x-y| < 2^{-j}r} \frac{\Omega(x-y)}{|x-y|^{n-\alpha(x)}} f(y)dy \\ &\leq \sum_{j=0}^{\infty} \frac{1}{(2^{-j-1}r)^{n-\alpha(x)}} \int_{|x-y|<2^{-j}r} \Omega(x-y)f(y)dy \\ &\leq 2^{n-\alpha(x)} M_\Omega f(x) \sum_{j=0}^{\infty} \frac{\left|B(x,2^{-j}r)\right|}{(2^{-j}r)^{n-\alpha(x)}}. \end{aligned}$$

Hence, by $\left|B(x,2^{-j}r)\right| \lesssim (2^{-j}r)^n$, we obtain

$$\mathcal{F}(x,r) \lesssim 2^{n-\alpha(x)} r^{\alpha(x)} M_\Omega f(x) \sum_{j=0}^{\infty} \left(2^{-j\alpha(x)}\right),$$

which gives estimate (4.16). Then by (4.16):

$$|\mathcal{F}(x,r)| \lesssim r^{\alpha(x)} M_\Omega f(x).$$

For $\mathcal{G}(x,r)$, from Lemma 4.2 and the procedure of Theorem 3 in [1], we may show that

$$|\mathcal{G}(x,r)| \lesssim r^{\alpha(x) - \frac{n-\lambda(x)}{p(x)}}.$$

Then, from (4.15), we get

$$I_{\Omega,\alpha(\cdot)}f(x) \lesssim \left[r^{\alpha(x)}M_{\Omega}f(x) + r^{\alpha(x)-\frac{n-\lambda(x)}{p(x)}}\right]. \tag{4.17}$$

As usual in Hedberg approach, we choose

$$r = [M_{\Omega}f(x)]^{-\frac{p(x)}{n-\lambda(x)}}.$$

Substituting this into (4.17), we get

$$\left|I_{\Omega,\alpha(\cdot)}f(x)\right| \lesssim (M_{\Omega}f(x))^{\frac{p(x)}{q(x)}},$$

here, we need (4.3). Therefore, by Theorem 5.1 in [18], we know that

$$\int_{\tilde{B}(x,r)} \left|I_{\Omega,\alpha(\cdot)}f(y)\right|^{q(y)} dy \lesssim \int_{\tilde{B}(x,r)} |M_{\Omega}f(y)|^{p(y)} dy \lesssim r^{\lambda(x)},$$

which completes the proof of Lemma 4.3. □

Theorem 4.3. *Let E be a bounded open set, $\Omega \in L_s(S^{n-1})$ with $1 < s < \infty$, $p(x), q(x) \in \mathcal{P}^{\log}(E)$, $\alpha(x)$ satisfying (4.14) and $(p')_+ \leq s$. Define $q(x)$, $\mu(x)$ by $\frac{1}{q(\cdot)} = \frac{1}{p(\cdot)} - \frac{\alpha(\cdot)}{n}$, $\frac{n-\mu(\cdot)}{q(\cdot)} = \frac{n-\lambda(\cdot)}{p(\cdot)} - \alpha(\cdot)$, respectively. Then, the rough Riesz-type potential operator $I_{\Omega,\alpha(\cdot)}$ is $\left(L^{p(\cdot),\lambda(\cdot)}(E) \to L^{q(\cdot),\mu(\cdot)}(E)\right)$-bounded. Moreover,*

$$\left\|I_{\Omega,\alpha(\cdot)}f\right\|_{L^{q(\cdot),\mu(\cdot)}(E)} \lesssim \|f\|_{L^{p(\cdot),\lambda(\cdot)}(E)}, \tag{4.18}$$

where

$$1 \leq q(\cdot) \leq \frac{p(\cdot)(n-\lambda(\cdot))}{n-\lambda(\cdot)-\alpha(\cdot)p(\cdot)}.$$

Proof. Since

$$\frac{p(\cdot)(n-\lambda(\cdot))}{n-\lambda(\cdot)-\alpha(\cdot)p(\cdot)} < \frac{np(\cdot)}{n-\alpha(\cdot)p(\cdot)},$$

from Lemma 4.3 and Theorem 4.1, we obtain

$$\begin{aligned}
\left\|I_{\Omega,\alpha(\cdot)}f\right\|_{L^{q(\cdot),\mu(\cdot)}(E)} &= \sup_{x\in E, r>0} r^{-\frac{\mu(x)}{q(x)}} \left\|I_{\Omega,\alpha(\cdot)}f\chi_{\tilde{B}(x,r)}\right\|_{L^{q(\cdot)}(E)} \\
&\lesssim \sup_{x\in E, r>0} r^{-\frac{\lambda(x)}{p(x)}} \left\|f\chi_{\tilde{B}(x,r)}\right\|_{L^{p(\cdot)}(E)} \\
&= \|f\|_{L^{p(\cdot),\lambda(\cdot)}(E)}.
\end{aligned}$$

Clearly, Theorem 4.3 holds. □

Theorem 4.4. *Under the conditions of Theorem 4.3,*

$$\left\|M_{\Omega,\alpha(\cdot)}f\right\|_{L^{q(\cdot),\mu(\cdot)}(E)} \lesssim \|f\|_{L^{p(\cdot),\lambda(\cdot)}(E)}. \tag{4.19}$$

Proof. Similar to the proof of Theorem 4.2, conclusion (4.19) is a direct consequence of (4.12) and (4.18). Indeed, from the process proving (4.18) in Theorem 4.3, it is easy to see that conclusion (4.18) also holds for $\widetilde{T}_{|\Omega|,\alpha(\cdot)}$. Combining this with (4.12), we can immediately obtain (4.19), which completes the proof. □

4.2.3 VARIABLE EXPONENT VANISHING GENERALIZED MORREY SPACES

Having set down the boundedness properties of the rough operators on variable exponent Morrey spaces, in this section, we are now interested in the boundedness properties of the rough operators on variable exponent vanishing generalized Morrey spaces. For this, we first consider the generalized Morrey spaces $L^{p(\cdot),w(\cdot)}(E)$ with a variable exponent $p(x)$ and a general function $w(x,r) : \Pi \times (0, diam(E)) \to \mathbb{R}_+$, $\Pi \subset E \subset \mathbb{R}^n$, defining the Morrey-type norm on sets $E \subset \mathbb{R}^n$ which may be both bounded and unbounded; see the definition of the spaces $L^{p(\cdot),w(\cdot)}(E)$ in (4.21).

Everywhere in the sequel, the functions $w(x,r)$, $w_1(x,r)$, $w_2(x,r)$ used in the body of this paper are nonnegative measurable functions on $E \times (0,\infty)$, where $E \subset \mathbb{R}^n$ is an open set. We recall the definition of variable exponent-generalized Morrey space in the following.

Definition 4.3. *[10] Let $1 \le p(x) \le p_+ < \infty$, $\Pi \subset E \subset \mathbb{R}^n$, $x \in \Pi$, $w(x,r) : \Pi \times (0, diam(E)) \to \mathbb{R}_+$, where*

$$\inf_{x\in\Pi} w(x,r) > 0 \qquad r > 0. \tag{4.20}$$

Then, the variable exponent-generalized Morrey space $L_\Pi^{p(\cdot),w(\cdot)} \equiv L_\Pi^{p(\cdot),w(\cdot)}(E)$ is defined by

$$L_\Pi^{p(\cdot),w(\cdot)} \equiv L_\Pi^{p(\cdot),w(\cdot)}(E) = \left\{ \begin{array}{c} f \in L_{loc}^{p(\cdot)}(E): \\ \|f\|_{L_\Pi^{p(\cdot),w(\cdot)}} = \sup\limits_{x\in\Pi, r>0} w(x,r)^{-\frac{1}{p(x)}} \|f\|_{L^{p(\cdot)}(\tilde{B}(x,r))} < \infty \end{array} \right\}, \tag{4.21}$$

and one can also see that for bounded exponents p, there holds the following equivalence:

$$f \in L_\Pi^{p(\cdot),w(\cdot)} \text{ if and only if } \sup_{x\in\Pi, r>0} \int_{\tilde{B}(x,r)} \left|\frac{f(y)}{w(x,r)}\right|^{p(y)} dy < \infty.$$

On the other hand, the above definition recovers the definition of $L^{p(\cdot),\lambda(\cdot)}(E)$ if we choose $w(x,r) = r^{\frac{\lambda(x)}{p(x)}}$ and $\Pi = E$, i.e.,

$$L^{p(\cdot),\lambda(\cdot)}(E) = L_\Pi^{p(\cdot),w(\cdot)}(E) \,\Big|_{w(x,r)=r^{\frac{\lambda(x)}{p(x)}}}.$$

Also, when $\Pi = \{x_0\}$ and $\Pi = E$, $L_{\Pi}^{p(\cdot),w(\cdot)}$ turns into the local generalized Morrey space $L_{\{x_0\}}^{p(\cdot),w(\cdot)}(E)$ and the global generalized Morrey space $L_{E}^{p(\cdot),w(\cdot)}(E)$, respectively. Moreover, we point out that $w(x,r)$ is a measurable nonnegative function and no monotonicity-type condition is imposed on these spaces. Note that by the above definition of the norm in $L^{p(\cdot)}(E)$ (see 4.4), we can also write that

$$\|f\|_{L_{\Pi}^{p(\cdot),w(\cdot)}} = \sup_{x\in\Pi, r>0} \inf \left\{ \lambda = \lambda(x,r) : \int_{\tilde{B}(x,r)} \left| \frac{f(y)}{\lambda w(x,r)} \right|^{p(y)} dy \leq 1 \right\}.$$

Then, recall that the concept of the variable exponent vanishing generalized Morrey space $VL_{\Pi}^{p(\cdot),w(\cdot)}(E)$ has been introduced in [15] in the following form.

Definition 4.4. *[15] Let $1 \leq p(x) \leq p_+ < \infty$, $\Pi \subset E \subset \mathbb{R}^n$, $x \in \Pi$, $w(x,r)\colon \Pi \times (0, diam(E)) \to \mathbb{R}_+$. Then, the variable exponent vanishing generalized Morrey space $VL_{\Pi}^{p(\cdot),w(\cdot)} \equiv VL_{\Pi}^{p(\cdot),w(\cdot)}(E)$ is defined by*

$$\left\{ f \in L_{\Pi}^{p(\cdot),w(\cdot)}(E) : \lim_{r\to 0} \sup_{x\in\Pi} \mathfrak{M}_{p(\cdot),w(\cdot)}(f;x,r) = 0 \right\},$$

where

$$\mathfrak{M}_{p(\cdot),w(\cdot)}(f;x,r) := \frac{r^{-\frac{n}{p(x)}} \|f\|_{L^{p(\cdot)}(\tilde{B}(x,r))}}{w(x,r)^{\frac{1}{p(x)}}}.$$

Naturally, it is suitable to impose on $w(x,t)$ with the following conditions:

$$\limsup_{t\to 0} \sup_{x\in\Pi} \frac{t^{-\psi_p(x,t)}}{w(x,t)^{\frac{1}{p(x)}}} = 0 \tag{4.22}$$

and

$$\inf_{t>1} \sup_{x\in\Pi} w(x,t) > 0. \tag{4.23}$$

From (4.22) and (4.23), we easily know that the bounded functions with compact support belong to $VL_{\Pi}^{p(\cdot),w(\cdot)}(E)$, which make the spaces $VL_{\Pi}^{p(\cdot),w(\cdot)}(E)$ nontrivial.

The spaces $VL_{\Pi}^{p(\cdot),w(\cdot)}(E)$ are the Banach spaces with respect to the norm

$$\|f\|_{VL_{\Pi}^{p(\cdot),w(\cdot)}} \equiv \|f\|_{L_{\Pi}^{p(\cdot),w(\cdot)}} = \sup_{x\in\Pi, r>0} \mathfrak{M}_{p(\cdot),w(\cdot)}(f;x,r).$$

The spaces $VL_{\Pi}^{p(\cdot),w(\cdot)}(E)$ are also the closed subspaces of the Banach spaces $L_{\Pi}^{p(\cdot),w(\cdot)}(E)$, which may be shown by standard means.

Furthermore, we have the following embeddings:

$$VL_{\Pi}^{p(\cdot),w(\cdot)} \subset L_{\Pi}^{p(\cdot),w(\cdot)}, \qquad \|f\|_{L_{\Pi}^{p(\cdot),w(\cdot)}} \leq \|f\|_{VL_{\Pi}^{p(\cdot),w(\cdot)}}.$$

In 2016, for bounded or unbounded sets E, Long and Han [15] considered the Spanne-type boundedness of operators $M_{\alpha(\cdot)}$ and $I_{\alpha(\cdot)}$ on $VL_{\Pi}^{p(\cdot),w(\cdot)}(E)$.

Now, in this section, we extend Theorem 4.3. in [15] to rough kernel versions. In other words, Theorem 4.3. in [15] allows us to use the known results for the boundedness of the operators $I_{\alpha(\cdot)}$ and $M_{\alpha(\cdot)}$ in generalized variable exponent Morrey spaces to transfer them to the operators $I_{\Omega,\alpha(\cdot)}$ and $M_{\Omega,\alpha(\cdot)}$. We give two versions of such an extension, the one being a generalization of Spanne's result for rough potential operators with variable order and the other extending the corresponding Adams' result, respectively.

In this context, we will give some answers to the above explanations in the following.

Theorem 4.5. ***(Spanne-type result with variable*** $\alpha(x)$***)*** *(our main result) Let E be a bounded open set,* $\Omega \in L_s(S^{n-1})$, $1 < s \leq \infty$, $\Omega(\mu x) = \Omega(x)$ *for any* $\mu > 0$, $x \in \mathbb{R}^n \setminus \{0\}$, $p(x) \in \mathcal{P}^{\log}(E)$, $\alpha(x)$ *satisfying assumption (4.5). Define* $q(x)$ *by (4.7). Suppose that* $q(\cdot)$ *and* $\alpha(\cdot)$ *satisfy (4.1). For* $\frac{s}{s-1} < p^- \leq p(\cdot) < \frac{n}{\alpha(\cdot)}$, *the following pointwise estimate*

$$\left\|I_{\Omega,\alpha(\cdot)}f\right\|_{L^{q(\cdot)}(\tilde{B}(x,r))} \lesssim r^{\frac{n}{q(x)}} \int_r^{diam(E)} \|f\|_{L^{p(\cdot)}(\tilde{B}(x,t))} \frac{dt}{t^{\frac{n}{q(x)}+1}} \tag{4.24}$$

holds for any ball $\tilde{B}(x,r)$ *and for all* $f \in L_{loc}^{p(\cdot)}(E)$.

If the functions $w_1(x,r)$ *and* $w_2(x,r)$ *satisfy (4.20) as well as the following Zygmund condition*

$$\int_r^{diam(E)} \frac{w_1^{\frac{1}{p(x)}}(x,t)}{t^{1-\alpha(x)}} dt \lesssim w_2^{\frac{1}{q(x)}}(x,r), \qquad r \in (0, diam(E)] \tag{4.25}$$

and additionally these functions satisfy the conditions (4.22)-(4.23),

$$c_\delta := \int_\delta^{diam(E)} \sup_{x\in\Pi} \frac{w_1^{\frac{1}{p(x)}}(x,t)}{t^{1-\alpha(x)}} dt < \infty, \qquad \delta > 0 \tag{4.26}$$

then the operators $I_{\Omega,\alpha(\cdot)}$ *and* $M_{\Omega,\alpha(\cdot)}$ *are* $\left(VL_{\Pi}^{p(\cdot),w_1(\cdot)}(E) \to VL_{\Pi}^{q(\cdot),w_2(\cdot)}(E)\right)$-*bounded. Moreover,*

$$\left\|I_{\Omega,\alpha(\cdot)}f\right\|_{VL_{\Pi}^{q(\cdot),w_2(\cdot)}(E)} \lesssim \|f\|_{VL_{\Pi}^{p(\cdot),w_1(\cdot)}(E)}, \tag{4.27}$$

$$\left\|M_{\Omega,\alpha(\cdot)}f\right\|_{VL_{\Pi}^{q(\cdot),w_2(\cdot)}(E)} \lesssim \|f\|_{VL_{\Pi}^{p(\cdot),w_1(\cdot)}(E)}.$$

Proof. Since inequality (4.24) is the key of the proof of (4.27), we first prove (4.24).

For any $x \in E$, we write as

$$f(y) = f_1(y) + f_2(y), \tag{4.28}$$

where $f_1(y) = f(y)\chi_{\tilde{B}(x,2r)}(y)$, $r > 0$ such that

$$I_{\Omega,\alpha(\cdot)}f(y) = I_{\Omega,\alpha(\cdot)}f_1(y) + I_{\Omega,\alpha(\cdot)}f_2(y).$$

By using triangle inequality, we get

$$\left\|I_{\Omega,\alpha(\cdot)}f\right\|_{L^{q(\cdot)}(\tilde{B}(x,r))} \le \left\|I_{\Omega,\alpha(\cdot)}f_1\right\|_{L^{q(\cdot)}(\tilde{B}(x,r))} + \left\|I_{\Omega,\alpha(\cdot)}f_2\right\|_{L^{q(\cdot)}(\tilde{B}(x,r))}.$$

Now, let us estimate $\left\|I_{\Omega,\alpha(\cdot)}f_1\right\|_{L^{q(\cdot)}(\tilde{B}(x,r))}$ and $\left\|I_{\Omega,\alpha(\cdot)}f_2\right\|_{L^{q(\cdot)}(\tilde{B}(x,r))}$, respectively.

By Hardy-Littlewood-Sobolev-type inequality and Theorem 4.1, we obtain that

$$\begin{aligned}
\left\|I_{\Omega,\alpha(\cdot)}f_1\right\|_{L^{q(\cdot)}(\tilde{B}(x,r))} &\le \left\|I_{\Omega,\alpha(\cdot)}f_1\right\|_{L^{q(\cdot)}(E)} \lesssim \|f_1\|_{L^{p(\cdot)}(E)} = \|f\|_{L^{p(\cdot)}(\tilde{B}(x,2r))} \\
&\approx r^{\frac{n}{q(x)}} \|f\|_{L^{p(\cdot)}(\tilde{B}(x,2r))} \int_{2r}^{diam(E)} \frac{dt}{t^{\frac{n}{q(x)}+1}} \\
&\le r^{\frac{n}{q(x)}} \int_{r}^{diam(E)} \|f\|_{L^{p(\cdot)}(\tilde{B}(x,t))} \frac{dt}{t^{\frac{n}{q(x)}+1}},
\end{aligned}$$

where in the last inequality, we have used the following fact:

$$\|f\|_{L^{p(\cdot)}(\tilde{B}(x,2r))} \le \|f\|_{L^{p(\cdot)}(\tilde{B}(x,t))}, \text{ for } t > 2r.$$

Now, let us estimate the second part $\left(= \left\|I_{\Omega,\alpha(\cdot)}f_2\right\|_{L^{q(\cdot)}(\tilde{B}(x,r))}\right)$.

If $|x-z| \le r$ and $|z-y| \ge r$, then $|x-y| \le |x-z| + |y-z| \le 2|y-z|$. By the generalized Minkowski's inequality, we get

$$\begin{aligned}
\left\|I_{\Omega,\alpha(\cdot)}f_2\right\|_{L^{q(\cdot)}(\tilde{B}(x,r))} &= \left\| \int_{E\setminus\tilde{B}(x,2r)} \frac{\Omega(z-y)}{|z-y|^{n-\alpha(x)}} f(y)dy \right\|_{L^{q(\cdot)}(\tilde{B}(x,r))} \\
&\lesssim \int_{E\setminus\tilde{B}(x,2r)} \frac{|\Omega(z-y)|\,|f(y)|}{|x-y|^{n-\alpha(x)}} dy \left\|\chi_{\tilde{B}(x,r)}\right\|_{L^{q(\cdot)}(E)}.
\end{aligned}$$

Put $\gamma > \frac{n}{q(\cdot)}$. Provided that $1 < s' < p^- \le p^+ < \infty$, $\sup_{x\in E}(\alpha(x) + \gamma - n) < \infty$, and $\inf_{x\in E}\left(n + (\alpha(x) + \gamma - n)\left(\frac{p(\cdot)}{s'}\right)'\right) < \infty$, by generalized Hölder's inequality for $L^{p(\cdot)}(E)$,Fubini's theorem, and Lemma 4.2 and (4.3), we obtain

$$
\begin{aligned}
&\int\limits_{E\backslash\tilde{B}(x,2r)} \frac{|\Omega(z-y)|\,|f(y)|}{|x-y|^{n-\alpha(x)}}dy \\
&\lesssim \int\limits_{E\backslash\tilde{B}(x,2r)} \frac{|\Omega(z-y)|\,|f(y)|}{|x-y|^{n-\alpha(x)-\gamma}}dy \int\limits_{|x-y|}^{diam(E)} \frac{dt}{t^{\gamma+1}} \\
&= \int\limits_{2r}^{diam(E)} \frac{dt}{t^{\gamma+1}} \int\limits_{\{y\in E:2r\le|x-y|\le t\}} \frac{|\Omega(z-y)|\,|f(y)|}{|x-y|^{n-\alpha(x)-\gamma}}dy \\
&\lesssim \int\limits_{2r}^{diam(E)} \|f\|_{L^{p(\cdot)}(\tilde{B}(x,t))} \left\| x-\cdot\right|^{\alpha(x)+\gamma-n}\Big\|_{L^{v(\cdot)}(\tilde{B}(x,t))} \|\Omega\,(z-y)\|_{L_s(\tilde{B}(x,t))} \frac{dt}{t^{\gamma+1}}
\end{aligned}
$$

$$
\lesssim \int\limits_{r}^{diam(E)} \|f\|_{L^{p(\cdot)}(\tilde{B}(x,t))} \frac{dt}{t^{\frac{n}{q(x)}+1}} \tag{4.29}
$$

for $\frac{1}{p(\cdot)}+\frac{1}{s}+\frac{1}{v(\cdot)}=1$. Thus, by (4.13), we get

$$
\left\|I_{\Omega,\alpha(\cdot)}f_2\right\|_{L^{q(\cdot)}(\tilde{B}(x,r))} \lesssim r^{\frac{n}{q(x)}} \int\limits_{r}^{diam(E)} \|f\|_{L^{p(\cdot)}(\tilde{B}(x,t))} \frac{dt}{t^{\frac{n}{q(x)}+1}}.
$$

Combining all the estimates for $\left\|I_{\Omega,\alpha(\cdot)}f_1\right\|_{L^{q(\cdot)}(\tilde{B}(x,r))}$ and $\left\|I_{\Omega,\alpha(\cdot)}f_2\right\|_{L^{q(\cdot)}(\tilde{B}(x,r))}$, we get (4.24).

At last, by Definition 4.4, (4.24), and (4.25), we get

$$
\begin{aligned}
\left\|I_{\Omega,\alpha(\cdot)}f\right\|_{VL_{\Pi}^{q(\cdot),w_2(\cdot)}(E)} &= \sup_{x\in\Pi,r>0} \frac{r^{-\frac{n}{q(x)}}\left\|I_{\Omega,\alpha(\cdot)}f\right\|_{L^{q(\cdot)}(\tilde{B}(x,r))}}{w_2(x,r)^{\frac{1}{q(x)}}} \\
&\lesssim \sup_{x\in\Pi,r>0} \frac{1}{w_2(x,r)^{\frac{1}{q(x)}}} \int\limits_{r}^{diam(E)} \|f\|_{L^{p(\cdot)}(\tilde{B}(x,t))} \frac{dt}{t^{\frac{n}{q(x)}+1}} \\
&\lesssim \|f\|_{VL_{\Pi}^{p(\cdot),w_1(\cdot)}(E)} \sup_{x\in\Pi,r>0} \frac{1}{w_2(x,r)^{\frac{1}{q(x)}}} \int\limits_{r}^{diam(E)} \frac{w_1^{\frac{1}{p(x)}}(x,t)}{t^{1-\alpha(x)}}dt \\
&\lesssim \|f\|_{VL_{\Pi}^{p(\cdot),w_1(\cdot)}(E)}
\end{aligned}
$$

and

$$
\limsup_{r\to 0}\sup_{x\in\Pi} \frac{r^{-\frac{n}{q(x)}}\left\|I_{\Omega,\alpha(\cdot)}f\right\|_{L^{q(\cdot)}(\tilde{B}(x,r))}}{w_2(x,t)^{\frac{1}{q(x)}}} \lesssim \limsup_{r\to 0}\sup_{x\in\Pi} \frac{r^{-\frac{n}{p(x)}}\|f\|_{L^{p(\cdot)}(\tilde{B}(x,r))}}{w_1(x,t)^{\frac{1}{p(x)}}} = 0.
$$

Thus, (4.27) holds. On the other hand, since $M_{\Omega,\alpha(\cdot)}(f) \lesssim I_{|\Omega|,\alpha(\cdot)}(|f|)$ (see Lemma 4.1), we can also use the same method for $M_{\Omega,\alpha(\cdot)}$, so we omit the details. As a result, we complete the proof of Theorem 4.5. □

Definition 4.5. *[13] **(Rough (p,q)-admissible $T_{\Omega,\alpha(\cdot)}$-potential type operator with variable order)** Let $1 \leq p_-(E) \leq p(\cdot) \leq p_+(E) < \infty$. A rough sublinear operator with variable order $T_{\Omega,\alpha(\cdot)}$, i.e., $|T_{\Omega,\alpha(\cdot)}(f+g)| \leq |T_{\Omega,\alpha(\cdot)}(f)| + |T_{\Omega,\alpha(\cdot)}(g)|$ and for $\forall\lambda \in \mathbb{C}$ $|T_{\Omega,\alpha(\cdot)}(\lambda f)| = |\lambda|\,|T_{\Omega,\alpha(\cdot)}(f)|$, will be called rough (p,q)-admissible $T_{\Omega,\alpha(\cdot)}$-potential type operator with variable order if*

· $T_{\Omega,\alpha(\cdot)}$ fulfills the following size condition:

$$\chi_{B(z,r)}(x)\,|T_{\Omega,\alpha(\cdot)}(f\chi_{E\setminus B(z,2r)})(x)| \leq C\chi_{B(z,r)}(x) \int\limits_{E\setminus B(z,2r)} \frac{|\Omega(x-y)|}{|x-y|^{n-\alpha(\cdot)}}|f(y)|\,dy, \tag{4.30}$$

· $T_{\Omega,\alpha(\cdot)}$ is $\left(L^{p(\cdot)}(E) \to L^{q(\cdot)}(E)\right)$-bounded.

Remark 4.3. *Note that rough (p,q)-admissible potential type operators were introduced to study their boundedness on Morrey spaces with variable exponents in [13]. The operators $M_{\Omega,\alpha(\cdot)}$ and $I_{\Omega,\alpha(\cdot)}$ are also rough (p,q)-admissible potential type operators. Moreover, these operators satisfy (4.30).*

Corollary 4.3. *Obviously, under the conditions of Theorem 4.5, if the rough (p,q)-admissible $T_{\Omega,\alpha(\cdot)}$-potential type operator is $\left(L^{p(\cdot)}(E) \to L^{q(\cdot)}(E)\right)$-bounded and satisfies (4.30), the result in Theorem 4.5 still holds.*

For $\alpha(x) = 0$ in Theorem 4.5, we get the following new result.

Corollary 4.4. *Let E, Ω, $p(x)$ be the same as in Theorem 4.5. Then, for $\frac{s}{s-1} < p^- \leq p(\cdot) \leq p^+ < \infty$, the following pointwise estimate*

$$\|T_\Omega f\|_{L^{p(\cdot)}(\tilde{B}(x,r))} \lesssim r^{\frac{n}{p(x)}} \int\limits_r^{diam(E)} t^{-\frac{n}{p(x)}-1} \|f\|_{L^{p(\cdot)}(\tilde{B}(x,t))}\,dt$$

holds for any ball $\tilde{B}(x,r)$ and for all $f \in L^{p(\cdot)}_{loc}(E)$.

If the function $w(x,r)$ satisfies (4.20) as well as the following Zygmund condition

$$\int\limits_r^{diam(E)} \frac{w^{\frac{1}{p(x)}}(x,t)}{t}dt \lesssim w^{\frac{1}{p(x)}}(x,r), \qquad r \in (0, diam(E)]$$

and additionally this function satisfies conditions (4.22) and (4.23),

$$c_\delta := \int\limits_\delta^{diam(E)} \sup_{x\in\Pi} \frac{w^{\frac{1}{p(x)}}(x,t)}{t}dt < \infty, \qquad \delta > 0$$

then the operators T_Ω and M_Ω are bounded on $VL_\Pi^{p(\cdot),w(\cdot)}(E)$. Moreover,

$$\|T_\Omega f\|_{VL_\Pi^{p(\cdot),w(\cdot)}(E)} \lesssim \|f\|_{VL_\Pi^{p(\cdot),w(\cdot)}(E)},$$

$$\|M_\Omega f\|_{VL_\Pi^{p(\cdot),w(\cdot)}(E)} \lesssim \|f\|_{VL_\Pi^{p(\cdot),w(\cdot)}(E)}. \tag{4.31}$$

Theorem 4.6. *(**Adams-type result with variable** $\alpha(x)$) (our main result) Let E, Ω, $p(x)$, $q(x)$, $\alpha(x)$ be the same as in Theorem 4.5. Then, for $\frac{s}{s-1} < p^- \le p(\cdot) < \frac{n}{\alpha(\cdot)}$, the following pointwise estimate*

$$\left|I_{\Omega,\alpha(\cdot)}f(x)\right| \lesssim r^{\alpha(x)}M_\Omega f(x) + \int_r^{diam(E)} t^{\alpha(x)-\frac{n}{p(x)}-1}\|f\|_{L_p(\tilde{B}(x,t))}\,dt \tag{4.32}$$

holds for any ball $\tilde{B}(x,r)$ and for all $f \in L_{loc}^{p(\cdot)}(E)$.

The function $w(x,t)$ satisfies (4.20), (4.22)-(4.23) as well as the following conditions:

$$\int_r^{diam(E)} \frac{w^{\frac{1}{p(x)}}(x,t)}{t}dt \lesssim w^{\frac{1}{p(x)}}(x,r),$$

$$\int_r^{diam(E)} \frac{w^{\frac{1}{p(x)}}(x,t)}{t^{1-\alpha(x)}}dt \lesssim r^{-\frac{\alpha(x)p(x)}{q(x)-p(x)}}, \tag{4.33}$$

where $p(x) < q(x)$. Then, the operators $I_{\Omega,\alpha(\cdot)}$ and $M_{\Omega,\alpha(\cdot)}$ are $\left(VL_\Pi^{p(\cdot),w^{\frac{1}{p(\cdot)}}}(E) \to VL_\Pi^{q(\cdot),w^{\frac{1}{q(\cdot)}}}(E)\right)$-bounded. Moreover,

$$\left\|I_{\Omega,\alpha(\cdot)}f\right\|_{VL_\Pi^{q(\cdot),w^{\frac{1}{q(\cdot)}}}(E)} \lesssim \|f\|_{VL_\Pi^{p(\cdot),w^{\frac{1}{p(\cdot)}}}(E)},$$

$$\left\|M_{\Omega,\alpha(\cdot)}f\right\|_{VL_\Pi^{q(\cdot),w^{\frac{1}{q(\cdot)}}}(E)} \lesssim \|f\|_{VL_\Pi^{p(\cdot),w^{\frac{1}{p(\cdot)}}}(E)}.$$

Proof. As in the proof of Theorem 4.5, we represent the function f in the form (4.28) and have

$$I_{\Omega,\alpha(\cdot)}f(x) = I_{\Omega,\alpha(\cdot)}f_1(x) + I_{\Omega,\alpha(\cdot)}f_2(x).$$

For $I_{\Omega,\alpha(\cdot)}f_1(x)$, similar to the proof of (4.17), we obtain the following pointwise estimate:

$$\left|I_{\Omega,\alpha(\cdot)}f_1(x)\right| \lesssim t^{\alpha(x)}M_\Omega f(x). \tag{4.34}$$

For $I_{\Omega,\alpha(\cdot)}f_2(x)$, similar to the proof of (4.29), applying Fubini's theorem, Hölder's inequality, and (4.3), we get

$$\left|I_{\Omega,\alpha(\cdot)}f_2(x)\right| \lesssim \int_r^{diam(E)} t^{\alpha(x)-\frac{n}{p(x)}-1}\|f\|_{L_p(\tilde{B}(x,t))}\,dt \tag{4.35}$$

and by (4.34) and (4.35) complete the proof of (4.32).

Since $M_{\Omega,\alpha(\cdot)}(f) \lesssim I_{|\Omega|,\alpha(\cdot)}(|f|)$ (see Lemma 4.1), it suffices to treat only the case of the operator $I_{\Omega,\alpha(\cdot)}$. In this sense, by (4.32) and (4.33), we obtain

$$\left|I_{\Omega,\alpha(\cdot)}f(x)\right| \lesssim r^{\alpha(x)}M_{\Omega}f(x) + r^{-\frac{\alpha(x)p(x)}{q(x)-p(x)}}\|f\|_{VL_{\Pi}^{p(\cdot),w(\cdot)}(E)}.$$

Then, choosing $r = \left(\frac{\|f\|_{VL_{\Pi}^{p(\cdot),w(\cdot)}(E)}}{M_{\Omega}f(x)}\right)^{\frac{q(x)-p(x)}{\alpha(x)p(x)}}$ for every $x \in E$ supposing that f is not equal 0, thus we have

$$\left|I_{\Omega,\alpha(\cdot)}f(x)\right| \lesssim \left(M_{\Omega}f(x)\right)^{\frac{p(x)}{q(x)}}\|f\|_{VL_{\Pi}^{p(\cdot),w(\cdot)}(E)}^{1-\frac{p(x)}{q(x)}}. \tag{4.36}$$

Finally, by Definition 4.4, (4.36), and (4.31), we get

$$\begin{aligned}
\left\|I_{\Omega,\alpha(\cdot)}f\right\|_{VL_{\Pi}^{q(\cdot),w^{\frac{1}{q(\cdot)}}}(E)} &= \sup_{x\in\Pi,r>0}\frac{r^{-\frac{n}{q(x)}}\left\|I_{\Omega,\alpha(\cdot)}f\right\|_{L^{p(\cdot)}(\tilde{B}(x,r))}}{w(x,r)^{\frac{1}{q(x)}}} \\
&\lesssim \|f\|_{VL_{\Pi}^{p(\cdot),w(\cdot)}(E)}^{1-\frac{p(x)}{q(x)}}\sup_{x\in\Pi,r>0}\frac{r^{-\frac{n}{q(x)}}}{w(x,r)^{\frac{1}{q(x)}}}\|M_{\Omega}f\|_{L^{p(\cdot)}(\tilde{B}(x,r))}^{\frac{p(x)}{q(x)}} \\
&\lesssim \|f\|_{VL_{\Pi}^{p(\cdot),w(\cdot)}(E)}^{1-\frac{p(x)}{q(x)}}\left(\sup_{x\in\Pi,r>0}\frac{r^{-\frac{n}{p(x)}}}{w(x,r)^{\frac{1}{p(x)}}}\|M_{\Omega}f\|_{L^{p(\cdot)}(\tilde{B}(x,r))}\right)^{\frac{p(x)}{q(x)}} \\
&\lesssim \|f\|_{VL_{\Pi}^{p(\cdot),w(\cdot)}(E)}^{1-\frac{p(x)}{q(x)}}\|M_{\Omega}f\|_{VL_{\Pi}^{p(\cdot),w^{\frac{1}{p(\cdot)}}}(E)}^{\frac{p(x)}{q(x)}} \\
&\lesssim \|f\|_{VL_{\Pi}^{p(\cdot),w^{\frac{1}{p(\cdot)}}}(E)}
\end{aligned}$$

if $p(x) < q(x)$ and

$$\limsup_{r\to 0}\sup_{x\in\Pi}\frac{r^{-\frac{n}{q(x)}}\left\|I_{\Omega,\alpha(\cdot)}f\right\|_{L^{q(\cdot)}(\tilde{B}(x,r))}}{w_2(x,t)^{\frac{1}{q(x)}}} \lesssim \limsup_{r\to 0}\sup_{x\in\Pi}\frac{r^{-\frac{n}{p(x)}}\|f\|_{L^{p(\cdot)}(\tilde{B}(x,r))}}{w_1(x,t)^{\frac{1}{p(x)}}} = 0,$$

which completes the proof of Theorem 4.6. □

Corollary 4.5. *Obviously, under the conditions of Theorem 4.6, if the rough (p,q)-admissible $T_{\Omega,\alpha(\cdot)}$-potential type operator is $\left(L^{p(\cdot)}(E) \to L^{q(\cdot)}(E)\right)$-bounded and satisfies (4.30), the result in Theorem 4.6 still holds.*

Remark 4.4. *Let E be a bounded open set and $\lambda(x)$ be a measurable function on E with values in $[0,n]$. Then, the variable exponent vanishing Morrey space*

$VL_{\Pi}^{p(\cdot),\lambda(\cdot)} \equiv VL_{\Pi}^{p(\cdot),\lambda(\cdot)}(E)$ *is defined by*

$$VL_{\Pi}^{p(\cdot),\lambda(\cdot)} \equiv VL_{\Pi}^{p(\cdot),\lambda(\cdot)}(E) = \left\{ \begin{array}{c} f \in L^{p(\cdot),\lambda(\cdot)}(E): \\ \|f\|_{VL_{\Pi}^{p(\cdot),\lambda(\cdot)}} = \lim\limits_{r\to 0} \sup\limits_{\substack{x\in E \\ 0<t<r}} t^{-\frac{\lambda(x)}{p(x)}} \left\| f\chi_{\tilde{B}(x,t)} \right\|_{L^{p(\cdot)}(E)} = 0 \end{array} \right\}.$$

Corollary 4.6. *Let E, Ω, $p(x)$, $\alpha(x)$ be the same as in Theorem 4.5. Define $q(x)$ by $\frac{1}{q(x)} = \frac{1}{p(x)} - \frac{\alpha(x)}{n-\lambda(x)}$. Let also the following conditions hold:*

$$\lambda(x) \geq 0, \qquad \operatorname*{esssup}_{x\in E}[\lambda(x) + \alpha(x)p(x)] < n.$$

Then for $(p_-)' \leq s$, the operators $I_{\Omega,\alpha(\cdot)}$ and $M_{\Omega,\alpha(\cdot)}$ are $\left(VL_{\Pi}^{p(\cdot),\lambda(\cdot)}(E) \to VL_{\Pi}^{q(\cdot),\lambda(\cdot)}(E)\right)$-bounded. Moreover,

$$\left\|I_{\Omega,\alpha(\cdot)}f\right\|_{VL_{\Pi}^{q(\cdot),\lambda(\cdot)}(E)} \lesssim \|f\|_{VL_{\Pi}^{p(\cdot),\lambda(\cdot)}(E)},$$

$$\left\|M_{\Omega,\alpha(\cdot)}f\right\|_{VL_{\Pi}^{q(\cdot),\lambda(\cdot)}(E)} \lesssim \|f\|_{VL_{\Pi}^{p(\cdot),\lambda(\cdot)}(E)}.$$

In the case of $\lambda(x) \equiv 0$, for the spaces $L^{p(\cdot)}(E)$, from Corollary 4.6, we get the following.

Corollary 4.7. *Let E, Ω, $p(x)$, $q(x)$, $\alpha(x)$ be the same as in Theorem 4.5. Then, the operators $I_{\Omega,\alpha(\cdot)}$ and $M_{\Omega,\alpha(\cdot)}$ are $\left(L^{p(\cdot)}(E) \to L^{q(\cdot)}(E)\right)$-bounded. Moreover,*

$$\left\|I_{\Omega,\alpha(\cdot)}f\right\|_{L^{q(\cdot)}(E)} \lesssim \|f\|_{L^{p(\cdot)}(E)},$$

$$\left\|M_{\Omega,\alpha(\cdot)}f\right\|_{L^{q(\cdot)}(E)} \lesssim \|f\|_{L^{p(\cdot)}(E)}.$$

4.2.4 VARIABLE EXPONENT-GENERALIZED CAMPANATO SPACES $C_{\Pi}^{Q(\cdot),\gamma(\cdot)}$

In this section, we first introduce the variable exponent-generalized Campanato spaces and then obtain the boundedness of the commutators of the operators $I_{\Omega,\alpha(\cdot)}$, $M_{\Omega,\alpha(\cdot)}$, T_Ω, and M_Ω on the spaces $VL_{\Pi}^{p(\cdot),w(\cdot)}(E)$.

Definition 4.6. *Let $1 \leq q(\cdot) \leq q^+ < \infty$ and $0 \leq \gamma(\cdot) < \frac{1}{n}$. Define the generalized Campanato space $C_{\Pi}^{q(\cdot),\gamma(\cdot)}$ with variable exponents $q(\cdot)$, $\gamma(\cdot)$ as follows:*

$$C_{\Pi}^{q(\cdot),\gamma(\cdot)}(E) = \left\{ f \in L_{loc}^{q(\cdot)}\left(\tilde{B}(x,r)\right) : \|f\|_{C_{\Pi}^{q(\cdot),\gamma(\cdot)}(E)} < \infty \right\},$$

where

$$\|f\|_{C_{\Pi}^{q(\cdot),\gamma(\cdot)}(E)} = \sup_{x\in\Pi, r>0} |B(x,r)|^{-\frac{1}{q(x)}-\gamma(x)} \left\|f - f_{B(x,r)}\right\|_{L^{q(\cdot)}(\tilde{B}(x,r))}$$

such that

$$\left\|f - f_{B(x,r)}\right\|_{L^{q(\cdot)}\left(\tilde{B}(x,r)\right)} \lesssim r^{\frac{n}{q(x)}+n\gamma(x)} \left\|f\right\|_{C_{\Pi}^{q(\cdot),\gamma(\cdot)}(E)}. \tag{4.37}$$

When $\Pi = \{x_0\}$ and $\Pi = E$, $C_{\Pi}^{q(\cdot),\gamma(\cdot)}(E)$ turns into the local generalized Campanato space $C_{\{x_0\}}^{q(\cdot),\gamma(\cdot)}(E)$ and the global generalized Campanato space $C_{E}^{q(\cdot),\gamma(\cdot)}(E)$, respectively. If $q(\cdot)$, $\gamma(\cdot)$ are the constant functions and $\Pi = E$, then the variable exponent-generalized Campanato space $C_{\Pi}^{q(\cdot),\gamma(\cdot)}(E)$ is exactly the usual Campanato space $C^{q,\gamma}(E)$. If $\gamma(\cdot) \equiv 0$ and $q(\cdot) \equiv q$, the generalized Campanato space $C_{\Pi}^{q(\cdot),\gamma(\cdot)}(E)$ is just the central $BMO(E)$ (the local version of $BMO(E)$).

Theorem 4.7. *Let E, Ω, $p(x)$, $q(x)$, $\alpha(x)$ be the same as in Theorem 4.5. Let also $\frac{1}{p(\cdot)} = \frac{1}{p_1(\cdot)} + \frac{1}{p_2(\cdot)}$, $\frac{1}{q_1(\cdot)} = \frac{1}{p_1(\cdot)} - \frac{\alpha(\cdot)}{n}$ and $b \in C_{\Pi}^{p_2(\cdot),\gamma(\cdot)}(E)$. Suppose that $p_1(\cdot)$, $p_2(\cdot)$, $q(\cdot)$, $q_1(\cdot)$, and $\alpha(\cdot)$ satisfy (4.1). Then, for $\frac{s}{s-1} < p^- \le p(\cdot) < \frac{n}{\alpha(\cdot)}$, the following pointwise estimate*

$$\left\| \left[b, I_{\Omega,\alpha(\cdot)}\right] f \right\|_{L_q(\tilde{B}(x,r))} \lesssim \|b\|_{C_{\Pi}^{p_2(\cdot),\gamma(\cdot)}} r^{\frac{n}{q(x)}} \int_{2r}^{diam(E)} \left(1 + \ln\frac{t}{r}\right) t^{n\gamma(x) - \frac{n}{q_1(x)} - 1} \|f\|_{L^{p_1(\cdot)}(\tilde{B}(x,t))}\, dt \tag{4.38}$$

holds for any ball $\tilde{B}(x,r)$ and for all $f \in L_{loc}^{p_1(\cdot)}(E)$.

If the functions $w_1(x,r)$ and $w_2(x,r)$ satisfy (4.20) as well as the following Zygmund condition

$$\int_{r}^{diam(E)} \left(1 + \ln\frac{t}{r}\right) \frac{w_1^{\frac{1}{p_1(x)}}(x,t)}{t^{1-(\alpha(x)+n\gamma(x))}} dt \lesssim w_2^{\frac{1}{q(x)}}(x,r), \qquad r \in (0, diam(E)] \tag{4.39}$$

and additionally these functions satisfy conditions (4.22) and (4.23),

$$d_\delta := \int_{\delta}^{diam(E)} \sup_{x\in\Pi} \left(1 + \ln\frac{t}{r}\right) \frac{w_1^{\frac{1}{p_1(x)}}(x,t)}{t^{1-(\alpha(x)+n\gamma(x))}} dt < \infty, \qquad \delta > 0,$$

then the operators $\left[b, I_{\Omega,\alpha(\cdot)}\right]$ and $\left[b, M_{\Omega,\alpha(\cdot)}\right]$ are $\left(VL_{\Pi}^{p_1(\cdot),w_1(\cdot)}(E) \to VL_{\Pi}^{q(\cdot),w_2(\cdot)}(E)\right)$-bounded. Moreover,

$$\left\| \left[b, I_{\Omega,\alpha(\cdot)}\right] f \right\|_{VL_{\Pi}^{q(\cdot),w_2(\cdot)}(E)} \lesssim \|b\|_{C_{\Pi}^{p_2(\cdot),\gamma(\cdot)}} \|f\|_{VL_{\Pi}^{p_1(\cdot),w_1(\cdot)}(E)},$$

$$\left\| \left[b, M_{\Omega,\alpha(\cdot)}\right] f \right\|_{VL_{\Pi}^{q(\cdot),w_2(\cdot)}(E)} \lesssim \|b\|_{C_{\Pi}^{p_2(\cdot),\gamma(\cdot)}} \|f\|_{VL_{\Pi}^{p_1(\cdot),w_1(\cdot)}(E)}.$$

Proof. Since $\left[b, M_{\Omega,\alpha(\cdot)}\right](f) \lesssim \left[b, I_{|\Omega|,\alpha(\cdot)}\right](|f|)$, it suffices to treat only the case of the operator $\left[b, I_{\Omega,\alpha(\cdot)}\right]$. As in the proof of Theorem 4.5, we represent the function f in the form (4.28) and have

$$\left[b, I_{\Omega,\alpha(\cdot)}\right] f(x) = \left(b(x) - b_{B(x,r)}\right) I_{\Omega,\alpha(\cdot)} f_1(x) - I_{\Omega,\alpha(\cdot)}\left(\left(b(\cdot) - b_{B(x,r)}\right) f_1\right)(x)$$

$$+\left(b(x)-b_{B(x,r)}\right)I_{\Omega,\alpha(\cdot)}f_2(x)-I_{\Omega,\alpha(\cdot)}\left(\left(b(\cdot)-b_{B(x,r)}\right)f_2\right)(x)$$
$$\equiv F_1+F_2+F_3+F_4.$$

Hence, we get

$$\left\|\left[b,I_{\Omega,\alpha(\cdot)}\right]f\right\|_{L^{q(\cdot)}(\tilde{B}(x,r))}\leq\|F_1\|_{L^{q(\cdot)}(\tilde{B}(x,r))}+\|F_2\|_{L^{q(\cdot)}(\tilde{B}(x,r))}+\|F_3\|_{L^{q(\cdot)}(\tilde{B}(x,r))}$$
$$+\|F_4\|_{L^{q(\cdot)}(\tilde{B}(x,r))}. \tag{4.40}$$

First, we use the Hölder's inequality such that $\frac{1}{q(\cdot)}=\frac{1}{p_2(\cdot)}+\frac{1}{q_1(\cdot)}$, the boundedness of $I_{\Omega,\alpha(\cdot)}$ from $L^{p(\cdot)}$ into $L^{q(\cdot)}$ (see Theorem 4.1) and (4.37) to estimate $\|F_1\|_{L^{q(\cdot)}(\tilde{B}(x,r))}$, and we obtain

$$\begin{aligned}
\|F_1\|_{L^{q(\cdot)}(\tilde{B}(x,r))}&=\left\|(b(\cdot)-b_B)I_{\Omega,\alpha(\cdot)}f_1(\cdot)\right\|_{L^{q(\cdot)}(\tilde{B}(x,r))}\\
&\lesssim\|(b(\cdot)-b_B)\|_{L^{p_2(\cdot)}(\tilde{B}(x,r))}\left\|I_{\Omega,\alpha(\cdot)}f_1(\cdot)\right\|_{L^{q_1(\cdot)}(\tilde{B}(x,r))}\\
&\lesssim r^{\frac{n}{p_2(x)}+n\gamma(x)}\|b\|_{C_{\Pi}^{p_2(\cdot),\gamma(\cdot)}}\|f_1\|_{L^{p_1(\cdot)}(\tilde{B}(x,r))}\\
&=\|b\|_{C_{\Pi}^{p_2(\cdot),\gamma(\cdot)}}r^{\frac{n}{p_2(x)}+\frac{n}{q_1(x)}+n\gamma(x)}\|f\|_{L^{p_1(\cdot)}(\tilde{B}(x,2r))}\int_{2r}^{diam(E)}t^{-1-\frac{n}{q_1(x)}}dt\\
&\lesssim\|b\|_{C_{\Pi}^{p_2(\cdot),\gamma(\cdot)}}r^{\frac{n}{q(x)}}\int_{2r}^{diam(E)}\left(1+\ln\frac{t}{r}\right)\|f\|_{L^{p_1(\cdot)}(\tilde{B}(x,t))}t^{n\gamma(x)-\frac{n}{q_1(x)}-1}dt.
\end{aligned}$$

Second, for $\|F_2\|_{L^{q(\cdot)}(\tilde{B}(x,r))}$, applying the boundedness of $I_{\Omega,\alpha(\cdot)}$ from $L^{p(\cdot)}$ into $L^{q(\cdot)}$ (see Theorem 4.1), generalized Hölder's inequality such that $\frac{1}{p(\cdot)}=\frac{1}{p_1(\cdot)}+\frac{1}{p_2(\cdot)}$, $\frac{1}{q(\cdot)}=\frac{1}{p_2(\cdot)}+\frac{1}{q_1(\cdot)}$ and (4.37), we know that

$$\begin{aligned}
\|F_2\|_{L^{q(\cdot)}(\tilde{B}(x,r))}&=\left\|I_{\Omega,\alpha(\cdot)}\left(b(\cdot)-b_{B(x,r)}\right)f_1\right\|_{L^{q(\cdot)}(\tilde{B}(x,r))}\\
&\lesssim\|(b(\cdot)-b_B)f_1\|_{L^{p(\cdot)}(\tilde{B}(x,r))}\\
&\lesssim\|(b(\cdot)-b_B)\|_{L^{p_2(\cdot)}(\tilde{B}(x,r))}\|f_1\|_{L^{p_1(\cdot)}(\tilde{B}(x,r))}\\
&\lesssim\|f\|_{C_{\Pi}^{p_2(\cdot),\gamma(\cdot)}}r^{\frac{n}{p_2(x)}+\frac{n}{q_1(x)}+n\gamma(x)}\|f\|_{L^{p_1(\cdot)}(\tilde{B}(x,2r))}\int_{2r}^{diam(E)}t^{-1-\frac{n}{q_1(x)}}dt\\
&\lesssim\|b\|_{C_{\Pi}^{p_2(\cdot),\gamma(\cdot)}}r^{\frac{n}{q(x)}}\int_{2r}^{diam(E)}\left(1+\ln\frac{t}{r}\right)\|f\|_{L^{p_1(\cdot)}(\tilde{B}(x,t))}t^{n\gamma(x)-\frac{n}{q_1(x)}-1}dt.
\end{aligned}$$

Third, for $\|F_3\|_{L^{q(\cdot)}(\tilde{B}(x,r))}$, similar to the proof of (4.29), when $\frac{s}{s-1} \leq p_1(\cdot)$, by Fubini's theorem, generalized Hölder's inequality, and (4.3), we have

$$\begin{aligned} \left|I_{\Omega,\alpha(\cdot)} f_2(x)\right| &\lesssim \int\limits_{E \backslash \tilde{B}(x,2r)} \frac{|\Omega(z-y)||f(y)|}{|x-y|^{n-\alpha(x)}} dy \\ &\lesssim \int\limits_{2r}^{diam(E)} \|f\|_{L^{p_1(\cdot)}(\tilde{B}(x,t))} t^{-1-\frac{n}{q_1(x)}} dt. \end{aligned} \tag{4.41}$$

Thus, by generalized Hölder's inequality such that $\frac{1}{q(\cdot)} = \frac{1}{p_2(\cdot)} + \frac{1}{q_1(\cdot)}$, (4.37), and (4.41), we obtain

$$\begin{aligned} \|F_3\|_{L^{q(\cdot)}(\tilde{B}(x,r))} &= \left\|\left(b(\cdot)-b_{B(x,r)}\right) I_{\Omega,\alpha(\cdot)} f_2(\cdot)\right\|_{L^{q(\cdot)}(\tilde{B}(x,r))} \\ &\lesssim \left\|\left(b(\cdot)-b_{B(x,r)}\right)\right\|_{L^{p_2(\cdot)}(\tilde{B}(x,r))} \left\|I_{\Omega,\alpha(\cdot)} f_2(\cdot)\right\|_{L^{q_1(\cdot)}(\tilde{B}(x,r))} \\ &\lesssim r^{\frac{n}{p_2(x)}+n\gamma(x)} \|b\|_{C_{\Pi}^{p_2(\cdot),\gamma(\cdot)}} r^{\frac{n}{q_1(x)}} \int\limits_{2r}^{diam(E)} \|f\|_{L^{p_1(\cdot)}(\tilde{B}(x,t))} t^{-1-\frac{n}{q_1(x)}} dt \\ &\lesssim \|b\|_{C_{\Pi}^{p_2(\cdot),\gamma(\cdot)}} r^{\frac{n}{q(x)}} \int\limits_{2r}^{diam(E)} \left(1+\ln\frac{t}{r}\right) \|f\|_{L^{p_1(\cdot)}(\tilde{B}(x,t))} t^{n\gamma(x)-\frac{n}{q_1(x)}-1} dt. \end{aligned}$$

Finally, we consider the term $\|F_4\|_{L^{q(\cdot)}(\tilde{B}(x,r))} = \left\|I_{\Omega,\alpha(\cdot)}\left(\left(b(\cdot)-b_{B(x,r)}\right) f_2\right)(\cdot)\right\|_{L^{q(\cdot)}(\tilde{B}(x,r))}$. For $z \in B(x,r)$, when $\frac{s}{s-1} \leq p(\cdot)$, by the Fubini's theorem, applying the generalized Hölder's inequality, and from (4.3) and (4.37), we have

$$\begin{aligned} &\left|I_{\Omega,\alpha(\cdot)}\left(\left(b(\cdot)-b_{B(x,r)}\right) f_2\right)(z)\right| \\ &\lesssim \int\limits_{2r}^{diam(E)} \left|b(y)-b_{B(x,r)}\right| |\Omega(z-y)| \frac{|f(y)|}{|x-y|^{n-\alpha(x)}} dy \\ &\approx \int\limits_{2r}^{diam(E)} \int\limits_{2r<|x-y|<t} \left|b(y)-b_{B(x,r)}\right| |\Omega(z-y)| |f(y)| dy \frac{dt}{t^{n-\alpha(x)+1}} \\ &\lesssim \int\limits_{2r}^{diam(E)} \int\limits_{B(x,t)} \left|b(y)-b_{B(x,t)}\right| |\Omega(z-y)| |f(y)| dy \frac{dt}{t^{n-\alpha(x)+1}} \\ &+ \int\limits_{2r}^{diam(E)} \int\limits_{B(x,t)} \left|b_{B(x,r)}-b_{B(x,t)}\right| |\Omega(z-y)| |f(y)| dy \frac{dt}{t^{n-\alpha(x)+1}} \\ &\lesssim \int\limits_{2r}^{diam(E)} \left\|\left(b(\cdot)-b_{B(x,r)}\right)\right\|_{L^{p_2(\cdot)}(\tilde{B}(x,t))} \|f\|_{L^{p_1(\cdot)}(\tilde{B}(x,t))} \|\Omega(z-\cdot)\|_{L_s(\tilde{B}(x,t))} \\ &t^{-\frac{n}{q(x)}-\frac{n}{s}-1} dt \\ &+ \int\limits_{2r}^{diam(E)} \left\|\left(b_{B(x,r)}-b_{B(x,t)}\right)\right\|_{L^{p_2(\cdot)}(\tilde{B}(x,t))} \|f\|_{L^{p_1(\cdot)}(\tilde{B}(x,t))} \|\Omega(z-\cdot)\|_{L_s(\tilde{B}(x,t))} \\ &t^{-\frac{n}{q(x)}-\frac{n}{s}-1} dt \end{aligned}$$

$$\lesssim \int_{2r}^{diam(E)} \left\|\left(b(\cdot)-b_{B(x,r)}\right)\right\|_{L^{p_2(\cdot)}(\tilde{B}(x,t))} \|f\|_{L^{p_1(\cdot)}(\tilde{B}(x,t))} t^{-\frac{n}{q(x)}-1} dt$$

$$+\|b\|_{C_{\Pi}^{p_2(\cdot),\gamma(\cdot)}} \int_{2r}^{diam(E)} \left(1+\ln\tfrac{t}{r}\right) \|f\|_{L^{p_1(\cdot)}(\tilde{B}(x,t))} t^{n\gamma(x)-\frac{n}{q_1(x)}-1} dt$$

$$\lesssim \|b\|_{C_{\Pi}^{p_2(\cdot),\gamma(\cdot)}} \int_{2r}^{diam(E)} \left(1+\ln\frac{t}{r}\right) \|f\|_{L^{p_1(\cdot)}(\tilde{B}(x,t))} t^{n\gamma(x)-\frac{n}{q_1(x)}-1} dt. \tag{4.42}$$

Then, by (4.42), we have

$$\|F_4\|_{L^{q(\cdot)}(\tilde{B}(x,r))} = \left\|I_{\Omega,\alpha(\cdot)}\left(\left(b(\cdot)-b_{B(x,r)}\right)f_2\right)(x)\right\|_{L^{q(\cdot)}(\tilde{B}(x,r))}$$

$$\lesssim \|b\|_{C_{\Pi}^{p_2(\cdot),\gamma(\cdot)}} r^{\frac{n}{q(x)}} \int_{2r}^{diam(E)} \left(1+\ln\frac{t}{r}\right) \|f\|_{L^{p_1(\cdot)}(\tilde{B}(x,t))} t^{n\gamma(x)-\frac{n}{q_1(x)}-1} dt.$$

Combining all the estimates of $\|F_1\|_{L^{q(\cdot)}(\tilde{B}(x,r))}$, $\|F_2\|_{L^{q(\cdot)}(\tilde{B}(x,r))}$, $\|F_3\|_{L^{q(\cdot)}(\tilde{B}(x,r))}$, $\|F_4\|_{L^{q(\cdot)}(\tilde{B}(x,r))}$, we get (4.38).

At last, by Definition 4.4, (4.38), and (4.39), we get

$$\left\|\left[b,I_{\Omega,\alpha(\cdot)}\right]f\right\|_{VL_{\Pi}^{q(\cdot),w_2(\cdot)}(E)} = \sup_{x\in\Pi,r>0} \frac{r^{-\frac{n}{q(x)}} \left\|\left[b,I_{\Omega,\alpha(\cdot)}\right]f\right\|_{L^{q(\cdot)}(\tilde{B}(x,r))}}{w_2(x,r)^{\frac{1}{q(x)}}}$$

$$\lesssim \|b\|_{C_{\Pi}^{p_2(\cdot),\gamma(\cdot)}} \sup_{x\in\Pi,r>0} \frac{1}{w_2(x,r)^{\frac{1}{q(x)}}} \int_{2r}^{diam(E)} \left(1+\ln\frac{t}{r}\right) t^{n\gamma(x)-\frac{n}{q_1(x)}-1}$$

$$\times \|f\|_{L^{p_1(\cdot)}(\tilde{B}(x,t))} dt$$

$$\lesssim \|b\|_{C_{\Pi}^{p_2(\cdot),\gamma(\cdot)}} \|f\|_{VL_{\Pi}^{p_1(\cdot),w_1(\cdot)}(E)} \sup_{x\in\Pi,r>0} \frac{1}{w_2(x,r)^{\frac{1}{q(x)}}} \int_{r}^{diam(E)} \left(1+\ln\frac{t}{r}\right)$$

$$\times t^{\alpha(x)+n\gamma(x)} w_1^{\frac{1}{p_1(x)}}(x,t)\frac{dt}{t}$$

$$\lesssim \|b\|_{C_{\Pi}^{p_2(\cdot),\gamma(\cdot)}} \|f\|_{VL_{\Pi}^{p(\cdot),w_1(\cdot)}(E)}$$

and

$$\limsup_{r\to 0}\sup_{x\in\Pi} \frac{r^{-\frac{n}{q(x)}} \left\|\left[b,I_{\Omega,\alpha(\cdot)}\right]f\right\|_{L^{q(\cdot)}(\tilde{B}(x,r))}}{w_2(x,t)^{\frac{1}{q(x)}}} \lesssim \limsup_{r\to 0}\sup_{x\in\Pi} \frac{r^{-\frac{n}{p(x)}} \|f\|_{L^{p(\cdot)}(\tilde{B}(x,r))}}{w_1(x,t)^{\frac{1}{p(x)}}} = 0,$$

which completes the proof of Theorem 4.7. □

Corollary 4.8. *Let E, Ω, $p(x)$, $q(x)$, $\alpha(x)$ be the same as in Theorem 4.5. Suppose that $q(\cdot)$ and $\alpha(\cdot)$ satisfy (4.1). Then, for $\frac{s}{s-1} < p^- \le p(\cdot) < \frac{n}{\alpha(\cdot)}$ and $b \in BMO(E)$, the following pointwise estimate*

$$\| [b, I_{\Omega,\alpha(\cdot)}] f\|_{L^{q(\cdot)}(\tilde{B}(x,r))} \lesssim \|b\|_{BMO} r^{\frac{n}{q(x)}} \int_{2r}^{diam(E)} \left(1+\ln\frac{t}{r}\right) t^{\frac{n}{q(x)}-1} \|f\|_{L^{p(\cdot)}(\tilde{B}(x,t))} dt$$

holds for any ball $\tilde{B}(x,r)$ and for all $f \in L^{p(\cdot)}_{loc}(E)$.

If the functions $w_1(x,r)$ and $w_2(x,r)$ satisfy (4.20) as well as the following Zygmund condition

$$\int_{r}^{diam(E)} \left(1+\ln\frac{t}{r}\right) \frac{w_1^{\frac{1}{p(x)}}(x,t)}{t^{1-\alpha(x)}} dt \lesssim w_2^{\frac{1}{q(x)}}(x,r), \qquad r \in (0, diam(E)]$$

and additionally these functions satisfy conditions (4.22) and (4.23),

$$d_\delta := \int_{\delta}^{diam(E)} \sup_{x\in\Pi} \left(1+\ln\frac{t}{r}\right) \frac{w_1^{\frac{1}{p(x)}}(x,t)}{t^{1-\alpha(x)}} dt < \infty, \qquad \delta > 0,$$

then the operators $[b, I_{\Omega,\alpha(\cdot)}]$ and $[b, M_{\Omega,\alpha(\cdot)}]$ are $\left(VL_{\Pi}^{p(\cdot),w_1(\cdot)}(E) \to VL_{\Pi}^{q(\cdot),w_2(\cdot)}(E)\right)$-bounded. Moreover,

$$\left\| [b, I_{\Omega,\alpha(\cdot)}] f\right\|_{VL_{\Pi}^{q(\cdot),w_2(\cdot)}(E)} \lesssim \|b\|_{BMO} \|f\|_{VL_{\Pi}^{p_1(\cdot),w_1(\cdot)}(E)},$$

$$\left\| [b, M_{\Omega,\alpha(\cdot)}] f\right\|_{VL_{\Pi}^{q(\cdot),w_2(\cdot)}(E)} \lesssim \|b\|_{BMO} \|f\|_{VL_{\Pi}^{p_1(\cdot),w_1(\cdot)}(E)}.$$

For $\alpha(x) = 0$ in Theorem 4.7, we get the following new result.

Corollary 4.9. *Let E, Ω, $p(x)$ be the same as in Theorem 4.5. Let also $\frac{1}{p(\cdot)} = \frac{1}{p_1(\cdot)} + \frac{1}{p_2(\cdot)}$ and $b \in C_{\Pi}^{p_2(\cdot),\gamma(\cdot)}(E)$. Suppose that $p_1(\cdot)$ and $p_2(\cdot)$ satisfy (4.1). Then, for $\frac{s}{s-1} < p^- \le p(\cdot) \le p^+ < \infty$, the following pointwise estimate*

$$\| [b, T_\Omega] f\|_{L^{p(\cdot)}(\tilde{B}(x,r))} \lesssim \|b\|_{C_{\Pi}^{p_2(\cdot),\gamma(\cdot)}} r^{\frac{n}{p(x)}} \int_{2r}^{diam(E)} \left(1+\ln\frac{t}{r}\right) t^{n\gamma(x)-\frac{n}{p_1(x)}-1} \|f\|_{L^{p_1(\cdot)}(\tilde{B}(x,t))} dt$$

holds for any ball $\tilde{B}(x,r)$ and for all $f \in L^{p_1(\cdot)}_{loc}(E)$.

If the function $w(x,r)$ satisfies (4.20) as well as the following Zygmund condition

$$\int_{r}^{diam(E)} \left(1+\ln\frac{t}{r}\right) \frac{w^{\frac{1}{p_1(x)}}(x,t)}{t^{1-n\gamma(x)}} dt \lesssim w^{\frac{1}{p(x)}}(x,r), \qquad r \in (0, diam(E)]$$

and additionally this function satisfies conditions (4.22) and (4.23),

$$d_\delta := \int_\delta^{diam(E)} \sup_{x\in\Pi}\left(1+\ln\frac{t}{r}\right)\frac{w^{\frac{1}{p_1(x)}}(x,t)}{t^{1-n\gamma(x)}}dt < \infty, \qquad \delta>0,$$

then the operators $[b,T_\Omega]$ *and* $[b,M_\Omega]$ *are* $\left(VL_\Pi^{p_1(\cdot),w(\cdot)}(E)\to VL_\Pi^{p(\cdot),w(\cdot)}(E)\right)$*-bounded. Moreover,*

$$\|[b,T_\Omega]f\|_{VL_\Pi^{p(\cdot),w(\cdot)}(E)} \lesssim \|b\|_{C_\Pi^{p_2(\cdot),\gamma(\cdot)}}\|f\|_{VL_\Pi^{p_1(\cdot),w(\cdot)}(E)},$$

$$\|[b,M_\Omega]f\|_{VL_\Pi^{p(\cdot),w(\cdot)}(E)} \lesssim \|b\|_{C_\Pi^{p_2(\cdot),\gamma(\cdot)}}\|f\|_{VL_\Pi^{p_1(\cdot),w(\cdot)}(E)}.$$

From Corollary 4.9, we get the following.

Corollary 4.10. *Let* E, Ω, $p(x)$ *be the same as in Theorem 4.5. Then, for* $\frac{s}{s-1} < p^- \le p(\cdot) \le p^+ < \infty$ *and* $b\in BMO(E)$*, the following pointwise estimate*

$$\|[b,T_\Omega]f\|_{L^{p(\cdot)}(\tilde{B}(x,r))} \lesssim \|b\|_{BMO} r^{\frac{n}{p(x)}} \int_{2r}^{diam(E)} \left(1+\ln\frac{t}{r}\right) t^{\frac{n}{p(x)}-1}\|f\|_{L^{p(\cdot)}(\tilde{B}(x,t))}dt$$

holds for any ball $\tilde{B}(x,r)$ *and for all* $f\in L_{loc}^{p(\cdot)}(E)$.

If the function $w(x,r)$ *satisfies (4.20) as well as the following Zygmund condition*

$$\int_r^{diam(E)} \left(1+\ln\frac{t}{r}\right)\frac{w^{\frac{1}{p(x)}}(x,t)}{t}dt \lesssim w^{\frac{1}{p(x)}}(x,r), \qquad r\in(0,diam(E)]$$

and additionally this function satisfies conditions (4.22) and (4.23),

$$d_\delta := \int_\delta^{diam(E)} \sup_{x\in\Pi}\left(1+\ln\frac{t}{r}\right)\frac{w^{\frac{1}{p(x)}}(x,t)}{t}dt < \infty, \qquad \delta>0,$$

then the operators $[b,T_\Omega]$ *and* $[b,M_\Omega]$ *are bounded on* $VL_\Pi^{p(\cdot),w(\cdot)}(E)$*. Moreover,*

$$\|[b,T_\Omega]f\|_{VL_\Pi^{p(\cdot),w(\cdot)}(E)} \lesssim \|b\|_{BMO}\|f\|_{VL_\Pi^{p(\cdot),w(\cdot)}(E)},$$

$$\|[b,M_\Omega]f\|_{VL_\Pi^{p(\cdot),w(\cdot)}(E)} \lesssim \|b\|_{BMO}\|f\|_{VL_\Pi^{p(\cdot),w(\cdot)}(E)}.$$

4.3 CONCLUSION

It is well known that, for the purpose of researching non-smoothness partial differential equation, mathematicians pay more attention to the singular integrals with rough kernel. Moreover, the fractional-type operators and their weighted boundedness theory play important roles in harmonic analysis and other fields, and the multilinear operators arise in numerous situations involving product-like operations. Moreover, the multilinear operators are natural generalizations of linear case. In recent years, these topics gain the attention from a vast number of researchers in different function spaces. Especially, more and more researches focus on function spaces based on variable exponent Morrey spaces to fill in some gaps in the theory of Morrey-type spaces. Moreover, these spaces are useful in harmonic analysis and PDEs. But this topic exceeds the scope of this paper. Thus, we omit the details here. In this paper, we have shown that various classical operators (such as rough maximal, potential, and singular integral operators) and their commutators are bounded in different proper closed subspaces of variable exponent-generalized Morrey spaces. On the other hand, all these subspaces together provide an explicit description of the closure of nice functions in variable exponent Morrey norm in terms of vanishing properties of Morrey functions. Therefore, our results give a contribution for the development of harmonic analysis on variable exponent-generalized Morrey spaces.

As is well known, variable exponent Morrey-type spaces attract a lot of attention from the applications side. By this reason, we are convinced that the results exhibited in this paper can also be useful for regularity results in the theory of certain partial differential equations.

Finally, the results presented here are sure to be new and potentially useful. Since the research subject here and its related ones are so popular, the content of this paper may attract interested readers who have been interested in this and related research subjects. Therefore, the results in this paper are worthwhile to record.

FUNDING

This work is funded by Hakkari University Scientific Research Project (Grant No. FM18BAP1) under the research project "Some estimates for rough Riesz-type potential operator with variable order and rough fractional maximal operator with variable order both on generalized variable exponent Morrey spaces and on vanishing generalized variable exponent Morrey spaces," Institution of Higher Education Scientific Research Project in Ningxia (Grant No. NGY2017011) and Natural Science Foundation of China (Grant Nos. 11461053 and 11762016).

REFERENCES

[1] A. Almeida, J.J. Hasanov and S.G. Samko, Maximal and potential operators in variable exponent Morrey spaces, *Georgian Math. J.*, 2008, 15(2): 195–208.

[2] S.N. Antontsev and J.F. Rodrigues, On stationary thermorheological viscous flows, *Ann. Univ. Ferrara, Sez. VII, Sci. Mat.*, 2006, 52(1): 19–36.

[3] Y. Chen, S.E. Levine and M. Rao, Variable exponent, linear growth functionals in image restoration, *SIAM J. Appl. Math.*, 2006, 66(4): 1383–1406.

[4] D. Cruz-Uribe, A. Fiorenza, C.J. Neugebauer, The maximal function on variable L_p spaces, *Ann. Acad. Sci Fenn. Math.*, 2003, 28(1): 223–238.

[5] L. Diening, P. Harjulehto, P. Hästö and M. Růžička, *Lebesgue and Sobolev Space with Variable Exponents,* Springer Lecture Notes, Vol. 2017, Berlin: Springer-Verlag, 2011.

[6] L. Diening, P. Hästö and A. Nekvinda, Open problems in variable exponent Lebesgue and Sobolev spaces, in Proceedings of the Function Spaces, Differential Operators and Nonlinear Analysis (FSDONA'04), pp. 38–58, Milovy, Czech, 2004.

[7] L. Diening and M. Rüžička, Calderón-Zygmund operators on generalized Lebesgue Spaces $L^{p(x)}(\Omega)$ and problems related to fluid dynamics, *J. Reine Angew. Math.*, 2003, 563: 197–220.

[8] X.L. Fan and D. Zhao, On the spaces $L^{p(x)}(\Omega)$ and $W^{m,p(x)}(\Omega)$, *J. Math. Anal. Appl.*, 2001, 263(2): 424–446.

[9] M. Giaquinta, *Multiple integrals in the calculus of variations and non-linear elliptic systems.* Princeton, New Jersey: Princeton Univ. Press, 1983.

[10] V.S. Guliyev, J.J. Hasanov and S.G. Samko, Boundedness of the maximal, potential and singular integral operators in generalized variable exponent Morrey spaces, *Math. Scand.*, 2010, 107(2): 285–304.

[11] P. Harjulehto, P. Hästö, Út V. Lê and M. Nuortio, Overview of differential equations with nonstandard growth, *Nonlinear. Anal.*, 2010, 72(12): 4551–4574.

[12] K. Ho, The fractional integral operators on Morrey spaces with variable exponent on unbounded domains, *Math. Inequal. Appl.*, 2013, 16(2): 363–373.

[13] K. Ho, Fractional integral operators with homogeneous kernels on Morrey spaces with variable exponents, *J. Math. Soc. Japan*, 2017, 69(3): 1059–1077.

[14] O. Kováčik and J. Rákosník, On spaces $L^{p(x)}$ and $W^{k,p(x)}$, *Czechoslovak Math. J.*, 1991, 41(11): 592–618.

[15] P. Long and H. Han, Characterizations of some operators on the vanishing generalized Morrey spaces with variable exponent, *J. Math. Anal. Appl.*, 2016, 437(1): 419–430.

[16] C.B. Morrey, On the solutions of quasi-linear elliptic partial differential equations, *Trans. Amer. Math. Soc.*, 1938, 43: 126–166.

[17] W. Orlicz, Über konjugierte Exponentenfolgen, *Studia Math.*, 1931, 3: 200–211.

[18] H. Rafeiro and S.G. Samko, On maximal and potential operators with rough kernels in variable exponent spaces, *Rend. Lincei Mat. Appl.*, 2016,27(3): 309–325.

[19] M. Ružička, *Electrorheological fluids: modeling and mathematical theory,* Berlin: Springer, 2013.

[20] Y. Sawano, K. Ho, D.C. Yang and S. Yang, Hardy spaces for ball quasi-Banach function spaces, *Dissertationes Math.*, 2017, 525: 102 pp.

[21] M.E. Taylor, *Tools for PDE: Pseudodifferential Operators, Paradifferential Operators, and Layer Potentials,* Volume 81 of Math. Surveys and Monogr. AMS, Providence, R.I., 2000.

[22] J. Wu, Boundedness for Riesz-type potential operators on Herz-Morrey spaces with variable exponent, *Math. Inequal. Appl.*, 2015, 18(2): 471–484.

[23] T. Wunderli, On time flows of minimizers of general convex functionals of linear growth with variable exponent in BV space and stability of pseudosolutions, *J. Math. Anal. Appl.*, 2010, 364(2): 591–598.

[24] L.J. Wang and Sh. P. Tao, Boundedness of Littlewood-Paley operators and their commutators on Herz-Morrey spaces with variable exponent, *J. Inequal. Appl.*, 2014, 2014: 227.

5 Compact-Like Operators in Vector Lattices Normed by Locally Solid Lattices

Abdullah Aydın
Muş Alparslan University

CONTENTS

5.1 INTRODUCTION

Lattice-valued norms on vector lattices play an important and effective role in the functional analysis. It is well known that order convergence in vector lattice is not topological unless dimension is finite. However, thanks to the order convergence, some properties like order continuous in vector lattice can be defined. We refer the reader for an exposition on basic notions and properties of vector lattice (or, Riesz space) to [1–3,13,16,25–27,31,32] and on the theory of lattice-normed vector lattices to [9–11,16,23,24] and to locally solid vector lattices [2–6,20,32]. In this chapter, we introduce and study continuous and bounded operators with respect to the p_τ-convergence, which was introduced and investigated by Aydın in [5], and we also introduce the compact operators in vector lattice normed by locally solid lattices.

The concept of the unbounded convergence is crucial for the concept of p_τ-convergence. The *uo-convergence* was introduced in [28] under the name *individual convergence* and the name "unbounded order convergence" was first proposed by DeMarr in [14]. The relation between weak and *uo*-convergence in Banach lattices was studied by Wickstead in [30], and Kaplan established two characterizations of *uo*-convergence on Dedekind complete Riesz spaces with the weak unit in [22], and the *un-convergence* was introduced in [29] under the name *d-convergence*. We refer the reader for an exposition on *uo*-convergence to [15,17–19], on *un*-convergence to

[15,21,29], on the unbounded p-convergence to [10–12], and on the p_τ-convergence to [5,6]. For applications of uo-convergence, we refer to [4,7,8,18,19].

The structure of this chapter is as follows. In Section 5.2, we give some basic notions and properties of vector lattices, lattice-normed spaces, locally solid Riesz spaces, unbounded convergence, and unbounded p_τ-convergence.

In Section 5.3, we introduce the definitions of the p_τ-continuous and the p_τ-bounded operators between lattice-normed locally solid vector lattices. We show that p-boundedness coincides with p_τ-boundedness, and also, we prove that a dominated operator is p_τ-bounded; see Propositions 5.1 and 5.2, respectively. Also, we get a relation between the order continuity and p_τ-continuity; see Proposition 5.3. We show that a sequentially p_τ-continuous operator is norm continuous; see Proposition 5.4.

In Section 5.4, we introduce the notions of up_τ-continuous and sequentially up_τ-continuous operator between lattice-normed locally solid vector lattices. We prove that a dominated surjective lattice homomorphism operator is sequentially up_τ-continuous; see Theorem 5.1. We give some kind of up_τ-continuous operator; see Propositions 5.6 and 5.7.

In the last section, we introduce the notions of sequentially p_τ-compact and p_τ-compact operators, and we give some basic properties of them. We show that a sequence of order-bounded sequentially p_τ-compact operators is sequentially p_τ-compact (see Theorem 5.2). Also, it holds for the equicontinuously and uniformly convergence (see Theorem 5.3). We prove that a sequentially p_τ-compact operator is compact (see Proposition 5.9). Also, a GAM-compact operator is sequentially p_τ-compact (see Proposition 5.12).

5.2 PRELIMINARIES

First of all, let us recall some notations and terminologies used in this paper. In this chapter, all vector spaces are supposed to be real. Let E be a vector space. Then, E is called *ordered vector space* if it has an order relation "$\leq$" (i.e, it is reflexive, antisymmetric, and transitive) that is compatible with the algebraic structure of E, it means that $y \leq x$ implies $y+z \leq x+z$ for all $z \in E$ and $\lambda y \leq \lambda x$ for each positive scalar $\lambda \geq 0$.

An ordered vector E is said to be *vector lattice* (or, *Riesz space*) if, for each pair of vectors $x, y \in E$, the supremum $x \vee y = \sup\{x,y\}$ and the infimum $x \wedge y = \inf\{x,y\}$ both exist in E. Then, $x^+ := x \vee 0$, $x^- := (-x) \vee 0$, and $|x| := x \vee (-x)$ are called the *positive* part, the *negative* part, and the *absolute value* of $x \in E$, respectively. Also, two vectors x and y in a vector lattice are said to be *disjoint* whenever $|x| \wedge |y| = 0$. A vector lattice is called *order complete* if every nonempty bounded above subset has a supremum (or, equivalently, whenever every nonempty bounded below subset has an infimum).

A partially ordered set I is called *directed* if, for each $a_1, a_2 \in I$, there is another $a \in I$ such that $a \geq a_1$ and $a \geq a_2$ (or, $a \leq a_1$ and $a \leq a_2$). A function from a directed set I into a set E is called a *net* in E. A vector lattice is order complete if $0 \leq x_\alpha \uparrow \leq x$ implies the existence of sup x_α. A net $(x_\alpha)_{\alpha \in A}$ in a vector lattice X is called *order convergent* (or shortly, *o-convergent*) to $x \in X$ if there exists another net $(y_\beta)_{\beta \in B}$

satisfying $y_\beta \downarrow 0$, and for any $\beta \in B$, there exists $\alpha_\beta \in A$ such that $|x_\alpha - x| \leq y_\beta$ for all $\alpha \geq \alpha_\beta$. In this case, we write $x_\alpha \xrightarrow{o} x$; for more details, see, for example, [3,31,32]. In a vector lattice X, a net (x_α) is *unbounded order convergent* to $x \in X$ if $|x_\alpha - x| \wedge u \xrightarrow{o} 0$ for every $u \in X_+$; see, for example, [15,17–19,28].

Recall that every linear topology τ on a vector space E has a base $\mathcal{N}$ for the zero neighborhoods satisfying the following four properties: For each $V \in \mathcal{N}$, we have $\lambda V \subseteq V$ for all scalar $|\lambda| \leq 1$; for any $V_1, V_2 \in \mathcal{N}$, there is another $V \in \mathcal{N}$ such that $V \subseteq V_1 \cap V_2$; for each $V \in \mathcal{N}$, there exists another $U \in \mathcal{N}$ with $U + U \subseteq V$; and for any scalar λ and each $V \in \mathcal{N}$, the set λV is also in $\mathcal{N}$; for much more detail, see [2,3].

A subset A of a vector lattice is called *solid* whenever $|x| \leq |y|$ and $y \in A$ imply $x \in A$. A solid vector subspace is referred to as an *order ideal*. An order closed ideal is referred to as a *band*. A sublattice Y of a vector lattice is majorizing E if, for every $x \in E$, there exists $y \in Y$ with $x \leq y$; see, for example, [1–3,32]. Let E be a vector lattice and τ be a linear topology on E that has a base at zero consisting of solid sets. Then, the pair (E, τ) is said to be a *locally solid vector lattice* (or, *locally solid Riesz space*).

It should be noted that all topologies considered throughout this chapter are assumed to be Hausdorff. It follows from [2, Thm. 2.28.] that a linear topology τ on a vector lattice E is a locally solid if it is generated by a family of Riesz pseudonorms $\{\rho_j\}_{j \in J}$. Moreover, if a family of Riesz pseudonorms generates a locally solid topology τ on a vector lattice E, then $x_\alpha \xrightarrow{\tau} x$ if $\rho_j(x_\alpha - x) \to 0$ in $\mathbb{R}$ for each $j \in J$. Since E is Hausdorff, the family $\{\rho_j\}_{j \in J}$ of Riesz pseudonorms is separating; i.e., if $\rho_j(x) = 0$ for all $j \in J$, then $x = 0$. A locally solid vector lattice is said to have *Lebesgue property* if $x_\alpha \xrightarrow{o} 0$ in E implies $x_\alpha \xrightarrow{\tau} 0$. In this chapter, unless otherwise, the pair (E, τ) refers to as a locally solid vector lattice, and the topologies in locally solid vector lattices are generated by a family of Riesz pseudonorms $\{\rho_j\}_{j \in J}$.

Let X be a vector space, E be a vector lattice, and $p : X \to E_+$ be a vector norm (i.e., $p(x) = 0 \Leftrightarrow x = 0$, $p(\lambda x) = |\lambda| p(x)$ for all $\lambda \in \mathbb{R}$, $x \in X$, and $p(x+y) \leq p(x) + p(y)$ for all $x, y \in X$), then the triple (X, p, E) is called a *lattice-normed space*, abbreviated as *LNS*. The lattice norm p in an LNS (X, p, E) is said to be *decomposable* if, for all $x \in X$ and $e_1, e_2 \in E_+$, it follows from $p(x) = e_1 + e_2$ that there exists $x_1, x_2 \in X$ such that $x = x_1 + x_2$ and $p(x_k) = e_k$ for $k = 1, 2$. We refer the reader for more information on *LNS*s to [13,23,24] and [10–12]. A linear operator T between two *LNS*s (X, p, E) and (Y, m, F) is said to be *dominated* if there is a positive operator $S : E \to F$ satisfying $m(Tx) \leq S(p(x))$ for all $x \in X$. In an *LNS* (X, p, E), a subset A of X is called *p-bounded* if there exists $e \in E$ such that $p(a) \leq e$ for all $a \in A$; see [11, Def. 2.].

If X is a vector lattice and the vector norm p is monotone (i.e., $|x| \leq |y| \Rightarrow p(x) \leq p(y)$), then the triple (X, p, E) is called a *lattice-normed vector lattice*, abbreviated as *LNVL*; see [10–12]. Let (X, p, E) be an LNVL with (E, τ), which is a locally solid vector lattice, then (X, p, E_τ) is said to be a *lattice-normed locally solid Riesz space* (or *lattice-normed locally solid vector lattice*), abbreviated as *LSNVL* in [5]. Throughout this chapter, we use X instead of (X, p, E_τ) and Y instead of $(Y, m, F_{\hat{\tau}})$.

Note also that $L(X,Y)$ denotes the space of all linear operators between vector spaces X and Y. If X is a normed space, then X^* denotes the topological dual of X and B_X denotes the closed unit ball of X.

We abbreviate the convergence $p(x_\alpha - x)\xrightarrow{\tau}0$ as $x_\alpha\xrightarrow{\mathrm{p}\tau}x$, and say in this case that (x_α) p_τ-converges to x. A net $(x_\alpha)_{\alpha\in A}$ in an *LSNVL* (X,p,E_τ) is said to be *p_τ-Cauchy* if the net $(x_\alpha - x_{\alpha'})_{(\alpha,\alpha')\in A\times A}$ p_τ-converges to 0. An *LSNVL* (X,p,E_τ) is called *(sequentially) p_τ-complete* if every p_τ-Cauchy (sequence) net in X is p_τ-convergent. In an *LSNVL* (X,p,E_τ), a subset A of X is called *p_τ-bounded* if $p(A)$ is τ-bounded in E. An *LSNVL* (X,p,E_τ) is called *op_τ-continuous* if $x_\alpha\xrightarrow{\mathrm{o}}0$ implies that $p(x_\alpha)\xrightarrow{\tau}0$. A net (x_α) in an *LSNVL* (X,p,E_τ) is said to be *unbounded p_τ-convergent* to $x\in X$ (shortly, (x_α) *up_τ*-converges to x or $x_\alpha\xrightarrow{\mathrm{up}\tau}x$) if $p(|x_\alpha - x|\wedge u)\xrightarrow{\tau}0$ for all $u\in X_+$; see for much more information [5].

Let (X,p,E) be an LNS and $(E,\|\cdot\|_E)$ be a normed vector lattice. The *mixed-norm* on X is defined by $p\text{-}\|x\|_E = \|p(x)\|_E$ for all $x\in X$. In this case, the normed space $(X,p\text{-}\|\cdot\|_E)$ is called a *mixed-normed space*; see [23, 7.1.1, p. 292].

Lastly, it should be noticed that the theory of lattice-normed spaces is well developed in the case of decomposable lattice norms. In this chapter, we usually do not assume lattice norms to be decomposable. On the other hand, throughout this chapter, all vector lattices are assumed to be Archimedean, and also, we frequently use the following lemmas and so we shall keep in mind them; see [5, Lem. 1.1.] and [11, Lem. 1.], respectively.

Lemma 5.1. *If $(x_\alpha)_{\alpha\in A}$ and $(y_\alpha)_{\alpha\in A}$ be two nets in a locally solid vector lattice (E,τ) such that $|x_\alpha|\leq|y_\alpha|$ for all $\alpha\in A$ and $y_\alpha\xrightarrow{\tau}0$, then $x_\alpha\xrightarrow{\tau}0$.*

Lemma 5.2. *Let (X,p,E) be an LNS such that $(E,\|\cdot\|_E)$ is a Banach space. If (X,p,E) is sequentially p-complete, then $(X,p\text{-}\|\cdot\|_E)$ is the Banach space.*

5.3 p_τ-CONTINUOUS AND p_τ-BOUNDED OPERATORS

In this section, we define the notions of p_τ-continuous and p_τ-bounded operators between LSNVLs. Recall that an operator T between two LNVLs X and Y is called *p-continuous* if $x_\alpha\xrightarrow{\mathrm{p}}0$ in X implies $Tx_\alpha\xrightarrow{\mathrm{p}}0$ in Y, and if the condition holds only for sequences, then T is called *sequentially p-continuous*. Moreover, T is also called *p-bounded* if it maps p-bounded sets in X to p-bounded sets in Y; see [11]. Motivated by these definitions, we give the following notions.

Definition 5.1. *Let X and Y be two LSNVLs and $T\in L(X,Y)$. Then, T is called*

1. *p_τ-continuous if $x_\alpha\xrightarrow{\mathrm{p}\tau}0$ in X implies $Tx_\alpha\xrightarrow{\mathrm{p}\tau}0$ in Y, and if the condition holds only for sequences, then T is called sequentially p_τ-continuous,*
2. *p_τ-bounded if it maps p_τ-bounded sets to p_τ-bounded sets.*

Remark 5.1.

i. *Let T be an operator between LSNVLs (X,p,E_τ) and $(Y,m,F_{\hat{\tau}})$ with (E,τ) and $(F,\hat{\tau})$ having order-bounded neighborhoods of zero. Then, by applying [2, Thm. 2.19(i)] and [20, Thm. 2.2.], one can see that T is p-bounded if it is p_τ-bounded. Moreover, $T:(E,|\cdot|,E_\tau)\to(F,|\cdot|,F_{\hat{\tau}})$ is p_τ-bounded if $T:X\to Y$ is order-bounded.*

ii. *Let X be a vector lattice and $(Y,\|\cdot\|_Y)$ be a normed space. Then, $T\in L(X,Y)$ is called order-to-norm continuous if $x_\alpha\xrightarrow{o}0$ in X implies $Tx_\alpha\xrightarrow{\|\cdot\|_Y}0$; see [27, Sec. 4., p. 468]. For a locally solid lattice (X,τ) with the Lebesgue property, the p_τ-continuity of $T:(X,|\cdot|,X_\tau)\to(Y,\|\cdot\|_Y,\mathbb{R})$ implies order-to-norm continuity of it.*

iii. *Let X be a vector lattice and $(Y,m,F_{\hat{\tau}})$ be an LSNVL, and $T:X\to Y$ be a strictly positive operator. Define $p:X\to F_+$, by $p(x)=(m\circ T)(|x|)$. Then, $(X,p,F_{\hat{\tau}})$ is an LSNVL, and also, the map $T:(X,p,F_{\hat{\tau}})\to(Y,m,F_{\hat{\tau}})$ is p_τ-continuous.*

In the following work, we show that the collection of all p_τ-continuous operators between *LSNVLs* is the vector space.

Lemma 5.3. *Let (X,p,E_τ) and $(Y,m,F_{\hat{\tau}})$ be two LSNVLs.*

i. *If $T,S:(X,p,E_\tau)\to(Y,m,F_{\hat{\tau}})$ are p_τ-continuous operators, then $\lambda S+\mu T$ is p_τ-continuous for any real numbers λ and μ. In particular, if $H=T-S$, then H is p_τ-continuous.*

ii. *If $-T_1\le T\le T_2$ with T_1 and T_2 positive and p_τ-continuous operators, then T is p_τ-continuous.*

Proof. (i) We show $x_\alpha\xrightarrow{p_\tau}0$ in X implies $m(Sx_\alpha+Tx_\alpha)\xrightarrow{\hat{\tau}}0$ in F. To do this, we consider the following inequality:

$$m(Sx_\alpha+Tx_\alpha)\le m(Sx_\alpha)+m(Tx_\alpha)\xrightarrow{\hat{\tau}}0.$$

So $S+T$ is also p_τ-continuous. Now, for arbitrary real numbers λ and μ, we have

$$\begin{aligned} m(\lambda Sx_\alpha+\mu Tx_\alpha) &\le |\lambda|m(Sx_\alpha)+|\mu|m(Tx_\alpha) \\ &\le (|\lambda|+|\mu|)\big(m(Sx_\alpha)+m(Tx_\alpha)\big)\xrightarrow{\hat{\tau}}0. \end{aligned}$$

Therefore, $\lambda S+\mu T$ is p_τ-continuous.

(ii) If $-T_1\le T\le T_2$ with T_1 *and* T_2 positive and p_τ-continuous, then we have

$$0\le T+T_1\le T_2+T_1$$

where $T+T_1$ is positive and p_τ-continuous. Thus, it follows from (i) that $T=(T+T_1)-T_1$ is p_τ-continuous. □

By the following result, we prove that there is a relation between p- and p_τ-bounded notions.

Proposition 5.1. *Let T be an operator between LSNVLs (X,p,E_τ) and $(Y,m,F_{\acute{\tau}})$ with (E,τ) and $(F,\acute{\tau})$ having order-bounded neighborhoods of zero. Then, T is p-bounded if it is p_τ-bounded.*

Proof. We show only one direction. Assume T is p-bounded. Suppose A is p_τ-bounded subset in X, and so $p(A)$ is τ-bounded in E. Thus, $p(A)$ is also order-bounded in E; see [20, Thm. 2.2.]. Since T is p-bounded, $T(A)$ is p-bounded in Y, and so $m\big(T(A)\big)$ is order-bounded in F. Therefore, $m\big(T(A)\big)$ is $\acute{\tau}$-bounded in F; see [2, Thm. 2.19(i)]. Hence, $T(A)$ is p_τ-bounded in Y. □

Proposition 5.2. *Any dominated operator T from an LSNVL (X,p,E_τ) with (E,τ) having order-bounded τ-neighborhood of zero to an LSNVL $(Y,m,F_{\acute{\tau}})$ is p_τ-bounded.*

Proof. Consider a p_τ-bounded subset A in X. That is, $p(A)$ is τ-bounded in E. So $p(A)$ is order-bounded in E; see [20, Thm. 2.2.]. Let S be the dominant of T. Since S is a positive operator, $S\big(p(A)\big)$ is order-bounded in F. Also, we know that $m\big(T(a)\big) \leq S\big(p(a)\big)$ for all $a \in A$, and so $m\big(T(A)\big)$ is order-bounded in F. Hence, by applying [2, Thm. 2.19(i)], $m\big(T(A)\big)$ is $\acute{\tau}$-bounded in F. Therefore, T is p_τ-bounded. □

The converse of Proposition 5.2 is not true in general. For instance, consider ℓ_∞ with the norm topology and $\mathbb{R}$ with the usual topology, and the identity operator $I : (\ell_\infty, |\cdot|, \ell_\infty) \to (\ell_\infty, \|\cdot\|, \mathbb{R})$. It is p_τ-bounded. Indeed, for any p_τ-bounded set A in ℓ_∞, $|A|$ is τ-bounded in ℓ_∞. Thus, $\big\||A|\big\| = \|A\|$ is bounded in $\mathbb{R}$. But it is not dominated; see [13, Rem., p. 388]. □

The next proposition gives a relation between the p_τ- and order continuity.

Proposition 5.3. *Let $(Y,m,F_{\acute{\tau}})$ be arbitrary and (X,p,E_τ) be op_τ-continuous LSNVLs and $T : (X,p,E_\tau) \to (Y,m,F_{\acute{\tau}})$ be a (sequentially) p_τ-continuous positive operator. Then, $T : X \to Y$ is (σ-) order continuous operator.*

Proof. Assume $x_\alpha \downarrow 0$ in X. Since X is op_τ-continuous, we have $p(x_\alpha) \xrightarrow{\tau} 0$, and so $x_\alpha \xrightarrow{p_\tau} 0$ in X. By the p_τ-continuity of T, $m(Tx_\alpha) \xrightarrow{\acute{\tau}} 0$ in F. It can be seen that $Tx_\alpha \downarrow$ because T is positive. Then, applying [5, Prop. 2.4.], we get $Tx_\alpha \downarrow 0$. Thus, T is order continuous. □

Corollary 5.1. *Let (X,p,E) be an op_τ-continuous LSNVL and $(Y,m,F_{\acute{\tau}})$ be an LSNVL with Y being order complete. If $T : (X,p,E) \to (Y,m,F)$ is p_τ-continuous and $T \in L^\sim(X,Y)$, then $T : X \to Y$ is order continuous.*

Proof. Since Y is order-complete and T is order-bounded, by Riesz-Kantorovich formula, we have $T = T^+ - T^-$. Now, Proposition 5.3 implies that T^+ and T^- are both order continuous, and so T is order continuous. □

In the following result, which is p_τ-version of [11, Prop. 3.], we give norm continuity of sequentially p_τ-continuous operator on the mixed-norms.

Proposition 5.4. *Let (X,p,E_τ) and $(Y,m,F_{\acute{\tau}})$ be two LSNVLs with $(E,\|\cdot\|_E)$ and $(F,\|\cdot\|_F)$ being normed vector lattices, and where τ and $\acute{\tau}$ are generated by the norms. If $T:(X,p,E_\tau)\to(Y,m,F_{\acute{\tau}})$ is sequentially p_τ-continuous, then $T:(X,p\text{-}\|\cdot\|_E)\to(Y,m\text{-}\|\cdot\|_F)$ is norm continuous.*

Proof. ($\Rightarrow$) Assume $T:(X,|\cdot|,X)\to(X,\|\cdot\|_X,\mathbb{R})$ to be sequentially p_τ-continuous, and a sequence $x_n\xrightarrow{\|\cdot\|}0$ in X. Thus, $x_n\xrightarrow{\tau}0$ in X, and since T is sequentially p_τ-continuous, $Tx_n\xrightarrow{\|\cdot\|}0$.

($\Leftarrow$) Assume that T is norm continuous, and let $x_n\xrightarrow{p_\tau}0$ in $(X,|\cdot|,X)$. Then, $|x_n|\xrightarrow{\tau}0$ or $x_n\xrightarrow{\|\cdot\|}0$ in X. Hence, $Tx_n\xrightarrow{\|\cdot\|}0$. Therefore, $T:(X,|\cdot|,X)\to(X,\|\cdot\|_X,\mathbb{R})$ is sequentially p-continuous. □ □

Remark 5.2. *By applying [2, Thm. 2.19(i)] and [11, Prop. 4.], one can see that every p_τ-continuous operator is p_τ-bounded. But a p_τ-continuous operator $T:(X,p,E_\tau)\to(Y,m,F_{\acute{\tau}})$ need not to be order-bounded from X to Y. Indeed, consider the classical "Fourier coefficients" operator $T:L_1[0,1]\to c_0$ defined by the formula*

$$T(f)=\left(\int_0^1 f(x)\sin x\,dx,\int_0^1 f(x)\sin 2x\,dx,\ldots\right).$$

Then, $T:L_1[0,1]\to c_0$ is norm-bounded, but it is not order-bounded; see [2, Exer. 10., p. 289]. So $T:(L_1[0,1],\|\cdot\|_{L_1},\mathbb{R})\to(c_0,\|\cdot\|_\infty,\mathbb{R})$ is p_τ-continuous and is not order-bounded.

Using [20, Thm. 2.2.] in Remark 5.2, it can be seen that p_τ-continuity implies order boundedness if $(F,\acute{\tau})$ has order-bounded $\acute{\tau}$-neighborhood of zero. Recall that an operator $T\in L(X,Y)$, where X and Y are the normed spaces, is called *Dunford-Pettis* if $x_n\xrightarrow{w}0$ in X implies $Tx_n\xrightarrow{\|\cdot\|}0$ in Y. We show the following result, which is p_τ-version of [11, Prop. 5.], so we omit its proof.

Proposition 5.5. *Let $(X,\|\cdot\|_X)$ be a normed vector lattice and $(Y,\|\cdot\|_Y)$ be a normed space. Put $E:=\mathbb{R}^{X^*}$ and define $p:X\to E_+$, by $p(x)[f]=|f|(|x|)$ for $f\in X^*$. It is easy to see that (X,p,E_τ), where τ is the topology generated by the norm $\|\cdot\|_{X^*}$, is an LSNVL. Then, the followings hold:*

i. If $T\in L(X,Y)$ is a Dunford-Pettis operator then $T:(X,p,E_\tau)\to(Y,\|\cdot\|_Y,\mathbb{R})$ is sequentially p_τ-continuous,
ii. The converse holds if the lattice operations of X are weakly sequentially continuous.

5.4 up_τ-CONTINUOUS OPERATORS

Recall that a net (x_α) in an *LSNVL* (X,p,E_τ) is said to be *unbounded p_τ-convergent* to x if $p(|x_\alpha - x| \wedge u) \xrightarrow{\tau} 0$ for all $u \in X_+$; see [5]. Motivated by *up*-continuous operators in [11] and *un*-continuous functionals in [21, p. 16], and by using the up_τ-convergence, we introduce the following notion.

Definition 5.2. *An operator T between two LSNVLs X and Y is called up_τ-continuous if it maps the up_τ-convergent net to up_τ-convergent nets. If it holds only for sequence, then T is called sequentially up_τ-continuous.*

It is clear that if T is (sequentially) p_τ-continuous operator, then T is (sequentially) up_τ-continuous. For an *LSNVL* (X,p,E_τ), a sublattice Y of X is called *up_τ-regular* if, for any net (y_α) in Y, the convergence $y_\alpha \xrightarrow{up_\tau} 0$ in Y implies $y_\alpha \xrightarrow{up_\tau} 0$ in X. The following is a more general extension of [21, Prop. 9.4.].

Theorem 5.1. *Let (X,p,E_τ) and $(Y,m,F_{\acute\tau})$ be LSNVLs with $(E,\|\cdot\|_E)$ being a Banach lattice and $(F,\|\cdot\|_F)$ being a normed vector lattice, and also τ and $\acute\tau$ are being generated by the norms. Then, the followings hold:*

i. *A dominated surjective lattice homomorphism operator $T \in L(X,Y)$ is sequentially up_τ-continuous;*
ii. *If $T \in L(X,Y)$ is a dominated lattice homomorphism operator and $T(X)$ is up_τ-regular in Y, then it is sequentially up_τ-continuous;*
iii. *If $T \in L(X,Y)$ is a dominated lattice homomorphism operator and $I_{T(X)}$ (the ideal generated by $T(X)$) is up_τ-regular in Y, then it is sequentially up_τ-continuous.*

Proof. (i) Let's fix a sequence $x_n \xrightarrow{up_\tau} 0$ in X and $u \in Y_+$. Since T is a surjective lattice homomorphism, we have some $v \in X_+$ such that $Tv = u$. So we have $p(|x_n| \wedge v) \xrightarrow{\tau} 0$ in E. Since T is dominated, there is a positive operator $S : E \to F$ such that

$$m\big(T(|x_n| \wedge v)\big) \leq S\big(p(|x_n| \wedge v)\big).$$

Taking into account that T is a lattice homomorphism and $Tv = u$, we get $m\big(|Tx_n| \wedge u\big) \leq S\big(p(|x_n| \wedge v)\big)$. By using [2, Thm. 4.3.], we know that every positive operator from a Banach lattice to the normed vector lattice is continuous, and so S is continuous. Hence, we get $S\big(p(|x_n| \wedge v)\big) \xrightarrow{\acute\tau} 0$ in F. That is, $m\big(|Tx_n| \wedge u\big) \xrightarrow{\acute\tau} 0$, and we get the desired result.

(ii) Since T is a lattice homomorphism, $T(X)$ is a vector sublattice of Y. So $\big(T(X),m,F_{\acute\tau}\big)$ is an *LSNVL*. Thus, by (i), we have $T : (X,p,E_\tau) \to \big(T(X),m,F_{\acute\tau}\big)$ is sequentially up_τ-continuous.

Next, we show that $T : (X,p,E_\tau) \to (Y,m,F_{\acute\tau})$ is sequentially up_τ-continuous. Consider an up_τ-convergent to zero sequence (x_n) in X. That is, $Tx_n \xrightarrow{up_\tau} 0$ in $T(X)$.

Since $T(X)$ is up_τ-regular in Y, $T(x_n) \xrightarrow{\text{up}_\tau} 0$ in Y. Therefore, T is sequentially up_τ-continuous.

(iii) Let $(x_n) \xrightarrow{\text{up}_\tau} 0$ sequence in X. Thus, $p(|x_n| \wedge u) \xrightarrow{\tau} 0$ in E for all $u \in X_+$. Fix $0 \leq w \in I_{T(X)}$. Then, there is $x \in X_+$ such that $0 \leq w \leq Tx$. For a dominant S, we have $m\big(T(|x_n| \wedge x)\big) \leq S\big(p(|x_n| \wedge x)\big)$, and so, by taking lattice homomorphism of T, we have

$$m\big((|Tx_n| \wedge Tx)\big) \leq S\big(p(|x_n| \wedge x)\big).$$

It follows from $0 \leq w \leq Tx$ that $m\big((|Tx_n| \wedge w)\big) \leq S\big(p(|Tx_n| \wedge x)\big)$. Now, the argument given in the proof of (i) can be repeated here as well. Thus, we see that $T : (X,p,E_\tau) \to \big(I_{T(X)},m,F_{\hat{\tau}}\big)$ is sequentially up_τ-continuous. Since $I_{T(X)}$ is up_τ-regular in Y, it can be easily seen by (ii) that $T : X \to Y$ is sequentially up_τ-continuous. □

It should be mentioned, by using Theorem 5.1, that an operator, surjective lattice homomorphism with order continuous dominant, is up_τ-continuous.

Proposition 5.6. *Let (X,p,E_τ) and $(Y,m,F_{\hat{\tau}})$ be two LSNVLs with Y being order-complete vector lattice. For a positive up_τ-continuous operator $T : (X,p,E_\tau) \to (Y,m,F_{\hat{\tau}})$, consider the operator $S : (X_+,p,E_\tau) \to (Y_+,m,F_{\hat{\tau}})$ defined by $S(x) = \sup\{T(x_\alpha \wedge x) : x_\alpha \in X_+, x_\alpha \xrightarrow{\text{up}_\tau} 0\}$ for each $x \in X_+$. Then, we have the followings:*

i. *S is the up_τ-continuous operator;*
ii. *The Kantorovich extension of S is the up_τ-continuous operator.*

Proof. (i) Suppose $X_+ \ni y_\beta \xrightarrow{\text{up}_\tau} 0$. Then, $Ty_\beta \xrightarrow{\text{up}_\tau} 0$ in Y, and so $m\big(Ty_\beta \wedge w\big) \xrightarrow{\hat{\tau}} 0$ in F for all $w \in Y_+$. For any net $x_\alpha \in X_+$ and fixed $w \in Y_+$, we have

$$T(x_\alpha \wedge y_\beta) \wedge w \leq T(y_\beta) \wedge w \xrightarrow{\hat{\tau}} 0.$$

Therefore, we get the desired result.

(ii) We show first that S has the Kantorovich extension. To show this, let's see additivity of it. By using [3, Lem. 1.4.], for any up_τ-null net (x_α) in X_+, we have

$$T\big((x+y) \wedge x_\alpha\big) \leq T(x \wedge x_\alpha) + T(y \wedge x_\alpha) \leq S(x) + S(y).$$

So by taking supremum, we get $S(x+y) \leq S(x) + S(y)$. On the other hand, for any two up_τ-null nets (x_α) and (y_β) in X_+, using the formula in the proof of [3, Thm. 1.28.], we get

$$T(x \wedge x_\alpha) + T(y \wedge y_\beta) = T(x \wedge x_\alpha + y \wedge y_\beta) \leq T\big((x+y) \wedge (x_\alpha + y_\beta)\big) \leq S(x+y).$$

So $S(x) + S(y) \leq S(x+y)$. By [3, Thm. 1.10.], S extends to a positive operator, denoted by $\hat{S} : (X,p,E_\tau) \to (Y,m,F_{\hat{\tau}})$. That is, $\hat{S}x = S(x^+) - S(x^-)$ for all $x \in X$. Now, we show up_τ-continuity of $\hat{S}$. Fix a net $w_\beta \xrightarrow{\text{up}_\tau} 0$ in X. Then, $w_\beta^+ \xrightarrow{\text{up}_\tau} 0$ and $w_\beta^- \xrightarrow{\text{up}_\tau} 0$ in X, and so $S(w_\beta^+) \xrightarrow{\text{up}_\tau} 0$ and $S(w_\beta^-) \xrightarrow{\text{up}_\tau} 0$ in Y. Hence, $\hat{S}w_\beta = S(w_\beta^+) - S(w_\beta^-) \xrightarrow{\text{up}_\tau} 0$ in Y. □

We complete this section with the following technical work.

Proposition 5.7. *Consider a positive up_τ-continuous operator T between LSNVLs X and Y, and an ideal A in X. Then, an operator $S : (X, p, E_\tau) \to (Y, m, F_{\hat{\tau}})$ defined by $S(x) = \sup_{a \in A} T(|x| \wedge a)$ for each $x \in X$ is up_τ-continuous operator.*

Proof. Let $x_\alpha \xrightarrow{up_\tau} 0$ be a net in X. Then, $|x_\alpha| \xrightarrow{up_\tau} 0$, and so $T(|x_\alpha|) \xrightarrow{up_\tau} 0$ in Y. Thus, for each $u \in Y_+$, we have

$$|S(x_\alpha)| \wedge u = \left| \sup_{a \in A} T(|x_\alpha| \wedge a) \right| \wedge u \leq \left| T(|x_\alpha|) \right| \wedge u \leq T(|x_\alpha|) \wedge u \xrightarrow{up_\tau} 0.$$

Therefore, $S(x_\alpha) \xrightarrow{up_\tau} 0$ in Y. □

5.5 THE COMPACT-LIKE OPERATORS

In this section, we define the notions of p_τ-compact and sequentially p_τ-compact operators in *LSNVLs* and study their properties. A linear operator T between normed spaces is called compact if $T(B_X)$ is relatively compact, or equivalently, T is compact if, for any norm-bounded sequence (x_n), there is a subsequence (x_{n_k}) such that the sequence (Tx_{n_k}) is convergent. Similarly, we introduce the following notions.

Definition 5.3. *Let X and Y be two LSNVLs and $T \in L(X,Y)$. Then, T is called p_τ-compact if, for any p_τ-bounded net (x_α) in X, there is a subnet (x_{α_β}) such that $Tx_{\alpha_\beta} \xrightarrow{p_\tau} y$ in Y for some $y \in Y$. If it holds only for sequence, then T is called sequentially p_τ-compact.*

Example 5.1.

i. *Let $(X, \|\cdot\|_X)$ and $(Y, \|\cdot\|_Y)$ be the normed spaces. Then, $T : (X, \|\cdot\|_X, \mathbb{R}) \to (Y, \|\cdot\|_Y, \mathbb{R})$ is (sequentially) p_τ-compact if $T : X \to Y$ is compact.*
ii. *Let X be a vector lattice and Y be a normed space. An operator $T \in L(X,Y)$ is said to be AM-compact if $T[-x,x]$ is relatively compact for every $x \in X_+$; see [26, Def. 3.7.1.]. Let (X, τ) be a locally solid vector lattice with order-bounded τ-neighborhood and $(Y, \|\cdot\|_Y)$ be a normed vector lattice. Then, $T \in L(X,Y)$ is the AM-compact operator if $T : (X, |\cdot|, X_\tau) \to (Y, \|\cdot\|_Y, \mathbb{R})$ is p_τ-compact; apply [20, Thm. 2.2.] and [2, Thm. 2.19(i)].*
iii. *Consider a topology τ in ℓ_2 with the norm $\|\cdot\|_2$, and so $(\ell_2, |\cdot|, \ell_{2\tau})$ and $(\ell_2, \|\cdot\|_2, \mathbb{R})$ are two LSNVLs. Thus, a linear operator $T_n : (\ell_2, |\cdot|, \ell_{2\tau}) \to (\ell_2, \|\cdot\|_2, \mathbb{R})$ defined by $T_n x = (x_1, x_2, \cdots, x_n, 0, 0 \cdots)$ for all $x \in \ell_2$ is p_τ-compact. Indeed, let (x_n) be a p_τ-bounded, or $|(x_n)|$ is $\|\cdot\|_2$-bounded sequence in ℓ_2. Hence, (Tx_n) is $\|\cdot\|_2$-bounded in ℓ_2. Since every bounded sequence in $\mathbb{R}^n$ has a convergent subsequence, it follows that T_n is p_τ-compact.*

Lemma 5.4. *If S and T are (sequentially) p_τ-compact operators between LSNVLs, then $T + S$ and λT, for any real number λ, are also (sequentially) p_τ-compact operators.*

Proof. We show that only $T+S$ is p_τ-compact, and the other cases are analogous. Let (x_α) be a p_τ-bounded net in X. Since T is p_τ-compact, there is a subnet (x_{α_β}) such that $m(Tx_{\alpha_\beta}-y)\xrightarrow{\hat{\tau}}0$ in F for some $y\in Y$. On the other hand, (x_{α_β}) is also p_τ-bounded in X and S is p_τ-compact, and then there is a subnet $(x_{\alpha_{\beta_\gamma}})$ such that $m(Sx_{\alpha_{\beta_\gamma}}-z)\xrightarrow{\hat{\tau}}0$ in F for some $z\in Y$. Also, we have $m(Tx_{\alpha_{\beta_\gamma}}-y)\xrightarrow{\hat{\tau}}0$ in F. Thus, $m[(S+T)(x_{\alpha_{\beta_\gamma}})-y-z)]\xrightarrow{\hat{\tau}}0$. Hence, $S+T$ is p_τ-compact. □

Proposition 5.8. *Let (X,p,E_τ) be an LSNVL and $R,T,S\in L(X)$.*

i. If T is a (sequentially) p_τ-compact and S is a (sequentially) p_τ-continuous operator, then $S\circ T$ is (sequentially) p_τ-compact.
ii. If T is a (sequentially) p_τ-compact and R is a p_τ-bounded operator, then $T\circ R$ is (sequentially) p_τ-compact.

Proof. We give the only p_τ-compactness, and the sequential case is analogous.

i. Consider a p_τ-bounded net (x_α) in X. Since T is p_τ-compact, there are a subnet (x_{α_β}) and $x\in X$ such that $Tx_{\alpha_\beta}\xrightarrow{p_\tau}x$. By p_τ-continuity of S, we have $S(Tx_{\alpha_\beta})\xrightarrow{p_\tau}Sx$ in E. Therefore, $S\circ T$ is p_τ-compact.
ii. Suppose (x_α) is a p_τ-bounded net in X. Since R is p_τ-bounded, (Rx_α) is p_τ-bounded. Then, by p_τ-compact of T, there are a subnet (x_{α_β}) and $y\in X$ such that $T(Rx_{\alpha_\beta})\xrightarrow{p_\tau}y$. Therefore, $T\circ R$ is p_τ-compact.

□

Remark 5.3.

i. Let X be an LSNVL and $(Y,\hat{\tau})$ be a locally solid vector lattice with Y being compact. Then, each operator $T:(X,p,E_\tau)\to(Y,|\cdot|,Y_{\hat{\tau}})$ is (sequentially) p_τ-compact.
ii. Let X be an LSNVL and $(Y,\|\cdot\|_Y)$ be a finite dimensional normed space, and $\hat{\tau}$ be the topology generated by this norm. If $T:(X,p,E_\tau)\to(Y,|\cdot|,Y_{\hat{\tau}})$ is p_τ-bounded operator, then it is sequentially p_τ-compact.
iii. Let (X,τ) be a locally solid vector lattice with an order-bounded τ-neighborhood of zero and $(Y,m,F_{\hat{\tau}})$ be an op_τ-continuous LSNVL with Y being an atomic KB-space. If $T:X\to Y$ is an order-bounded operator, then $T:(X,|\cdot|,X_\tau)\to(Y,m,F_{\hat{\tau}})$ is p_τ-compact; see [20, Thm. 2.2.] and [11, Rem. 6.].

Remark 5.4. *Let (T_m) be a sequence of sequentially p_τ-compact operators from X to Y. For a given p_τ-bounded sequence (x_n) in X, by a standard diagonal argument, there exists a subsequence (x_{n_k}) such that, for any $m\in\mathbb{N}$, $T_mx_{n_k}\xrightarrow{p_\tau}y_m$ for some $y_m\in Y$.*

Theorem 5.2. *Let (T_m) be a sequence of order-bounded sequentially p_τ-compact operators from (X,p,E_τ) to a sequentially p_τ-complete op_τ-continuous $(Y,q,F_{\hat{\tau}})$ with Y being order complete. If $T_m \xrightarrow{o} T$ in $L_b(X,Y)$, then T is sequentially p_τ-compact.*

Proof. Let (x_n) be a p_τ-bounded sequence in X. By Remark 5.4, there exists a subsequence (x_{n_k}) such that, for any $m \in \mathbb{N}$, $T_m x_{n_k} \xrightarrow{p_\tau} y_m$ for some $y_m \in Y$. We show that (y_m) is a p_τ-Cauchy sequence. Consider the following formula

$$q(y_m - y_j) \leq q(y_m - T_m x_{n_k}) + q(T_m x_{n_k} - T_j x_{n_k}) + q(T_j x_{n_k} - y_j). \tag{5.1}$$

The first and the third terms in the last inequality both $\hat{\tau}$-converge to zero as $m \to \infty$ and $j \to \infty$, respectively. Since $T_m \xrightarrow{o} T$, we have $T_m x_{n_k} \xrightarrow{o} T x_{n_k}$ for all x_{n_k}; see [31, Thm. VIII.2.3.]. Then, for a fixed index k, we have

$$|T_m x_{n_k} - T_j x_{n_k}| \leq |T_m x_{n_k} - T x_{n_k}| + |T x_{n_k} - T_j x_{n_k}| \xrightarrow{o} 0$$

as $m, j \to \infty$, and so $(T_m - T_j)x_{n_k} \xrightarrow{o} 0$ in Y. Hence, by op_τ-continuity of $(Y,q,F_{\hat{\tau}})$, we get $q(T_m x_{n_k} - T_j x_{n_k}) \xrightarrow{\hat{\tau}} 0$ in F. By formula (1), (y_m) is p_τ-Cauchy. Since Y is sequentially p_τ-complete, there is $y \in Y$ such that $q(y_m - y) \xrightarrow{\hat{\tau}} 0$ in F as $m \to \infty$. So, for arbitrary m, if we take $\hat{\tau}$-limit with k in the following formula

$$q(T x_{n_k} - y) \leq q(T x_{n_k} - T_m x_{n_k}) + q(T_m x_{n_k} - y_m) + q(y_m - y),$$

we get $\hat{\tau} - \lim q(T x_{n_k} - y) \leq q(T x_{n_k} - T_m x_{n_k}) + q(y_m - y)$ because of $q(T_m x_{n_k} - y_m) \xrightarrow{\hat{\tau}} 0$. Since m is arbitrary, $\hat{\tau} - \lim q(T x_{n_k} - y) \xrightarrow{\hat{\tau}} 0$. Thus, T is sequentially p_τ-compact. □

Similar to Theorem 5.2, we give the following theorem by using equicontinuously and uniformly convergence.

Theorem 5.3. *Let (T_m) be a sequence of sequentially p_τ-compact operators from $(X,|\cdot|,X_\tau)$ to a sequentially p_τ-complete LSNVL $(Y,|\cdot|,Y_\tau)$. Then, the followings hold:*

i. *If (T_m) converges equicontinuously to an operator $T : (X,|\cdot|,X_\tau) \to (Y,|\cdot|,Y_{\hat{\tau}})$, then T is sequentially p_τ-compact,*
ii. *If (T_m) uniformly converges on zero neighborhoods to an operator $T : (X,|\cdot|,X_\tau) \to (Y,|\cdot|,Y_{\hat{\tau}})$, then T is sequentially p_τ-compact.*

Let (X,E) be a decomposable *LNS* and (Y,F) be an *LNS* with F being order complete. Then, each dominated operator $T : X \to Y$ has the exact dominant $|T| : E \to F$; see [23, 4.1.2., p. 142]. For a sequence (T_n) in the set of dominated operators $M(X,Y)$, we call $T_n \to T$ in $M(X,Y)$ whenever $|T_n - T|(e) \xrightarrow{\hat{\tau}} 0$ in F for each $e \in E$.

Theorem 5.4. *Let (X,p,E_τ) be a decomposable and $(Y,q,F_{\hat{\tau}})$ be a sequentially p_τ-complete LSNVLs with F being order complete. If (T_m) is a sequence of sequentially p_τ-compact operators and $T_m \to T$ in $M(X,Y)$ then T is sequentially p_τ-compact.*

Proof. Let (x_n) be a p_τ-bounded sequence in X. By Remark 5.4, there exists a subsequence (x_{n_k}) and a sequence (y_m) in Y such that, for any $m \in \mathbb{N}$, $T_m x_{n_k} \xrightarrow{p_\tau} y_m$. We show that (y_m) is p_τ-Cauchy sequence in Y. Consider formula (1) of Theorem 5.2. Similarly, the first and the third terms in the last inequality of (1) both $\acute{\tau}$-converge to zero as $m \to \infty$ and $j \to \infty$, respectively. Since $T_m \in M(X,Y)$ for all $m \in \mathbb{N}$,

$$q(T_m x_{n_k} - T_j x_{n_k}) \leq |T_m - T_j|(p(x_{n_k})) \leq |T_m - T|(p(x_{n_k})) + |T - T_j|(p(x_{n_k})) \xrightarrow{\acute{\tau}} 0$$

as $m, j \to \infty$. Thus, $q(y_m - y_j) \xrightarrow{\acute{\tau}} 0$ in F as $m, j \to \infty$. Therefore, (y_m) is p_τ-Cauchy. Since Y is sequentially p_τ-complete, there is $y \in Y$ such that $q(y_m - y) \xrightarrow{\acute{\tau}} 0$ in F as $m \to \infty$. By the following formula

$$\begin{aligned} q(Tx_{n_k} - y) &\leq q(Tx_{n_k} - T_m x_{n_k}) + q(T_m x_{n_k} - y_m) + q(y_m - y) \\ &\leq |T_m - T|(p(x_{n_k})) + q(T_m x_{n_k} - y_m) + q(y_m - y) \end{aligned}$$

and by repeating the same of last part of Theorem 5.2, we get $q(Tx_{n_k} - y) \xrightarrow{\acute{\tau}} 0$. Therefore, T is sequentially p_τ-compact. □

Proposition 5.9. *Let (X,p,E_τ) be an LSNVL, where $(E, \|\cdot\|_E)$ is an AM-space with a strong unit, and $(Y,m,F_{\acute{\tau}})$ be an LSNVL, where $(F, \|\cdot\|_F)$ is the normed vector lattice and $\acute{\tau}$ is generated by the norm $\|\cdot\|_F$. If $T : (X,p,E_\tau) \to (Y,m,F_{\acute{\tau}})$ is sequentially p_τ-compact, then $T : (X, p\text{-}\|\cdot\|_E) \to (Y, m\text{-}\|\cdot\|_F)$ is compact.*

Proof. Let (x_n) be a normed bounded sequence in $(X, p\text{-}\|\cdot\|_E)$. That is, $p\text{-}\|x_n\|_E = \|p(x_n)\|_E < \infty$ for all $n \in N$. Since $(E, \|\cdot\|_E)$ is an *AM*-space with a strong unit, $p(x_n)$ is order-bounded in E. Thus, $p(x_n)$ is τ-bounded in E; see [2, Thm. 2.19(i)]. So (x_n) is a p_τ-bounded sequence in (X,p,E_τ). Since T is sequentially p_τ-compact, there are a subsequence x_{n_k} and $y \in Y$ such that $m(Tx_{n_k} - y) \xrightarrow{\acute{\tau}} 0$ in F. Then, $\|m(Tx_{n_k} - y)\|_F \to 0$ or $m\text{-}\|Tx_{n_k} - y\|_F \to 0$ in F. Thus, the operator $T : (X, p\text{-}\|\cdot\|_E) \to (Y, m\text{-}\|\cdot\|_F)$ is compact. □

It is known that a finite rank operator is compact. Similarly, we give the following result.

Proposition 5.10. *Let (X,p,E_τ) and $(Y,m,F_{\acute{\tau}})$ be LSNVLs with $(F,\acute{\tau})$ having the Lebesgue property. Consider an operator $T : (X,p,E_\tau) \to (Y,m,F_{\acute{\tau}})$ defined by $Tx = f(x)y_0$, where $y_0 \in Y$ and f is a linear functional on X. If $f : (X,p,E_\tau) \to (\mathbb{R}, |\cdot|, \mathbb{R})$ is p_τ-bounded, then T is (sequentially) p_τ-compact.*

Proof. Suppose (x_α) is a p_τ-bounded net in X. Since f is p_τ-bounded, $f(x_\alpha)$ is bounded in $\mathbb{R}$. Then, there is a subnet (x_{α_β}) such that $f(x_{\alpha_\beta}) \to \lambda$ for some $\lambda \in \mathbb{R}$. For $y_0 \in Y$, we have the following formula:

$$m(Tx_{\alpha_\beta} - \lambda y_0) = m(f(x_{\alpha_\beta})y_0 - \lambda y_0) = m\big((f(x_{\alpha_\beta}) - \lambda)y_0\big) = |f(x_{\alpha_\beta}) - \lambda| m(y_0) \xrightarrow{o} 0.$$

By the Lebesgue property of F, we get $m(Tx_{\alpha_\beta} - \lambda y) \xrightarrow{\tau} 0$ in E. Thus, T is p_τ-compact. □

Proposition 5.11. *Let (X, p, E_τ) be an LSNVL with (E, τ) having an order-bounded τ-neighborhood and $(Y, m, F_{\acute{\tau}})$ be an LSNVL, where $(Y, \|\cdot\|_Y)$ is an order continuous atomic KB-space and $\acute{\tau}$ is generated by $\|\cdot\|_Y$. If $T : (X, p, E_\tau) \to (Y, |\cdot|, Y_{\acute{\tau}})$ is p-bounded or dominated operator, then it is p_τ-compact.*

Recall that a linear operator T from an LNS (X, E) to a Banach space $(Y, \|\cdot\|_Y)$ is called the *generalized AM-compact* or *GAM-compact* if, for any p-bounded set A in X, $T(A)$ is relatively compact in $(Y, \|\cdot\|_Y)$.

Proposition 5.12. *Let (X, p, E_τ) be an LSNVL with (E, τ) having an order-bounded τ-neighborhood and $(Y, m, F_{\acute{\tau}})$ be an op_τ-continuous LSNVL with a Banach lattice $(Y, \|\cdot\|_Y)$. If $T : (X, p, E_\tau) \to (Y, \|\cdot\|_Y)$ is GAM-compact, then $T : (X, p, E_\tau) \to (Y, m, F_{\acute{\tau}})$ is sequentially p_τ-compact.*

Proof. Let (x_n) be a p_τ-bounded sequence in X. By [20, Thm. 2.2.], (x_n) is p-bounded in (X, p, E_τ). Since T is *GAM*-compact, there are a subsequence (x_{n_k}) and some $y \in Y$ such that $\|Tx_{n_k} - y\|_Y \to 0$. Since $(Y, \|\cdot\|_Y)$ is Banach lattice then, by [31, Thm. VII.2.1.], there is a further subsequence $(x_{n_{k_j}})$ such that $Tx_{n_{k_j}} \xrightarrow{o} y$ in Y. Then, by op_τ-continuity of $(Y, m, F_{\acute{\tau}})$, we get $Tx_{n_{k_j}} \xrightarrow{p_\tau} y$ in Y. Hence, T is sequentially p_τ-compact. □

Proposition 5.13. *Let $(X, \|\cdot\|_X)$ be a normed lattice and $(Y, \|\cdot\|_Y)$ be a Banach lattice. If $T : (X, \|\cdot\|_X, \mathbb{R}) \to (Y, |\cdot|, Y_{\acute{\tau}})$ is sequentially p_τ-compact and p-bounded, and $f : Y \to \mathbb{R}$ is σ-order continuous, then $(f \circ T) : X \to \mathbb{R}$ is compact.*

Proof. Assume (x_n) be a norm-bounded sequence in X. Since T is sequentially p_τ-compact, there are a subsequence (x_{n_k}) and $y \in Y$ such that $Tx_{n_k} \xrightarrow{p_\tau} y$ or $|Tx_{n_k} - y| \xrightarrow{\acute{\tau}} 0$ or $Tx_{n_k} \xrightarrow{\|\cdot\|_Y} y$ in Y. Since $(Y, \|\cdot\|_Y)$ be a Banach lattice, there is a further subsequence $(x_{n_{k_j}})$ such that $Tx_{n_{k_j}} \xrightarrow{o} y$ in Y; see [31, Thm. VII.2.1.]. By σ-order continuity of f, we have $(f \circ T)x_{n_{k_j}} \to f(y)$ in $\mathbb{R}$. □

We now turn our attention to the up_τ-compact operators.

Definition 5.4. *Let X and Y be two LSNVLs and $T \in L(X, Y)$. Then, T is called up_τ-compact if, for any p_τ-bounded net (x_α) in X, there is a subnet (x_{α_β}) such that $Tx_{\alpha_\beta} \xrightarrow{up_\tau} y$ in Y for some $y \in Y$. If the condition holds only for sequences, then T is called sequentially-up_τ-compact.*

It is clear that a p_τ-compact operator is up_τ-compact, and similar to Lemma 5.4, linear properties hold for up_τ-compact operators. Moreover, an operator $T \in L(X, Y)$

is (sequentially) un-compact if $T:(X,\|\cdot\|_X,\mathbb{R})\to(Y,\|\cdot\|_Y,\mathbb{R})$ is (sequentially) up_τ-compact; see [21, Sec. 9., p. 28]. Similar to Proposition 5.8, we give the following results.

Proposition 5.14. *Let (X,p,E_τ) be an LSNVL and $R,T,S,H\in L(X)$.*

i. If T is an (sequentially) up_τ-compact and S is a (sequentially) p_τ-continuous, then $S\circ T$ is (sequentially) up_τ-compact.
ii. If T is an (sequentially) up_τ-compact and R is a p_τ-bounded, then $T\circ R$ is (sequentially) up_τ-compact.

Now, we investigate a relation between sequentially up_τ-compact operators and dominated lattice homomorphisms. The following is a more general extension of [21, Prop. 9.4.] and [11, Thm. 8.], and its proof is similar to Theorem 5.1.

Theorem 5.5. *Let (X,p,E_τ), $(Y,m,F_{\acute{\tau}})$, and $(Z,q,G_{\hat{\tau}})$ be LSNVLs with $(F,\|\cdot\|_F)$ being Banach lattice and $(G,\|\cdot\|_G)$ normed lattice, and $\acute{\tau}$ and $\hat{\tau}$ are being generated by the norms. Then, the followings hold:*

i. If $T\in L(X,Y)$ is a sequentially up_τ-compact operator and $S\in L(Y,Z)$ is a dominated surjective lattice homomorphism, then $S\circ T$ is sequentially up_τ-compact;
ii. If $T\in L(X,Y)$ is a sequentially up_τ-compact, $S\in L(Y,Z)$ is a dominated lattice homomorphism, and $S(Y)$ is up_τ-regular in Z, then $S\circ T$ is sequentially up_τ-compact;
iii. If $T\in L(X,Y)$ is a sequentially up_τ-compact, $S\in L(Y,Z)$ is a dominated lattice homomorphism operator, and $I_{S(Y)}$ (the ideal generated by $S(Y)$) is up_τ-regular in Z, then $S\circ T$ is sequentially up_τ-compact.

Proposition 5.15. *Let (X,p,E_τ) be an LSNVL and $(Y,m,F_{\acute{\tau}})$ be an up_τ-complete LSNVL, and S, $T:(X,p,E_\tau)\to(Y,m,F_{\acute{\tau}})$ be operators with $0\le S\le T$. If T is a lattice homomorphism and (sequentially) up_τ-compact, then S is (sequentially) up_τ-compact.*

Proof. We will prove the sequential case; the other case is similar. Let (x_n) be a p_τ-bounded sequence in X. So there are a subsequence (x_{n_k}) and some $y\in Y$ such that $Tx_{n_k}\xrightarrow{up_\tau}y$ in Y. In particular, it is up_τ-Cauchy. Fix $u\in Y_+$, and note that

$$|Sx_{n_k}-Sx_{n_j}|\wedge u\le(S|x_{n_k}-x_{n_j}|)\wedge u\le(T|x_{n_k}-x_{n_j}|)\wedge u=|Tx_{n_k}-Tx_{n_j}|\wedge u\xrightarrow{\acute{\tau}}0$$

as $k,j\to\infty$. Thus, we get (Sx_{n_k}), which is a up_τ-Cauchy sequence in Y. Therefore, it follows from up_τ-complete of Y. □

Lemma 5.5. *Let (X,p,E_τ) and $(Y,m,F_{\acute{\tau}})$ be two LSNVLs with Y being order-complete vector lattice. If $T:(X,p,E_\tau)\to(Y,m,F_{\acute{\tau}})$ is a positive up_τ-compact operator, then the operator $S:(X_+,p,E_\tau)\to(Y_+,m,F_{\acute{\tau}})$ defined by $S(x)=\sup\{T(u\wedge x):u\in X_+\}$ for each $x\in X_+$ is also up_τ-compact operator.*

Proof. Suppose (y_β) is a p_τ-bounded net in X_+. Then, there is a subnet (y_{β_γ}) such that $Ty_{\beta_\gamma} \xrightarrow{\text{up}_\tau} y$ for some $y \in Y$, and so $m(|Ty_{\beta_\gamma} - y| \wedge w) \xrightarrow{\acute{t}} 0$ in F for all $w \in Y_+$. For $u \in X_+$ and fixed $w \in Y_+$, we have $0 \leq T(u \wedge y_{\beta_\gamma}) \leq T(y_{\beta_\gamma})$, and so $|T(u \wedge y_{\beta_\gamma}) - y| \wedge w \leq |T(y_{\beta_\gamma}) - y| \wedge w$. By taking supremum over $u \in X_+$, we get $|Sy_{\beta_\gamma} - y| \wedge w \leq |T(y_{\beta_\gamma}) - y| \wedge w \xrightarrow{\acute{t}} 0$, and so we get the desired result. □

Remark 5.5. *The sum of two p_τ-bounded subsets is also p_τ-bounded since the sum of two solid subsets is solid. Moreover, for a p_τ-bounded net (x_α) in an LSNVL (X, p, E_τ), the nets (x_α^+) and (x_α^-) are p_τ-bounded.*

The following theorem is up_τ-compact version of Proposition 5.6, so we omit its proof.

Theorem 5.6. *Let (X, p, E_τ) and $(Y, m, F_{\acute{t}})$ be two LSNVLs with Y being order-complete vector lattice. If $T : (X, p, E_\tau) \to (Y, m, F_{\acute{t}})$ is a positive up_τ-compact operator, then the Kantorovich extension of $S : (X_+, p, E_\tau) \to (Y_+, m, F_{\acute{t}})$ defined by $S(x) = \sup\{T(x_\alpha \wedge x) : x_\alpha \in X_+$ is p_τ-bounded$\}$ for each $x \in X_+$ is also up_τ-compact.*

BIBLIOGRAPHY

[1] Y.A. Abramovich and C.D. Aliprantis, *An invitation to operator theory*, Graduate Studies in Mathematics, Vol. 50, American Mathematical Society, Rhoda Island, (2002).

[2] C.D. Aliprantis and O. Burkinshaw, *Locally solid Riesz spaces with applications to economics*, Pure and Applied Mathematics, Vol. 105, American Mathematical Society, Indianapolish, (2003).

[3] C.D. Aliprantis and O. Burkinshaw, *Positive operators*, Springer, Vol. 119, Dordrecht, (2006).

[4] A. Aydın, Topological algebras of bounded operators with locally solid Riesz spaces, *Journal of Science and Technology of Erzincan University,* Vol. 11, 543–549, (2018).

[5] A. Aydın, Unbounded p_τ-convergence in vector lattices normed by locally solid lattices, Academic studies in mathematic and natural sciences, IVPE, Cetinje-Montenegro, 118–134, (2019).

[6] A. Aydın, *Convergence via filter in locally solid Riesz spaces*, *International Journal of Science and Research,* Vol. 8, 351–353, (2019).

[7] A. Aydın, Multiplicative order convergence in f-algebras, *Hacettepe Journal of Mathematics and Statistics,* Vol. 49, 998–1005, (2020).

[8] A. Aydın, Multiplicative norm convergence in Banach lattice f-algebras, *Hacettepe Journal of Mathematics and Statistics,* in press, (2020).

[9] A. Aydın and M. Çınar, Multiplicative norm compact operators on Banach lattice f-algebras, *International Journal on Mathematics, Engineering and Natural Science,* Vol. 9, 8–13, (2019).

[10] A. Aydın, E.Y. Emel'yanov, N.E. Özcan and M.A.A. Marabeh, Compact-like operators in lattice-normed spaces, *Indagationes Mathematicae,* Vol. 2, 633–656, (2018).

[11] A. Aydın, E.Y. Emel'yanov, N.E. Özcan and M.A.A. Marabeh, Unbounded p-convergence in lattice-normed vector lattices, *Siberian Advances in Mathematics,* Vol. 29, 164–182, (2019).

[12] A. Aydın, S.G. Gorokhova and H. Gül, Nonstandard hulls of lattice-normed ordered vector spaces, *Turkish Journal of Mathematics,* Vol. 42, 155–163, (2018).

[13] A.V. Bukhvalov, A.E. Gutman, V.B. Korotkov, A.G. Kusraev, S.S. Kutateladze and B.M. Makarov, *Vector lattices and integral operators*, Mathematics and its Applications, Vol. 358, Kluwer Academic Publishers Group, Dordrecht, (1996).

[14] R. DeMarr, Partially ordered linear spaces and locally convex linear topological spaces, *Illinois Journal of Mathematics,* Vol. 8, 601–606, (1964).

[15] Y. Deng, M. O'Brien and V.G. Troitsky, Unbounded norm convergence in Banach lattices, *Positivity,* Vol. 21, 963–974, (2017).

[16] E.Y. Emel'yanov, Infinitesimal analysis and vector lattices, *Siberian Advances in Mathematics,* Vol. 6, 19–70, (1996).

[17] N. Gao, Unbounded order convergence in dual spaces, *Journal of Mathematical Analysis and Applications,* Vol. 419, 347–354, (2014).

[18] N. Gao and F. Xanthos, Unbounded order convergence and application to martingales without probability, *Journal of Mathematical Analysis and Applications,* Vol. 415, 931–947, (2014).

[19] N. Gao, V.G. Troitsky and F. Xanthos, Uo-convergence and its applications to Cesáro means in Banach lattices, *Israel Journal of Mathematics,* Vol. 220, 649–689, (2017).

[20] L. Hong, On order bounded subsets of locally solid Riesz spaces, *Quaestiones Mathematicae,* Vol. 39, 381–389, (2016).

[21] M. Kandić, M.A.A. Marabeh and V.G. Troitsky, *Unbounded norm topology in Banach lattices, Mathematical Analysis and Applications,* Vol. 451, 259–279, (2017).

[22] S. Kaplan, On unbounded order convergence, *Real Analysis Exchange,* Vol. 23, 175–184, (1998).

[23] A.G. Kusraev, *Dominated operators*, Mathematics and its Applications, (2000).

[24] A.G. Kusraev and S.S. Kutateladze, *Boolean valued analysis*, Mathematics and its Applications, (1999).

[25] W.A.J. Luxemburg and A.C. Zaanen, *Riesz spaces I*, North-Holland, Amsterdam, (1971).

[26] P. Meyer-Nieberg, *Banach lattices*, Universitext, Springer-Verlag, Berlin, (1991).

[27] O.V. Maslyuchenko, V.V. Mykhaylyuk and M.M. Popov, A lattice approach to narrow operators, *Positivity,* Vol. 13, 459–495, (2009).

[28] H. Nakano, Ergodic theorems in semi-ordered linear spaces, *Annals of Mathematics,* Vol. 49, 538–556, (1948).

[29] V.G. Troitsky, Measures of non-compactness of operators on Banach lattices, *Positivity,* Vol. 8, 165–178, (2004).

[30] A.W. Wickstead, Weak and unbounded order convergence in Banach lattices, *Journal of the Australian Mathematical Society,* Vol. 24, 312–319, (1977).

[31] B.Z. Vulikh, *Introduction to the theory of partially ordered spaces*, Wolters-Noordhoff Scientific Publications, Ltd., Groningen, (1967).

[32] A.C. Zaanen, *Riesz spaces II*, North-Holland Mathematical Library, Vol. 30, North-Holland Publishing Co., Amsterdam, (1983).

6 On Indexed Product Summability of an Infinite Series

B. P. Padhy
KIIT Deemed to be University

P. Baliarsingh
Gangadhar Meher University

CONTENTS

6.1 INTRODUCTION

6.1.1 HISTORICAL BACKGROUND

Initially, in 1952, Szasz [10] published some results on products of summability methods. Subsequently, Rajgopal [7] in 1954, Parameswaran [6] in 1957, Ramanujan [8] in 1958, etc. published some more results on the products of summability methods. Later on, Das [2] in 1969 proved some related results on the absolute product summability. In 2008, Sulaiman [9] provided a result on the indexed product summability of an infinite series. The result found by Sulaiman was then extended by Paikray et al. [5] in 2010. Also, we can see some more developments on this idea in Ref. [1,3,4] in the recent past.

6.1.2 NOTATIONS AND DEFINITIONS

Let $\sum a_n$ be an infinite series with the sequence of partial sums $\{s_n\}$. Let $\{p_n\}$ be a sequence of positive real constants such that

$$P_n = p_0 + p_1 + \cdots + p_n \to \infty \; as \; n \to \infty \,(P_{-i} = p_{-1} = 0). \tag{6.1}$$

The sequence-to-sequence transformation

$$t_n = \frac{1}{P_n} \sum_{v=0}^{n} p_v \, s_v \tag{6.2}$$

defines $(R,\ p_n)$ transform of $\{s_n\}$ generated by $\{p_n\}$.

The series $\sum a_n$ is said to be summable $|R, p_n|_k$, $k \geq 1$, if

$$\sum_{n=1}^{\infty} n^{k-1} \left| t_n - t_{n-1} \right|^k < \infty. \tag{6.3}$$

Similarly, the sequence-to-sequence transformation

$$T_n = \frac{1}{P_n} \sum_{v=0}^{n} p_{n-v} \, s_v$$

defines the (N, p_n) transform of $\{s_n\}$ generated by $\{p_n\}$.

Let $\{\tau_n\}$ be the sequence of (N, q_n) transform of the (N, p_n) transform of $\{s_n\}$, generated by the sequences $\{q_n\}$ and $\{p_n\}$, respectively. Then, the series $\sum a_n$ is said to be summable $|(N, q_n)\ (N, p_n)|_k$, $k \geq 1$, if

$$\sum_{n=1}^{\infty} n^{k-1} \left| \tau_n - \tau_{n-1} \right|^k < \infty, \tag{6.4}$$

and the series $\sum a_n$ is said to be summable $|(N, q_n)(N, p_n), \delta|_k, k \geq 1, 1 \geq \delta k \geq 0$, if

$$\sum_{n=1}^{\infty} n^{(\delta k + k - 1)} \left| \tau_n - \tau_{n-1} \right|^k < \infty. \tag{6.5}$$

Similarly, if $\{\alpha_n\}$ is any sequence of positive numbers, then the series $\sum a_n$ is said to be summable $|(N, q_n)\ (N, p_n), \alpha_n|_k$, $k \geq 1$, if

$$\sum_{n=1}^{\infty} \alpha_n^{k-1} \left| \tau_n - \tau_{n-1} \right|^k < \infty, \tag{6.6}$$

and the series $\sum a_n$ is said to be summable $|(N, q_n)(N, p_n), \alpha_n; \delta|_k, k \geq 1, 1 \geq k\delta \geq 0$, if

$$\sum_{n=1}^{\infty} \alpha_n^{\delta k + k - 1} \left| \tau_n - \tau_{n-1} \right|^k < \infty. \tag{6.7}$$

Let f be a function of α_n, if

$$\sum_{n=1}^{\infty} \{f(\alpha_n)\}^k (\alpha_n)^{k-1} \left| \tau_n - \tau_{n-1} \right|^k < \infty, \tag{6.8}$$

then the series $\sum a_n$ is said to be $|(N, q_n)\ (N, p_n), \alpha_n; f|_k, k \geq 1$ summable.

Clearly for $f(\alpha_n) = \alpha_n^{\delta}, \delta \geq 0$, $|(N, q_n)\ (N, p_n), \alpha_n; f|_k = |(N, q_n)\ (N, p_n), \alpha_n; \delta|_k$., and for $\delta = 0$, $|(N, q_n)\ (N, p_n), \alpha_n; f|_k = |(N, q_n)\ (N, p_n), \alpha_n|_k$.

We assume throughout this chapter that $Q_n = q_0 + \cdots + q_n \to \infty$ as $n \to \infty$ and $P_n = p_0 + \cdots + p_n \to \infty$ as $n \to \infty$

6.2 KNOWN RESULTS

In 2008, Sulaiman [9] has proved the following theorem.

Theorem 6.1. *Let $k \geq 1$ and (λ_m) be a sequence of constants. Let us define*

$$f_v = \sum_{r=v}^{n} \frac{q_r}{P_r}, \; F_v = \sum_{r=v}^{n} p_r \, f_r \tag{6.9}$$

Let $p_n \, Q_n = O(P_n)$ such that

$$\sum_{n=v+1}^{\infty} \frac{n^{k-1} q_n^k}{Q_n^k Q_{n-1}} = O\left(\frac{\left(v q_v^{k-1}\right)}{Q_v^k}\right). \tag{6.10}$$

Then, the sufficient conditions for the implication that $\sum a_n$ is summable $|R, r_n|_k \Rightarrow \sum a_n \, \lambda_n$ is summable $|(R, q_n) \; (R, p_n)|_k$ are

$$|\lambda_v| \; F_v = O(Q_v), \tag{6.11}$$

$$|\lambda_v| = O(Q_v), \tag{6.12}$$

$$p_v \, R_v \; |\lambda_v| = O(Q_v), \tag{6.13}$$

$$p_v \, q_v \, R_v \; |\lambda_v| = O(Q_v \, Q_{v-1} r_v), \tag{6.14}$$

$$p_n \, q_n \, R_n \; |\lambda_n| = O(P_n \, Q_n \, r_n), \tag{6.15}$$

$$R_{v-1} \, |\Delta \lambda_v| \; F_{v+1} = O(Q_v \, r_v), \tag{6.16}$$

and

$$R_{v-1} \, |\Delta \lambda_v| = O(Q_v \, r_v). \tag{6.17}$$

Subsequently, Paikray et al. [5] generalized the above theorem by replacing the (R, p_n) summability by A-summability and stated the results as follows.

Theorem 6.2. *Let $k \geq 1, \{\lambda_n\}$ be a sequence of constants. Let us define*

$$f_v = \sum_{r=v}^{n} q_r \, a_{rv}, \quad F_v = \sum_{r=v}^{n} f_r. \tag{6.18}$$

Then, the sufficient conditions for the implication that $\sum a_n$ is summable $|R, r_n|_k \Rightarrow \sum a_n \, \lambda_n$ is summable $|(R, q_n) \, (A)|_k$ are

$$\sum_{n=v+1}^{m+1} \frac{n^{k-1} q_n^k}{Q_n^k Q_{n-1}} = O\left(\frac{1}{\lambda_v^k}\right), \tag{6.19}$$

$$\left(\sum_{r=v}^{n} q_r^{\frac{k}{k-1}}\right) = O(q_v), \tag{6.20}$$

$$\left(\sum_{r=v}^{n} a_{r,v}^{k}\right) = O\left(v^{k-1}\right), \tag{6.21}$$

$$R_v = O(r_v), \tag{6.22}$$

$$\frac{q_n}{Q_n} = O(1), \tag{6.23}$$

$$\frac{q_n \lambda_n a_{n,n}}{Q_{n-1}} = O(1), \tag{6.24}$$

$$\frac{(\Delta\lambda_v)^k}{q_v^{k-1}} = O\left(v^{k-1}\right), \tag{6.25}$$

$$\frac{\Delta\lambda_v}{\lambda_v} = O(1), \tag{6.26}$$

and

$$\frac{\lambda_v^k}{q_v^{k-1}} = O\left(v^{k-1}\right). \tag{6.27}$$

In this chapter, we established the following theorems on the product summability of the infinite series $\sum a_n \lambda_n$.

6.3 MAIN RESULTS

Theorem 6.3. *For the sequences of real constants $\{p_n\}$ and $\{q_n\}$, define*

$$f_v = \sum_{i=v}^{n} \frac{q_{n-i} p_{i-v}}{P_i} \text{ and } F_v = \sum_{i=v}^{n} f_i. \tag{6.28}$$

Let

$$Q_n = O(q_n P_n) \tag{6.29}$$

and

$$\sum_{n=v+1}^{m+1} \frac{n^{\delta k+k-1} q_n^k}{Q_n^k Q_{n-1}} = O\left(\frac{(v q_v)^{k-1}}{Q_v^k}\right) \text{ as } m \to \infty. \tag{6.30}$$

Then for any sequences $\{r_n\}$ and $\{\lambda_n\}$, the sufficient conditions for the implication that $\sum a_n$ is summable $|R, r_n|_k \Rightarrow \sum a_n \lambda_n$ is summable $|(N, q_n)\ (N, p_n); \delta|_k$, $k \geq 1$, $1 \geq k\delta \geq 0$, are

$$|\lambda_n|\ F_v = O(Q_v), \tag{6.31}$$

$$|\lambda_n| = O(Q_n), \tag{6.32}$$

$$R_v\, F_v\, |\lambda_v| = O(Q_v\, r_v), \tag{6.33}$$

$$q_n\, R_n\, F_n\, |\lambda_n| = O(Q_n\, Q_{n-1} r_n), \tag{6.34}$$

$$R_{v-1} F_{v+1}\, |\Delta\lambda_v| = O(Q_v\, r_v), \tag{6.35}$$

$$R_{v-1}\, |\Delta\lambda_v| = O(Q_v\, r_v), \tag{6.36}$$

$$q_n\, R_n\, |\lambda_n| = O(Q_n\, Q_{n-1} r_n), \tag{6.37}$$

where $R_n = r_1 + r_2 + \ldots\ldots + r_n$.

Theorem 6.4. *For the sequences of real constants $\{p_n\}$ and $\{q_n\}$ and the sequence of positive numbers $\{\alpha_n\}$, define*

$$f_v = \sum_{i=v}^{n} \frac{q_{n-i} p_{i-v}}{P_i} \text{ and } F_v = \sum_{i=v}^{n} f_i. \tag{6.38}$$

Let

$$Q_n = O(q_n P_n) \tag{6.39}$$

and

$$\sum_{n=v+1}^{m+1} \frac{\alpha_n^{k-1} q_n^k}{Q_n^k Q_{n-1}} = O\left(\frac{(v q_v)^{k-1}}{Q_v^k}\right) \text{ as } m \to \infty. \tag{6.40}$$

Then, for any sequences $\{r_n\}$ and $\{\lambda_n\}$, the sufficient conditions for the implication that $\sum a_n$ is summable $|R, r_n|_k \Rightarrow \sum a_n \lambda_n$ is summable $|(N,q_n)\ (N,p_n), \alpha_n|_k$, $k \geq 1$, are

$$|\lambda_v|\ F_v = O(Q_v), \tag{6.41}$$

$$|\lambda_n| = O(Q_n), \tag{6.42}$$

$$R_v\ F_v\ |\lambda_v| = O(Q_v\ r_v), \tag{6.43}$$

$$q_n\ R_n\ F_n\ |\lambda_n| = O(Q_n\ Q_{n-1} r_n), \tag{6.44}$$

$$R_{v-1} F_{v+1}\ |\Delta\lambda_v| = O(Q_v\ r_v), \tag{6.45}$$

$$R_{v-1}\ |\Delta\lambda_v| = O(Q_v\ r_v), \tag{6.46}$$

$$q_n\ R_n\ |\lambda_n| = O(Q_n\ Q_{n-1} r_n), \tag{6.47}$$

$$\sum_{n=1}^{\infty} n^{k-1}\ |t_n|^k = O(1), \tag{6.48}$$

and

$$\sum_{n=2}^{\infty} \alpha_n^{k-1}\ |t_n|^k = O(1), \tag{6.49}$$

where $R_n = r_1 + r_2 + \ldots\ldots + r_n$ $t_n = \Delta t_n^{'}$ and $t_n^{'}$ is the n-th (R, r_n) transform of $\sum a_n$.

Theorem 6.5. *For the sequences of real constants $\{p_n\}$ and $\{q_n\}$ and the sequence of positive numbers $\{\alpha_n\}$, define*

$$f_v = \sum_{i=v}^{n} \frac{q_{n-i} p_{i-v}}{P_i} \text{ and } F_v = \sum_{i=v}^{n} f_i \tag{6.50}$$

Let

$$Q_n = O(q_n P_n) \tag{6.51}$$

and

$$\sum_{n=v+1}^{m+1} \frac{\alpha_n^{\delta k+k-1} q_n^k}{Q_n^k Q_{n-1}} = O\left(\frac{(v q_v)^{k-1}}{Q_v^k}\right) \text{ as } m \to \infty. \tag{6.52}$$

Then for any sequences $\{r_n\}$ *and* $\{\lambda_n\}$, *the sufficient conditions for the implication* $\sum a_n$ *is summable* $|R, r_n|_k \Rightarrow \sum a_n\lambda_n$ *is summable* $|(N,q_n)(N,p_n),\alpha_n;\delta|_k, k \geq 1, 1 \geq k\delta \geq 0$, *are*

$$|\lambda_n|\ F_v = O(Q_v), \tag{6.53}$$

$$|\lambda_n| = O(Q_n), \tag{6.54}$$

$$R_v\ F_v\ |\lambda_v| = O(Q_v\ r_v), \tag{6.55}$$

$$q_n\ R_n\ F_n\ |\lambda_n|\ \alpha_n^{\delta} = O(Q_n\ Q_{n-1} r_n), \tag{6.56}$$

$$R_{v-1} F_{v+1}\ |\Delta\lambda_v| = O(Q_v\ r_v), \tag{6.57}$$

$$R_{v-1}\ |\Delta\lambda_v| = O(Q_v\ r_v), \tag{6.58}$$

$$q_n\ R_n\ |\lambda_n|\ \alpha_n^{\delta} = O(Q_n\ Q_{n-1} r_n), \tag{6.59}$$

$$\sum_{n=1}^{\infty} n^{k-1}\ |t_n|^k = O(1), \tag{6.60}$$

and

$$\sum_{n=2}^{\infty} \alpha_n^{k-1}\ |t_n|^k = O(1), \tag{6.61}$$

where $R_n = r_1 + r_2 + \ldots\ldots + r_n$.

Theorem 6.6. *For the sequences of real constants* $\{p_n\}$ *and* $\{q_n\}$ *and the sequence of positive numbers* $\{\alpha_n\}$, *we define*

$$f_v = \sum_{i=v}^{n} \frac{q_{n-i} p_{i-v}}{P_i} \text{ and } F_v = \sum_{i=v}^{n} f_i. \tag{6.62}$$

Let

$$Q_n = O(q_n P_n) \tag{6.63}$$

and

$$\sum_{n=v+1}^{m+1} \frac{\{f(\alpha_n)\}^k (\alpha_n)^{k-1} q_n^k}{Q_n^k Q_{n-1}} = O\left(\frac{(vq_v)^{k-1}}{Q_v^k}\right) \text{ as } m \to \infty. \tag{6.64}$$

Then for any sequences $\{r_n\}$ *and* $\{\lambda_n\}$, *the sufficient conditions for the implication* $\sum a_n$ *is summable* $|R, r_n|_k \Rightarrow \sum a_n\lambda_n$ *is* $|(N,q_n)\ (N,p_n), \alpha_n; f|_k, k \geq 1$ *is summable, are*

$$|\lambda_n|\ F_v = O(Q_v), \tag{6.65}$$

$$|\lambda_n| = O(Q_n), \tag{6.66}$$

$$R_v\ F_v\ |\lambda_v| = O(Q_v\ r_v), \tag{6.67}$$

$$q_n\ R_n\ F_n\ |\lambda_n| = O(Q_n\ Q_{n-1} r_n), \tag{6.68}$$

$$R_{v-1} F_{v+1}\ |\Delta\lambda_v| = O(Q_v\ r_v), \tag{6.69}$$

$$R_{v-1}\ |\Delta\lambda_v| = O(Q_v\ r_v), \tag{6.70}$$

$$q_n\, R_n\, |\lambda_n| = O(Q_n\, Q_{n-1} r_n), \tag{6.71}$$

$$\sum_{n=1}^{\infty} n^{k-1}\, |t_n|^k = O(1), \tag{6.72}$$

and

$$\sum_{n=2}^{\infty} \{f(\alpha_n)\}^k (\alpha_n)^{k-1}\, |t_n|^k = O(1), \tag{6.73}$$

where $R_n = r_1 + r_2 + \ldots\ldots + r_n$.

6.4 PROOF OF MAIN RESULTS

In this section, we provide the detailed proof of all theorems as stated above.

Proof of Theorem 6.3. Let $\{t'_n\}$ be the (R, r_n) transform of the series $\sum a_n$. Then,

$$t'_n = \frac{1}{R} \sum_{v=0}^{n} r_v s_v .$$

Then

$$t_n = t'_n - t'_{n-1} = \frac{r_n}{R_n R_{n-1}} \sum_{v=1}^{n} R_{v-1} a_v .$$

Let $\{s_n\}$ be the sequence of partial sums of the series $\sum a_n\, \lambda_n$ and $\{\tau_n\}$ the sequence of $(N, q_n)\ (N, p_n)$ transform of the series $\sum a_n\, \lambda_n$. Then,

$$\begin{aligned}
\tau_n &= \frac{1}{Q_n} \sum_{r=0}^{n} q_{n-r} \frac{1}{P_r} \sum_{v=0}^{r} p_{r=v}\, s_v, \\
&\frac{1}{Q_n} \sum_{v=0}^{n} s_v \sum_{r=v}^{n} \frac{q_{n-v} p_{r-v}}{P_r}, \\
&= \frac{1}{Q_n} \sum_{v=0}^{n} f_v\, s_v .
\end{aligned} \tag{6.74}$$

Hence,

$$\begin{aligned}
T_n &= \tau_n - \tau_{n-1} \\
&= \frac{1}{Q_n} \sum_{v=0}^{n} f_v\, s_v - \frac{1}{Q_{n-1}} \sum_{v=0}^{n-1} f_v\, s_v \\
&= -\frac{q_n}{Q_n Q_{n-1}} \sum_{v=0}^{n} f_v\, s_v + \frac{f_n s_n}{Q_{n-1}} \\
&= -\frac{q_n}{Q_n Q_{n-1}} \sum_{r=0}^{n} f_r \sum_{v=0}^{r} a_v\, \lambda_v + \frac{f_n}{Q_{n-1}} \sum_{v=0}^{n} a_v\, \lambda_v \\
&= -\frac{q_n}{Q_n Q_{n-1}} \sum_{r=0}^{n} a_v\, \lambda_v \sum_{v=0}^{r} f_r + \frac{f_n}{Q_{n-1}} \sum_{v=0}^{n} a_v\, \lambda_v
\end{aligned} \tag{6.75}$$

$$
\begin{aligned}
&= -\frac{q_n}{Q_nQ_{n-1}}\sum_{v=1}^{n} R_{v-1}a_v\left(\frac{\lambda_v}{R_{v-1}}\sum_{r=v}^{n} f_r\right)+\frac{q_0p_0}{P_nQ_{n-1}}\sum_{v=1}^{n} R_{v-1}a_v\left(\frac{\lambda_v}{R_{v-1}}\right)\\
&= -\frac{q_n}{Q_nQ_{n-1}}\left[\sum_{v=1}^{n-1}\left(\sum_{r=1}^{v} R_{r-1}a_r\right)\Delta\left(\frac{\lambda_v}{R_{v-1}}\sum_{r=v}^{n} f_r\right)+\left(\sum_{v=1}^{n} R_{v-1}a_v\right)\frac{\lambda_n}{R_{n-1}}f_n\right]\\
&\quad+\frac{p_0q_0}{P_nQ_{n-1}}\left[\sum_{v=1}^{n-1}\left(\sum_{r=1}^{v} R_{r-1}a_v\right)\Delta\left(\frac{\lambda_v}{R_{v+1}}\right)+\left(\sum_{v=1}^{n} R_{v-1}a_v\right)\frac{\lambda_n}{R_{n-1}}\right]\\
&= -\frac{q_n}{Q_nQ_{n-1}}\left[\sum_{v=1}^{n-1}\left[\lambda_vF_vt_v+\frac{R_{v-1}}{r_v}f_v\,\lambda_v\,t_v+\frac{R_{v-1}}{r_v}(\Delta\lambda_v)\,F_{v+1}t_v\right]+\frac{R_n}{r_n}\lambda_n\,F_n\,t_n\right]\\
&\quad+\frac{p_0q_0}{P_nQ_{n-1}}\left[\sum_{v=1}^{n-1}\left(\lambda_v\,t_v+\frac{R_{v-1}}{r_v}(\Delta\lambda_v)t_v\right)+\frac{R_n}{r_n}\lambda_n\,t_n\right]\\
&= \sum_{i=1}^{7} T_{n,i}, \quad \text{say.}
\end{aligned}
\tag{6.76}
$$

In order to prove the theorem, using Minkowski's inequality, it is sufficient to show that

$$\sum_{n=1}^{\infty} n^{\delta k+k-1}\,|T_{n,i}|^k<\infty, \quad \text{for } i=1,2,3,4,5,6,7.$$

Now, on applying Holder's inequality, we have

$$
\begin{aligned}
\sum_{n=2}^{m+1} n^{\delta k+k-1}\,|T_{n,1}|^k &= \sum_{n=2}^{m+1} n^{\delta k+k-1}\left|\frac{q_n}{Q_nQ_{n-1}}\sum_{v=1}^{n-1}\lambda_v\,F_v\,t_v\right|^k\\
&\le \sum_{n=2}^{m+1} n^{\delta k+k-1}\frac{q_n^k}{Q_n^kQ_{n-1}}\sum_{v=1}^{n-1}\frac{|\lambda_v|^k F_v^k|t_v|^k}{q_v^{k-1}}\left(\frac{1}{Q_{n-1}}\sum_{v=1}^{n-1} q_v\right)^{k-1}\\
&= O(1)\sum_{v=1}^{m}\frac{1}{q_v^{k-1}}|\lambda_v|^k\,F_v^k\,|t_v|^k\sum_{n=v+1}^{m+1}\frac{n^{\delta k+k-1}q_n^k}{Q_n^kQ_{n-1}}\\
&= O(1)\sum_{v-1}^{m}\frac{1}{q_v^{k-1}}|\lambda_v|^k\,F_v^k\,|t_v|^k\frac{(vq_v)^{k-1}}{Q_v^k}, \text{ using (6.30)}\\
&= O(1)\sum_{v=1}^{m} v^{k-1}\,|t_v|^k\left(\frac{|\lambda_v|\,F_v}{Q_v}\right)^k\\
&= O(1)\sum_{v=1}^{m} v^{k-1}\,|t_v|^k, \quad \text{using (6.31)}\\
&= O(1),\ on\ m\to\infty.
\end{aligned}
$$

Next,

$$\sum_{n=2}^{m+1} n^{\delta k+k-1}\,|T_{n2}|^k=\sum_{n=2}^{m+1} n^{k-1}\left|\frac{q_n}{Q_nQ_{n-1}}\sum_{v=1}^{n-1}\frac{R_{v-1}}{r_v}f_v\,\lambda_v\,t_v\right|^k$$

$$\leq \sum_{n=2}^{m+1} n^{\delta k+k-1} \frac{q_n^k}{Q_n^k Q_{n-1}} \sum_{v=1}^{n-1} \frac{R_v^k F_v^k |\lambda_v|^k}{q_v^{k-1} r_v^k} \left(\frac{1}{Q_{n-1}} \sum_{v=1}^{n-1} q_v \right)^{k-1}$$

$$= O(1) \sum_{v=1}^{m} \frac{R_v^k F_v^k |\lambda_v|^k |t_v|^k}{q_v^{k-1} r_v^k} \sum_{n=v+1}^{m+1} \frac{n^{\delta k+k-1} q_n^k}{Q_n^k Q_{n-1}}$$

$$= O(1) \sum_{v=1}^{m} v^{k-1} |t_v|^k \left(\frac{R_v F_v |\lambda_v|}{r_v Q_v} \right)^k$$

$$= O(1) \sum_{v=1}^{m} v^{k-1} |t_v|^k, \text{ using } (6.33)$$

$$= O(1), \text{ as } m \to \infty.$$

$$\sum_{n=2}^{m+1} n^{\delta k+k-1} |T_{n,3}|^k = \sum_{n=2}^{m+1} n^{\delta k+k-1} \left| \frac{q_n}{Q_n Q_{n-1}} \sum_{v=1}^{n-1} \frac{R_{v-1}}{r_v} (\Delta\lambda_v) F_{v+1} t_v \right|^k$$

$$\leq \sum_{n=2}^{m+1} n^{\delta k+k-1} \frac{q_n^k}{Q_n^k Q_{n-1}} \sum_{v=1}^{n-1} \frac{R_{v-1}^k}{r_v^k q_v^{k-1}} |\Delta\lambda_v|^k \; F_{v+1}^k \; |t_v|^k \left(\frac{1}{Q_{n-1}} \sum_{v=1}^{n-1} q_v \right)^{k-1}$$

$$= O(1) \sum_{v=1}^{m} \frac{R_{v-1}^k |\Delta\lambda_r|^k}{r_v^k q_v^{k-1}} F_{v+1}^k \; |t_v|^k \sum_{n=v+1}^{m+1} \frac{n^{\delta k+k-1} q_n^k}{Q_n^k Q_{n-1}}, \text{ by } (6.30)$$

$$= O(1) \sum_{v=1}^{m} v^{k-1} |t_v|^k \left(\frac{R_{v-1} F_{v+1} |\Delta\lambda_v|}{r_v Q_v} \right)^k$$

$$= O(1) \sum_{v=1}^{m} v^{k-1} |t_v|^k, \text{ using } (6.35)$$

$$= O(1), \text{ as } m \to \infty.$$

$$\sum_{n=2}^{m+1} n^{\delta k+k-1} |T_{n,4}|^k = \sum_{n=2}^{m+1} n^{\delta k+k-1} \left| \frac{q_n}{Q_n Q_{n-1}} \frac{R_n \lambda_n f_n t_n}{r_n} \right|^k$$

$$\leq \sum_{n=2}^{m+1} n^{k-1} |t_n|^k \left(\frac{q_n R_n F_n |\lambda_n|}{Q_n Q_{n-1} r_n} \right)^k$$

$$= O(1) \sum_{n=2}^{m+1} n^{k-1} |t_n|^k, \text{ using } (6.34)$$

$$= O(1), \text{ as } m \to \infty.$$

$$\sum_{n=2}^{m+1} n^{\delta k+k-1} \left| T_{n,5} \right|^k = \sum_{n=2}^{m+1} n^{\delta k+k-1} \left| \frac{p_0 q_0}{P_n Q_{n-1}} \sum_{v=1}^{n-1} \lambda_v \, t_v \right|^k$$

$$\leq O(1) \sum_{n=2}^{m+1} n^{\delta k+k-1} \frac{1}{P_n^k Q_{n-1}} \sum_{v=1}^{n-1} \frac{|\lambda_v|^k}{q_v^{k-1}} |t_v|^k \left(\frac{1}{Q_{n-1}} \sum_{v=1}^{n-1} q_v \right)^{k-1}$$

$$
\begin{aligned}
&= O(1)\sum_{v=1}^{m}\frac{|\lambda_v|^k|t_v|^k}{q_v^{k-1}}\sum_{n=v+1}^{m+1} n^{\delta k+k-1}\cdot\frac{1}{P_n^kQ_{n-1}}\\
&= O(1)\sum_{v=1}^{m}\frac{|\lambda_v|^k|t_v|^k}{q_v^{k-1}}\sum_{n=v+1}^{m+1}\frac{n^{k-1}q_n^k}{Q_n^kQ_{n-1}},\ \text{using (6.29)}\\
&= O(1)\sum_{v=1}^{m} v^k\ |t_v|^k\left(\frac{|\lambda_n|^k}{Q_v}\right)^k\\
&= O(1)\sum_{v=1}^{m} v^k\ |t_v|^k,\ \text{using (6.34)}\\
&= O(1),\ as\ m\to\infty.
\end{aligned}
$$

$$
\begin{aligned}
\sum_{n=2}^{m+1} n^{\delta k+k-1}\left|T_{n,6}\right|^k &= \sum_{n=2}^{m+1} n^{\delta k+k-1}\left|\frac{p_0q_0}{P_nQ_{n-1}}\sum_{v=1}^{n-1}\frac{R_{v-1}}{r_v}(\Delta\lambda_v)\ t_v\right|^k\\
&\le O(1)\sum_{n=2}^{m+1} n^{\delta k+k-1}\cdot\frac{1}{P_n^kQ_{n-1}}\sum_{v=1}^{n-1}\frac{R_{v-1}^k}{r_v^kq_v^{k-1}}|\Delta\lambda_v|^k\ |t_v|^k\left(\frac{1}{Q_{n-1}}\sum_{v=1}^{n-1}q_v\right)^{n-1}\\
&= O(1)\sum_{v=1}^{m}\frac{R_{v-1}^k|\Delta\lambda_v|^k|t_v|^k}{r_v^kq_v^{k-1}}\sum_{n=v+1}^{m+1} n^{\delta k+k-1}\cdot\frac{1}{P_n^kQ_{n-1}}\\
&= O(1)\sum_{v=1}^{m} v^{k-1}|t_v|^k\left(\frac{R_{v-1}|\Delta\lambda_v|}{r_vQ_v}\right)^k\\
&= O(1)\sum_{v=1}^{m} v^{k-1}|t_v|^k,\ \text{using (6.36)}\\
&= O(1),\ as\ m\to\infty.
\end{aligned}
$$

Finally,

$$
\begin{aligned}
\sum_{n=2}^{m+1} n^{\delta k+k-1}|T_{n,7}|^k &= \sum_{n=2}^{m+1} n^{\delta k+k-1}\left|\frac{p_0q_0}{P_nQ_{n-1}}\frac{R_n\lambda_nt_n}{r_n}\right|^k\\
&= O(1)\sum_{n=2}^{m+1} n^{k-1}|t_n|^k\left(\frac{R_n|\lambda_n|}{P_nQ_{n-1}r_n}\right)^k\\
&= O(1)\sum_{n=2}^{m+1} n^{k-1}|t_n|^k\left(\frac{q_nR_n|\lambda_n|}{Q_nQ_{n-1}r_n}\right)^k\\
&= O(1)\sum_{n=2}^{m+1} n^{k-1}|t_n|^k,\ \text{using (6.37)}\\
&= O(1),\ as\ m\to\infty.
\end{aligned}
$$

This completes the Proof of Theorem 6.3.

Proof of Theorem 6.4. In order to prove this theorem, using (6.76) and Minkowski's inequality, it is sufficient to show that

$$\sum_{n=1}^{\infty} \alpha_n^{k-1} |T_{n,i}|^k < \infty, \text{ for } i = 1,2,3,4,5,6,7.$$

Now, on applying Holder's inequality, we have

$$\begin{aligned}
\sum_{n=2}^{m+1} \alpha_n^{k-1} |T_{n,1}|^k &= \sum_{n=2}^{m+1} \alpha_n^{k-1} \left| \frac{q_n}{Q_n Q_{n-1}} \sum_{v=1}^{n-1} \lambda_v F_v t_v \right|^k \\
&\leq \sum_{n=2}^{m+1} \alpha_n^{k-1} \frac{q_n^k}{Q_n^k Q_{n-1}} \sum_{v=1}^{n-1} \frac{|\lambda_v|^k F_v^k |t_v|^k}{q_v^{k-1}} \left(\frac{1}{Q_{n-1}} \sum_{v=1}^{n-1} q_v \right)^{k-1} \\
&= O(1) \sum_{v=1}^{m} \frac{1}{q_v^{k-1}} |\lambda_v|^k F_v^k |t_v|^k \sum_{n=v+1}^{m+1} \frac{\alpha_n^{k-1} q_n^k}{Q_n^k Q_{n-1}} \\
&= O(1) \sum_{v-1}^{m} \frac{1}{q_v^{k-1}} |\lambda_v|^k F_v^k |t_v|^k \frac{(v q_v)^{k-1}}{Q_v^k}, \text{ using (6.40)} \\
&= O(1) \sum_{v=1}^{m} v^{k-1} |t_v|^k \left(\frac{|\lambda_v| F_v}{Q_v} \right)^k \\
&= O(1) \sum_{v=1}^{m} v^{k-1} |t_v|^k, \text{ using (6.41)} \\
&= O(1), \; as\; m \to \infty.
\end{aligned}$$

Next,

$$\begin{aligned}
\sum_{n=2}^{m+1} \alpha_n^{k-1} |T_{n2}|^k &= \sum_{n=2}^{m+1} \alpha_n^{k-1} \left| \frac{q_n}{Q_n Q_{n-1}} \sum_{v=1}^{n-1} \frac{R_{v-1}}{r_v} f_v \lambda_v t_v \right|^k \\
&\leq \sum_{n=2}^{m+1} \alpha_n^{k-1} \frac{q_n^k}{Q_n^k Q_{n-1}} \sum_{v=1}^{n-1} \frac{R_v^k F_v^k |\lambda_v|^k}{q_v^{k-1} r_v^k} \left(\frac{1}{Q_{n-1}} \sum_{v=1}^{n-1} q_v \right)^{k-1} \\
&= O(1) \sum_{v=1}^{m} \frac{R_v^k F_v^k |\lambda_v|^k |t_v|^k}{q_v^{k-1} r_v^k} \sum_{n=v+1}^{m+1} \frac{\alpha_n^{k-1} q_n^k}{Q_n^k Q_{n-1}} \\
&= O(1) \sum_{v=1}^{m} v^{k-1} |t_v|^k \left(\frac{R_v F_v |\lambda_v|}{r_v Q_v} \right)^k \\
&= O(1) \sum_{v=1}^{m} v^{k-1} |t_v|^k, \text{ using (6.43)} \\
&= O(1), \; as\; m \to \infty.
\end{aligned}$$

Further,

$$\begin{aligned}
\sum_{n=2}^{m+1} \alpha_n^{k-1} |T_{n,3}|^k &= \sum_{n=2}^{m+1} \alpha_n^{k-1} \left| \frac{q_n}{Q_n Q_{n-1}} \sum_{v=1}^{n-1} \frac{R_{v-1}}{r_v} (\Delta\lambda_v) F_{v+1} t_v \right|^k \\
&\leq \sum_{n=2}^{m+1} \alpha_n^{k-1} \frac{q_n^k}{Q_n^k Q_{n-1}} \sum_{v=1}^{n-1} \frac{R_{v-1}^k}{r_v^k q_v^{k-1}} |\Delta\lambda_v|^k \; F_{v+1}^k \; |t_v|^k \left(\frac{1}{Q_{n-1}} \sum_{v=1}^{n-1} q_v \right)^{k-1} \\
&= O(1) \sum_{v=1}^{m} \frac{R_{v-1}^k |\Delta\lambda_r|^k}{r_v^k q_v^{k-1}} F_{v+1}^k \; |t_v|^k \sum_{n=v+1}^{m+1} \frac{\alpha_n^{k-1} q_n^k}{Q_n^k Q_{n-1}} \\
&= O(1) \sum_{v=1}^{m} v^{k-1} |t_v|^k \left(\frac{R_{v-1} F_{v+1} |\Delta\lambda_v|}{r_v Q_v} \right)^k, \text{ by (6.40)} \\
&= O(1) \sum_{v=1}^{m} v^{k-1} |t_v|^k \text{ , using (6.45)} \\
&= O(1), \; as \; m \to \infty.
\end{aligned}$$

Again,

$$\begin{aligned}
\sum_{n=2}^{m+1} \alpha_n^{k-1} |T_{n,4}|^k &= \sum_{n=2}^{m+1} \alpha_n^{k-1} \left| \frac{q_n}{Q_n Q_{n-1}} \frac{R_n \lambda_n f_n t_n}{r_n} \right|^k \\
&\leq \sum_{n=2}^{m+1} \alpha_n^{k-1} |t_n|^k \left(\frac{q_n R_n F_n |\lambda_n|}{Q_n Q_{n-1} r_n} \right)^k \\
&= O(1) \sum_{n=2}^{m+1} \alpha_n^{k-1} |t_n|^k \text{ , using (6.44)} \\
&= O(1), \; as \; m \to \infty.
\end{aligned}$$

Next,

$$\begin{aligned}
\sum_{n=2}^{m+1} \alpha_n^{k-1} |T_{n,5}|^k &= \sum_{n=2}^{m+1} \alpha_n^{k-1} \left| \frac{p_0 q_0}{P_n Q_{n-1}} \sum_{v=1}^{n-1} \lambda_v \, t_v \right|^k \\
&\leq O(1) \sum_{n=2}^{m+1} \alpha_n^{k-1} \frac{1}{P_n^k Q_{n-1}} \sum_{v=1}^{n-1} \frac{|\lambda_v|^k}{q_v^{k-1}} |t_v|^k \left(\frac{1}{Q_{n-1}} \sum_{v=1}^{n-1} q_v \right)^{k-1} \\
&= O(1) \sum_{v=1}^{m} \frac{|\lambda_v|^k |t_v|^k}{q_v^{k-1}} \sum_{n=v+1}^{m+1} \alpha_n^{k-1} \cdot \frac{1}{P_n^k Q_{n-1}} \\
&= O(1) \sum_{v=1}^{m} \frac{|\lambda_v|^k |t_v|^k}{q_v^{k-1}} \sum_{n=v+1}^{m+1} \frac{\alpha_n^{k-1} q_n^k}{Q_n^k Q_{n-1}}, \text{ using (6.39)} \\
&= O(1) \sum_{v=1}^{m} v^k \; |t_v|^k \left(\frac{|\lambda_n|^k}{Q_v} \right)^k
\end{aligned}$$

$$= O(1) \sum_{v=1}^{m} v^k \, |t_v|^k \text{, using (6.44)}$$
$$= O(1), \text{ as } m \to \infty.$$

Again,

$$\begin{aligned}
\sum_{n=2}^{m+1} \alpha_n^{k-1} |T_{n,6}|^k &= \sum_{n=2}^{m+1} \alpha_n^{k-1} \left| \frac{p_0 q_0}{P_n Q_{n-1}} \sum_{v=1}^{n-1} \frac{R_{v-1}}{r_v} (\Delta\lambda_v) \, t_v \right|^k \\
&\leq O(1) \sum_{n=2}^{m+1} \alpha_n^{k-1} \frac{1}{P_n^k Q_{n-1}} \sum_{v=1}^{n-1} \frac{R_{v-1}^k}{r_v^k q_v^{k-1}} |\Delta\lambda_v|^k \, |t_v|^k \left(\frac{1}{Q_{n-1}} \sum_{v=1}^{n-1} q_v \right)^{n-1} \\
&= O(1) \sum_{v=1}^{m} \frac{R_{v-1}^k |\Delta\lambda_v|^k |t_v|^k}{r_v^k q_v^{k-1}} \sum_{n=v+1}^{m+1} \alpha_n^{k-1} \frac{1}{P_n^k Q_{n-1}} \\
&= O(1) \sum_{v=1}^{m} v^{k-1} |t_v|^k \left(\frac{R_{v-1} |\Delta\lambda_v|}{r_v Q_v} \right)^k \\
&= O(1) \sum_{v=1}^{m} v^{k-1} |t_v|^k \text{, using (6.46)} \\
&= O(1), \text{ as } m \to \infty.
\end{aligned}$$

Finally,

$$\begin{aligned}
\sum_{n=2}^{m+1} \alpha_n^{k-1} |T_{n,7}|^k &= \sum_{n=2}^{m+1} \alpha_n^{k-1} \left| \frac{p_0 q_0}{P_n Q_{n-1}} \frac{R_n \lambda_n t_n}{r_n} \right|^k \\
&= O(1) \sum_{n=2}^{m+1} \alpha_n^{k-1} |t_n|^k \left(\frac{R_n |\lambda_n|}{P_n Q_{n-1} r_n} \right)^k \\
&= O(1) \sum_{n=2}^{m+1} \alpha_n^{k-1} |t_n|^k \left(\frac{q_n R_n |\lambda_n|}{Q_n Q_{n-1} r_n} \right)^k \\
&= O(1) \sum_{n=2}^{m+1} \alpha_n^{k-1} |t_n|^k \text{, using (6.47)} \\
&= O(1), \text{ as } m \to \infty.
\end{aligned}$$

This completes the Proof of Theorem 6.4.

Proof of Theorem 6.5. In order to prove this theorem, using (6.76) and Minkowski's inequality, it is sufficient to show that

$$\sum_{n=1}^{\infty} \alpha_n^{\delta k+k-1} |T_{n,i}|^k < \infty, \text{for } i = 1,2,3,4,5,6,7.$$

On applying Holder's inequality, we have

$$
\begin{aligned}
\sum_{n=2}^{m+1} \alpha_n^{\delta k+k-1} |T_{n,1}|^k &= \sum_{n=2}^{m+1} \alpha_n^{\delta k+k-1} \left| \frac{q_n}{Q_n Q_{n-1}} \sum_{v=1}^{n-1} \lambda_v F_v t_v \right|^k \\
&\leq \sum_{n=2}^{m+1} \alpha_n^{\delta k+k-1} \frac{q_n^k}{Q_n^k Q_{n-1}} \sum_{v=1}^{n-1} \frac{|\lambda_v|^k F_v^k |t_v|^k}{q_v^{k-1}} \left(\frac{1}{Q_{n-1}} \sum_{v=1}^{n-1} q_v \right)^{k-1} \\
&= O(1) \sum_{v=1}^{m} \frac{1}{q_v^{k-1}} |\lambda_v|^k F_v^k |t_v|^k \sum_{n=v+1}^{m+1} \frac{\alpha_n^{\delta k+k-1} q_n^k}{Q_n^k Q_{n-1}} \\
&= O(1) \sum_{v-1}^{m} \frac{1}{q_v^{k-1}} |\lambda_v|^k F_v^k |t_v|^k \frac{(vq_v)^{k-1}}{Q_v^k}, \text{ using (6.51)} \\
&= O(1) \sum_{v=1}^{m} v^{k-1} |t_v|^k \left(\frac{|\lambda_v| F_v}{Q_v} \right)^k \\
&= O(1) \sum_{v=1}^{m} v^{k-1} |t_v|^k, \text{ using (6.53)} \\
&= O(1), \; as \; m \to \infty.
\end{aligned}
$$

Next,

$$
\begin{aligned}
\sum_{n=2}^{m+1} \alpha_n^{\delta k+k-1} |T_{n,2}|^k &= \sum_{n=2}^{m+1} \alpha_n^{\delta k+k-1} \left| \frac{q_n}{Q_n Q_{n-1}} \sum_{v=1}^{n-1} \frac{R_{v-1}}{r_v} f_v \lambda_v t_v \right|^k \\
&\leq \sum_{n=2}^{m+1} \alpha_n^{\delta k+k-1} \frac{q_n^k}{Q_n^k Q_{n-1}} \sum_{v=1}^{n-1} \frac{R_v^k F_v^k |\lambda_v|^k}{q_v^{k-1} r_v^k} \left(\frac{1}{Q_{n-1}} \sum_{v=1}^{n-1} q_v \right)^{k-1} \\
&= O(1) \sum_{v=1}^{m} \frac{R_v^k F_v^k |\lambda_v|^k |t_v|^k}{q_v^{k-1} r_v^k} \sum_{n=v+1}^{m+1} \frac{\alpha_n^{\delta k+k-1} q_n^k}{Q_n^k Q_{n-1}} \\
&= O(1) \sum_{v=1}^{m} v^{k-1} |t_v|^k \left(\frac{R_v F_v |\lambda_v|}{r_v Q_v} \right)^k \\
&= O(1) \sum_{v=1}^{m} v^{k-1} |t_v|^k, \text{ using (6.55)} \\
&= O(1), \; as \; m \to \infty.
\end{aligned}
$$

Further,

$$
\begin{aligned}
\sum_{n=2}^{m+1} \alpha_n{}^{\delta k+k-1} |T_{n,3}|^k &= \sum_{n=2}^{m+1} \alpha_n^{\delta k+k-1} \left| \frac{q_n}{Q_n Q_{n-1}} \sum_{v=1}^{n-1} \frac{R_{v-1}}{r_v} (\Delta \lambda_v) F_{v+1} t_v \right|^k \\
&\leq \sum_{n=2}^{m+1} \alpha_n^{\delta k+k-1} \frac{q_n^k}{Q_n^k Q_{n-1}} \sum_{v=1}^{n-1} \frac{R_{v-1}^k}{r_v^k q_v^{k-1}} |\Delta \lambda_v|^k F_{v+1}^k |t_v|^k \left(\frac{1}{Q_{n-1}} \sum_{v=1}^{n-1} q_v \right)^{k-1}
\end{aligned}
$$

$$= O(1) \sum_{v=1}^{m} \frac{R_{v-1}^{k} |\Delta\lambda_r|^k}{r_v^k q_v^{k-1}} F_{v+1}^{k} \, |t_v|^k \sum_{n=v+1}^{m+1} \frac{\alpha_n^{\delta k+k-1} q_n^k}{Q_n^k Q_{n-1}}, \text{ by (6.52)}$$

$$= O(1) \sum_{v=1}^{m} v^{k-1} |t_v|^k \left(\frac{R_{v-1} F_{v+1} |\Delta\lambda_v|}{r_v Q_v} \right)^k$$

$$= O(1) \sum_{v=1}^{m} v^{k-1} |t_v|^k \text{ , using (6.56)}$$

$$= O(1), \ as\ m \to \infty.$$

Again,

$$\begin{aligned}
\sum_{n=2}^{m+1} \alpha_n^{\delta k+k-1} |T_{n,4}|^k &= \sum_{n=2}^{m+1} \alpha_n^{\delta k+k-1} \left| \frac{q_n}{Q_n Q_{n-1}} \frac{R_n \lambda_n f_n t_n}{r_n} \right|^k \\
&\leq \sum_{n=2}^{m+1} \alpha_n^{\delta k+k-1} |t_n|^k \left(\frac{q_n R_n F_n |\lambda_n|}{Q_n Q_{n-1} r_n} \right)^k \\
&= \sum_{n=2}^{m+1} \alpha_n^{k-1} |t_n|^k \left(\frac{q_n R_n F_n |\lambda_n| \alpha_n^{\delta}}{Q_n Q_{n-1} r_n} \right)^k \\
&= O(1) \sum_{n=2}^{m+1} \alpha_n^{k-1} |t_n|^k \text{ , using (6.56)} \\
&= O(1), \ as\ m \to \infty.
\end{aligned}$$

Next,

$$\begin{aligned}
\sum_{n=2}^{m+1} \alpha_n^{\delta k+k-1} |T_{n,5}|^k &= \sum_{n=2}^{m+1} \alpha_n^{\delta k+k-1} \left| \frac{p_0 q_0}{P_n Q_{n-1}} \sum_{v=1}^{n-1} \lambda_v \, t_v \right|^k \\
&\leq O(1) \sum_{n=2}^{m+1} \alpha_n^{\delta k+k-1} \frac{1}{P_n^k Q_{n-1}} \sum_{v=1}^{n-1} \frac{|\lambda_v|^k}{q_v^{k-1}} |t_v|^k \left(\frac{1}{Q_{n-1}} \sum_{v=1}^{n-1} q_v \right)^{k-1} \\
&= O(1) \sum_{v=1}^{m} \frac{|\lambda_v|^k |t_v|^k}{q_v^{k-1}} \sum_{n=v+1}^{m+1} \alpha_n^{\delta k+k-1} \frac{1}{P_n^k Q_{n-1}} \\
&= O(1) \sum_{v=1}^{m} \frac{|\lambda_v|^k |t_v|^k}{q_v^{k-1}} \sum_{n=v+1}^{m+1} \frac{\alpha_n^{\delta k+k-1} q_n^k}{Q_n^k Q_{n-1}}, \text{ using (6.51)} \\
&= O(1) \sum_{v=1}^{m} v^k \, |t_v|^k \left(\frac{|\lambda_n|^k}{Q_v} \right)^k \\
&= O(1) \sum_{v=1}^{m} v^k \, |t_v|^k \text{ , using (6.55)} \\
&= O(1), \ \ as\ m \to \infty.
\end{aligned}$$

Again,

$$\begin{aligned}
\sum_{n=2}^{m+1} \alpha_n{}^{\delta k+k-1} \left|T_{n,6}\right|^k &= \sum_{n=2}^{m+1} \alpha_n^{\delta k+k-1} \left| \frac{p_0 q_0}{P_n Q_{n-1}} \sum_{v=1}^{n-1} \frac{R_{v-1}}{r_v} (\Delta\lambda_v)\, t_v \right|^k \\
&\le O(1) \sum_{n=2}^{m+1} \alpha_n^{\delta k+k-1} \frac{1}{P_n^k Q_{n-1}} \sum_{v=1}^{n-1} \frac{R_{v-1}^k}{r_v^k q_v^{k-1}} |\Delta\lambda_v|^k\, |t_v|^k \left(\frac{1}{Q_{n-1}} \sum_{v=1}^{n-1} q_v \right)^{n-1} \\
&= O(1) \sum_{v=1}^{m} \frac{R_{v-1}^k |\Delta\lambda_v|^k |t_v|^k}{r_v^k q_v^{k-1}} \sum_{n=v+1}^{m+1} \alpha_n^{\delta k+k-1} \frac{1}{P_n^k Q_{n-1}} \\
&= O(1) \sum_{v=1}^{m} v^{k-1} |t_v|^k \left(\frac{R_{v-1} |\Delta\lambda_v|}{r_v Q_v} \right)^k \\
&= O(1) \sum_{v=1}^{m} v^{k-1} |t_v|^k, \text{ using } (6.58) \\
&= O(1), \; as\; m \to \infty.
\end{aligned}$$

Finally,

$$\begin{aligned}
\sum_{n=2}^{m+1} \alpha_n^{\delta k+k-1} |T_{n,7}|^k &= \sum_{n=2}^{m+1} \alpha_n^{\delta k+k-1} \left| \frac{p_0 q_0}{P_n Q_{n-1}} \frac{R_n \lambda_n t_n}{r_n} \right|^k \\
&= O(1) \sum_{n=2}^{m+1} \alpha_n^{\delta k+k-1} |t_n|^k \left(\frac{R_n |\lambda_n|}{P_n Q_{n-1} r_n} \right)^k \\
&= O(1) \sum_{n=2}^{m+1} \alpha_n^{\delta k+k-1} |t_n|^k \left(\frac{q_n R_n |\lambda_n|}{Q_n Q_{n-1} r_n} \right)^k \\
&= O(1) \sum_{n=2}^{m+1} \alpha_n^{k-1} |t_n|^k \left(\frac{q_n R_n |\lambda_n| \alpha_n^{\delta}}{Q_n Q_{n-1} r_n} \right)^k \\
&= O(1) \sum_{n=2}^{m+1} \alpha_n^{k-1} |t_n|^k, \text{ using } (6.59) \\
&= O(1), \; as\; m \to \infty.
\end{aligned}$$

This completes the Proof of the Theorem 6.5.

Proof of Theorem 6.6. In order to prove this theorem, using (6.76) and Minkowski's inequality, it is sufficient to show that

$$\sum_{n=1}^{\infty} \{f(\alpha_n)\}^k (\alpha_n)^{k-1} |T_{n,i}|^k < \infty, \text{ for } i = 1,2,3,4,5,6,7.$$

Now, on applying Holder's inequality, we have

$$
\begin{aligned}
&\sum_{n=2}^{m+1}\{f(\alpha_n)\}^k(\alpha_n)^{k-1}\,|T_{n,1}|^k \\
&= \sum_{n=2}^{m+1}\{f(\alpha_n)\}^k(\alpha_n)^{k-1}\left|\frac{q_n}{Q_nQ_{n-1}}\sum_{\nu=1}^{n-1}\lambda_\nu\,F_\nu\,t_\nu\right|^k \\
&\leq \sum_{n=2}^{m+1}\{f(\alpha_n)\}^k(\alpha_n)^{k-1}\frac{q_n^k}{Q_n^kQ_{n-1}}\sum_{\nu=1}^{n-1}\frac{|\lambda_\nu|^kF_\nu^k\,|t_\nu|^k}{q_\nu^{k-1}}\left(\frac{1}{Q_{n-1}}\sum_{\nu=1}^{n-1}q_\nu\right)^{k-1} \\
&= O(1)\sum_{\nu=1}^{m}\frac{1}{q_\nu^{k-1}}\,|\lambda_\nu|^k\;F_\nu^k\;|t_\nu|^k\sum_{n=\nu+1}^{m+1}\frac{\{f(\alpha_n)\}^k(\alpha_n)^{k-1}q_n^k}{Q_n^kQ_{n-1}} \\
&= O(1)\sum_{\nu-1}^{m}\frac{1}{q_\nu^{k-1}}\,|\lambda_\nu|^k\;F_\nu^k\;|t_\nu|^k\,\frac{(\nu q_\nu)^{k-1}}{Q_\nu^k},\ \text{using (6.64)} \\
&= O(1)\sum_{\nu=1}^{m}\nu^{k-1}\,|t_\nu|^k\left(\frac{|\lambda_\nu|\,F_\nu}{Q_\nu}\right)^k \\
&= O(1)\sum_{\nu=1}^{m}\nu^{k-1}\,|t_\nu|^k,\ \text{using (6.65)} \\
&= O(1),\ as\ m\to\infty.
\end{aligned}
$$

Next,

$$
\begin{aligned}
&\sum_{n=2}^{m+1}\{f(\alpha_n)\}^k(\alpha_n)^{k-1}\,|T_{n,2}|^k \\
&= \sum_{n=2}^{m+1}\{f(\alpha_n)\}^k(\alpha_n)^{k-1}\left|\frac{q_n}{Q_nQ_{n-1}}\sum_{\nu=1}^{n-1}\frac{R_{\nu-1}}{r_\nu}f_\nu\,\lambda_\nu\,t_\nu\right|^k \\
&\leq \sum_{n=2}^{m+1}\{f(\alpha_n)\}^k(\alpha_n)^{k-1}\frac{q_n^k}{Q_n^kQ_{n-1}}\sum_{\nu=1}^{n-1}\frac{R_\nu^kF_\nu^k\,|\lambda_\nu|^k}{q_\nu^{k-1}r_\nu^k}\left(\frac{1}{Q_{n-1}}\sum_{\nu=1}^{n-1}q_\nu\right)^{k-1} \\
&= O(1)\sum_{\nu=1}^{m}\frac{R_\nu^kF_\nu^k\,|\lambda_\nu|^k\,|t_\nu|^k}{q_\nu^{k-1}r_\nu^k}\sum_{n=\nu+1}^{m+1}\frac{\{f(\alpha_n)\}^k(\alpha_n)^{k-1}q_n^k}{Q_n^kQ_{n-1}} \\
&= O(1)\sum_{\nu=1}^{m}\nu^{k-1}\,|t_\nu|^k\left(\frac{R_\nu F_\nu\,|\lambda_\nu|}{r_\nu Q_\nu}\right)^k \\
&= O(1)\sum_{\nu=1}^{m}\nu^{k-1}\,|t_\nu|^k\ ,\ \text{using (6.67)} \\
&= O(1),\ as\ m\to\infty.
\end{aligned}
$$

Further,

$$\begin{aligned}
&\sum_{n=2}^{m+1}\{f(\alpha_n)\}^k(\alpha_n)^{k-1}\left|T_{n,3}\right|^k\\
&=\sum_{n=2}^{m+1}\{f(\alpha_n)\}^k(\alpha_n)^{k-1}\left|\frac{q_n}{Q_nQ_{n-1}}\sum_{v=1}^{n-1}\frac{R_{v-1}}{r_v}(\Delta\lambda_v)F_{v+1}t_v\right|^k\\
&\leq\sum_{n=2}^{m+1}\{f(\alpha_n)\}^k(\alpha_n)^{k-1}\frac{q_n^k}{Q_n^kQ_{n-1}}\sum_{v=1}^{n-1}\frac{R_{v-1}^k}{r_v^kq_v^{k-1}}\left|\Delta\lambda_v\right|^k F_{v+1}^k\left|t_v\right|^k\left(\frac{1}{Q_{n-1}}\sum_{v=1}^{n-1}q_v\right)^{k-1}\\
&=O(1)\sum_{v=1}^{m}\frac{R_{v-1}^k\left|\Delta\lambda_r\right|^k}{r_v^kq_v^{k-1}}F_{v+1}^k\left|t_v\right|^k\sum_{n=v+1}^{m+1}\frac{\{f(\alpha_n)\}^k(\alpha_n)^{k-1}q_n^k}{Q_n^kQ_{n-1}},\text{ by }(6.64)\\
&=O(1)\sum_{v=1}^{m}v^{k-1}\left|t_v\right|^k\left(\frac{R_{v-1}F_{v+1}\left|\Delta\lambda_v\right|}{r_vQ_v}\right)^k\\
&=O(1)\sum_{v=1}^{m}v^{k-1}\left|t_v\right|^k,\text{ using }(6.69)\\
&=O(1),\ \ as\ m\to\infty.
\end{aligned}$$

Again,

$$\begin{aligned}
&\sum_{n=2}^{m+1}\{f(\alpha_n)\}^k(\alpha_n)^{k-1}\left|T_{n,4}\right|^k\\
&=\sum_{n=2}^{m+1}\{f(\alpha_n)\}^k(\alpha_n)^{k-1}\left|\frac{q_n}{Q_nQ_{n-1}}\frac{R_n\lambda_nf_nt_n}{r_n}\right|^k\\
&\leq\sum_{n=2}^{m+1}\{f(\alpha_n)\}^k(\alpha_n)^{k-1}\left|t_n\right|^k\left(\frac{q_nR_nF_n\left|\lambda_n\right|}{Q_nQ_{n-1}r_n}\right)^k\\
&=\sum_{n=2}^{m+1}\{f(\alpha_n)\}^k(\alpha_n)^{k-1}\left|t_n\right|^k\left(\frac{q_nR_nF_n\left|\lambda_n\right|}{Q_nQ_{n-1}r_n}\right)^k\\
&=O(1)\sum_{n=2}^{m+1}\{f(\alpha_n)\}^k(\alpha_n)^{k-1}\left|t_n\right|^k,\text{ using }(6.68)\\
&=O(1),\ as\ m\to\infty.
\end{aligned}$$

Next,

$$\begin{aligned}
&\sum_{n=2}^{m+1}\{f(\alpha_n)\}^k(\alpha_n)^{k-1}\left|T_{n,5}\right|^k\\
&=\sum_{n=2}^{m+1}\{f(\alpha_n)\}^k(\alpha_n)^{k-1}\left|\frac{p_0q_0}{P_nQ_{n-1}}\sum_{v=1}^{n-1}\lambda_v\,t_v\right|^k
\end{aligned}$$

$$\leq O(1)\sum_{n=2}^{m+1}\{f(\alpha_n)\}^k(\alpha_n)^{k-1}\frac{1}{P_n^kQ_{n-1}}\sum_{v=1}^{n-1}\frac{|\lambda_v|^k}{q_v^{k-1}}|t_v|^k\left(\frac{1}{Q_{n-1}}\sum_{v=1}^{n-1}q_v\right)^{k-1}$$

$$= O(1)\sum_{v=1}^{m}\frac{|\lambda_v|^k|t_v|^k}{q_v^{k-1}}\sum_{n=v+1}^{m+1}\frac{\{f(\alpha_n)\}^k(\alpha_n)^{k-1}}{P_n^kQ_{n-1}}$$

$$= O(1)\sum_{v=1}^{m}\frac{|\lambda_v|^k|t_v|^k}{q_v^{k-1}}\sum_{n=v+1}^{m+1}\frac{\{f(\alpha_n)\}^k(\alpha_n)^{k-1}q_n^k}{Q_n^kQ_{n-1}}, \text{ using (6.63)}$$

$$= O(1)\sum_{v=1}^{m}v^k\ |t_v|^k\left(\frac{|\lambda_n|^k}{Q_v}\right)^k$$

$$= O(1)\sum_{v=1}^{m}v^k\ |t_v|^k, \text{ using (6.68)}$$

$$= O(1),\ as\ m\to\infty.$$

Again,

$$\sum_{n=2}^{m+1}\{f(\alpha_n)\}^k(\alpha_n)^{k-1}\left|T_{n,6}\right|^k$$

$$= \sum_{n=2}^{m+1}\{f(\alpha_n)\}^k(\alpha_n)^{k-1}\left|\frac{p_0q_0}{P_nQ_{n-1}}\sum_{v=1}^{n-1}\frac{R_{v-1}}{r_v}(\Delta\lambda_v)\ t_v\right|^k$$

$$\leq O(1)\sum_{n=2}^{m+1}\{f(\alpha_n)\}^k(\alpha_n)^{k-1}\frac{1}{P_n^kQ_{n-1}}\sum_{v=1}^{n-1}\frac{R_{v-1}^k}{r_v^kq_v^{k-1}}|\Delta\lambda_v|^k\ |t_v|^k\left(\frac{1}{Q_{n-1}}\sum_{v=1}^{n-1}q_v\right)^{n-1}$$

$$= O(1)\sum_{v=1}^{m}\frac{R_{v-1}^k|\Delta\lambda_v|^k|t_v|^k}{r_v^kq_v^{k-1}}\sum_{n=v+1}^{m+1}\frac{\{f(\alpha_n)\}^k(\alpha_n)^{k-1}}{P_n^kQ_{n-1}}$$

$$= O(1)\sum_{v=1}^{m}v^{k-1}\,|t_v|^k\left(\frac{R_{v-1}|\Delta\lambda_v|}{r_vQ_v}\right)^k$$

$$= O(1)\sum_{v=1}^{m}v^{k-1}\,|t_v|^k, \text{ using (6.70)}$$

$$= O(1),\ as\ m\to\infty.$$

Finally,

$$\sum_{n=2}^{m+1}\{f(\alpha_n)\}^k(\alpha_n)^{k-1}|T_{n,7}|^k$$

$$= \sum_{n=2}^{m+1}\{f(\alpha_n)\}^k(\alpha_n)^{k-1}\left|\frac{p_0q_0}{P_nQ_{n-1}}\frac{R_n\lambda_nt_n}{r_n}\right|^k$$

$$= O(1)\sum_{n=2}^{m+1}\{f(\alpha_n)\}^k(\alpha_n)^{k-1}|t_n|^k\left(\frac{R_n|\lambda_n|}{P_nQ_{n-1}r_n}\right)^k$$

$$
\begin{aligned}
&= O(1)\sum_{n=2}^{m+1}\{f(\alpha_n)\}^k(\alpha_n)^{k-1}|t_n|^k\left(\frac{q_nR_n|\lambda_n|}{Q_nQ_{n-1}r_n}\right)^k\\
&= O(1)\sum_{n=2}^{m+1}\{f(\alpha_n)\}^k(\alpha_n)^{k-1}|t_n|^k\left(\frac{q_nR_n|\lambda_n|}{Q_nQ_{n-1}r_n}\right)^k\\
&= O(1)\sum_{n=2}^{m+1}\{f(\alpha_n)\}^k(\alpha_n)^{k-1}|t_n|^k, \text{ using } (6.71)\\
&= O(1), \text{ as } m\to\infty.
\end{aligned}
$$

This completes the Proof of the Theorem 6.6.

6.5 CONCLUSION

From the above results and discussions, we are in a conclusion that our results are more generalized and in particular, these generalize the results of Sulaiman [9], Paikray et al. [5], and Das [2]. We have also observed that the sufficient conditions for $\sum a_n\lambda_n$ using the absolute indexed product summability $|(N,q_n)(N,p_n),\alpha_n;\delta|_k$ with $k\geq 1, 1\geq k\delta\geq 0$ and $|(N,q_n)\ (N,p_n),\ \alpha_n;f|_k,\ k\geq 1$ generalize the sufficient conditions for $\sum a_n\lambda_n$ using the absolute indexed product summability $|(N,q_n)\ (N,p_n),\ \alpha_n|_k,\ k\geq 1$. Similarly, the sufficient conditions for $\sum a_n\lambda_n$ using the absolute indexed product summabilities $|(N,q_n)\ (N,p_n);\delta|_k$, $k\geq 1, 1\geq k\delta\geq 0$, and $|(N,q_n)\ (N,p_n),\alpha_n|_k,\ k\geq 1$ generalize the sufficient conditions for $\sum a_n\lambda_n$ using the absolute indexed product summability $|(N,q_n)(N,p_n)|_k$.

As a future scope of this work, one may approach in the similar way to find the sufficient conditions for

- $\sum a_n\lambda_n$ is summable $|(N,q_n)(N,p_n),\alpha_n,\delta,\mu|_k$ if $\sum a_n$ is $|R,r_n|_k$-summable, where μ is a real number.
- $\sum a_n\lambda_n$ is summable $|(R,q_n)(R,p_n),\delta|_k, |(R,q_n)(R,p_n),\alpha_n|_k$, $|(R,q_n)(R,p_n),\alpha_n;\delta|_k$ and $|(R,q_n)(R,p_n),\alpha_n,f|_k$ if $\sum a_n$ is $|R,r_n|_k$-summable.
- $\sum a_n\lambda_n$ is summable $|(R,q_n)(C,1),\delta|_k, |(C,1)(R,p_n),\delta|_k, |(R,q_n)(C,1),\alpha_n|_k$, $|(C,1)(R,p_n),\alpha_n|_k$, $|(C,1)(R,p_n),\alpha_n;\delta|_k, |(R,q_n)(C,1),\alpha_n;\delta|_k$ if $\sum a_n$ is $|C,1|_k$-summable.

REFERENCES

[1] Aasma, A., Dutta, H. and Natarajan, P. N., *An Introductory Course in Summability Theory,* First Edition, John Wiley & Sons, Inc., USA, (2017).

[2] Das, G., Tauberian theorems for absolute Norlund summability, *Proceedings of the London Mathematical Society,* Vol. 19(2), (1969), 357–384.

[3] Dutta, H. and Rhoades, B.E. (Eds.), *Current Topics in Summability Theory and Applications,* First Edition, Springer, Singapore, (2016).

[4] Padhy, B.P., Misra, U.K. and Misra, M., *Summability Methods and its Applications,* Lap Lambart Academic Publications, Germany, (2012).

[5] Paikray, S.K., Misra, U.K. and Sahoo, N.C., Product Summability of an Infinite Series, *International Journal of Computer and Mathematical Sciences,* Vol-1(7), (2010), 853–863.

[6] Parameswaran, M.R., Some product theorems in summability, *Mathematische Zeitscher,* Vol. 68, (1957), 19–26.

[7] Rajgopal, C.T., Theorems on product of two summability methods, The *Journal of Indian Mathematical Society,* Vol. 18(1), (1954) .

[8] Ramanujan, M.S., On products of summability methods, *Mathematische Zeitscher,* Vol. 69(1), (1958), 423–428.

[9] Sulaiman, W.T., A Note on product summability of an infinite series, *International Journal of Mathematical Sciences,* Hindawi publishing corporation, (2008), Article ID 372604.

[10] Szasz, O., On products of summability methods, *Proceedings of American Mathematical Society,* Vol. 3(2), (1952).

7 On Some Important Inequalities

Zlatko Pavić
University of Osijek

CONTENTS

7.1 CONCEPTS OF AFFINITY AND CONVEXITY

7.1.1 AFFINE AND CONVEX SETS AND FUNCTIONS

Throughout this chapter, we will use a vector space $\mathbb{X}$ over the field of real numbers $\mathbb{R}$. We think about a binomial linear combination of points $x_1, x_2 \in \mathbb{X}$ and coefficients $\lambda_1, \lambda_2 \in \mathbb{R}$ as the sum

$$\lambda_1 x_1 + \lambda_2 x_2. \tag{7.1}$$

We briefly say that the above sum is a linear combination of points x_1 and x_2.

A linear combination in formula (7.1) is said to be affine if $\lambda_1 + \lambda_2 = 1$. A set $A \subseteq \mathbb{X}$ is said to be affine if it contains each affine combination of each pair of its points. A function $h : A \to \mathbb{R}$ is said to be affine if the equality

$$h(\lambda_1 x_1 + \lambda_2 x_2) = \lambda_1 h(x_1) + \lambda_2 h(x_2) \tag{7.2}$$

holds for each affine combination $\lambda_1 x_1 + \lambda_2 x_2$ of each pair of points $x_1, x_2 \in A$.

A linear combination in formula (7.1) is said to be convex if $\lambda_1 + \lambda_2 = 1$ and $\lambda_1, \lambda_2 \geq 0$. Really, we have that $\lambda_1, \lambda_2 \in [0,1]$. A set $C \subseteq \mathbb{X}$ is said to be convex if it contains each convex combination of each pair of its points. A function $f : C \to \mathbb{R}$ is said to be convex if the inequality

$$f(\lambda_1 x_1 + \lambda_2 x_2) \leq \lambda_1 f(x_1) + \lambda_2 f(x_2) \tag{7.3}$$

holds for each convex combination $\lambda_1 x_1 + \lambda_2 x_2$ of each pair of points $x_1, x_2 \in C$. In relation to this, a function f is said to be concave if $-f$ is convex.

A convex combination is affine. An affine set is convex. The empty set and singleton are affine and convex.

Let $S \subseteq \mathbb{X}$ be a set. The affine (convex) hull aff$S$ (conv$S$) of the set S is defined as the set containing each binomial affine (convex) combination of each pair of points from S. The set aff$S$ (conv$S$) is the smallest affine (convex) set containing S.

Let $n \geq 2$ be an integer. By using the method of mathematical induction, it can be demonstrated that the set affinity (convexity), function affinity (convexity), and affine (convex) hull apply to n-member affine (convex) combinations.

Comprehensive presentation of convex sets, convex functions, and their inequalities can be found in books [12,18].

7.1.2 EFFECT OF AFFINE AND CONVEX COMBINATIONS IN $\mathbb{R}^n$

Let a and b be the distinct points in the line $\mathbb{R}$. Then, each point $x \in \mathbb{R}$ can be represented by the affine combination of points a and b as

$$x = \alpha(x)a + \beta(x)b, \tag{7.4}$$

where

$$\alpha(x) = \frac{x-b}{a-b} = \frac{\begin{vmatrix} x & 1 \\ b & 1 \end{vmatrix}}{\begin{vmatrix} a & 1 \\ b & 1 \end{vmatrix}}, \quad \beta(x) = \frac{a-x}{a-b} = \frac{\begin{vmatrix} a & 1 \\ x & 1 \end{vmatrix}}{\begin{vmatrix} a & 1 \\ b & 1 \end{vmatrix}}. \tag{7.5}$$

The convex combinations in formula (7.4) accentuating points x with $\alpha(x) \geq 0$ and $\beta(x) \geq 0$ delineate the closed interval with endpoints a and b as the set

$$\Delta_{ab} = \operatorname{conv}\{a,b\} = \{\overline{\alpha}a + \overline{\beta}b : \overline{\alpha}, \overline{\beta} \geq 0, \overline{\alpha} + \overline{\beta} = 1\}. \tag{7.6}$$

Let $a = (a_1, a_2)$, $b = (b_1, b_2)$, and $c = (c_1, c_2)$ be the noncollinear points in the plane $\mathbb{R}^2$. Then, each point $x = (x_1, x_2) \in \mathbb{R}^2$ can be represented by the affine combination of points a, b, and c as

$$x = \alpha(x)a + \beta(x)b + \gamma(x)c, \tag{7.7}$$

where

$$\alpha(x)=\frac{\begin{vmatrix} x_1 & x_2 & 1\\ b_1 & b_2 & 1\\ c_1 & c_2 & 1\end{vmatrix}}{\begin{vmatrix} a_1 & a_2 & 1\\ b_1 & b_2 & 1\\ c_1 & c_2 & 1\end{vmatrix}},\ \beta(x)=\frac{\begin{vmatrix} a_1 & a_2 & 1\\ x_1 & x_2 & 1\\ c_2 & c_2 & 1\end{vmatrix}}{\begin{vmatrix} a_1 & a_2 & 1\\ b_1 & b_2 & 1\\ c_1 & c_2 & 1\end{vmatrix}},\ \gamma(x)=\frac{\begin{vmatrix} a_1 & a_2 & 1\\ b_1 & b_2 & 1\\ x_1 & x_2 & 1\end{vmatrix}}{\begin{vmatrix} a_1 & a_2 & 1\\ b_1 & b_2 & 1\\ c_1 & c_2 & 1\end{vmatrix}}. \tag{7.8}$$

The convex combinations in formula (7.7) emphasizing points x with $\alpha(x)\geq 0$, $\beta(x)\geq 0$, and $\gamma(x)\geq 0$ designate the triangle with vertices a, b, and c as the set

$$\Delta_{abc}=\operatorname{conv}\{a,b,c\}=\{\overline{\alpha}a+\overline{\beta}b+\overline{\gamma}c:\overline{\alpha},\overline{\beta},\overline{\gamma}\geq 0,\ \overline{\alpha}+\overline{\beta}+\overline{\gamma}=1\}. \tag{7.9}$$

Let $a_0=(a_{01},\ldots,a_{0n}),\ldots,a_n=(a_{n1},\ldots,a_{nn})$ be the points in the space $\mathbb{R}^n$ such that differences $a_1-a_0,\ldots,a_n-a_0$ are linearly independent. Then, each point $x=(x_1,\ldots,x_n)\in\mathbb{R}^n$ can be represented by the affine combination of points a_i as

$$x=\sum_{i=0}^{n}\alpha_i(x)a_i, \tag{7.10}$$

where

$$\alpha_i(x)=\begin{vmatrix} a_{01} & \cdots & a_{0n} & 1\\ \vdots & \ddots & \vdots & \vdots\\ x_1 & \cdots & x_n & 1\\ \vdots & \ddots & \vdots & \vdots\\ a_{n1} & \cdots & a_{nn} & 1\end{vmatrix}\cdot\begin{vmatrix} a_{01} & \cdots & a_{0n} & 1\\ \vdots & \ddots & \vdots & \vdots\\ a_{i1} & \cdots & a_{in} & 1\\ \vdots & \ddots & \vdots & \vdots\\ a_{n1} & \cdots & a_{nn} & 1\end{vmatrix}^{-1}. \tag{7.11}$$

Since the differences $a_1-a_0,\ldots,a_n-a_0$ are linearly independent, the affine combination in formula (7.10) is unique for each $x\in\mathbb{R}^n$. The affine hull of the set of simplex vertices covers the space $\mathbb{R}^n$, i.e., $\mathbb{A}_{a_0\ldots a_n}=\operatorname{aff}\{a_0,\ldots,a_n\}=\mathbb{R}^n$.

The convex combinations in formula (7.10) highlighting points x with all $\alpha_i(x)\geq 0$ appoint the n-simplex with vertices a_i as the set

$$\Delta_{a_0\ldots a_n}=\operatorname{conv}\{a_0,\ldots,a_n\}=\left\{\textstyle\sum_{i=0}^{n}\overline{\alpha}_i a_i:\overline{\alpha}_i\geq 0,\ \sum_{i=0}^{n}\overline{\alpha}_i=1\right\}. \tag{7.12}$$

7.1.3 COEFFICIENTS OF AFFINE AND CONVEX COMBINATIONS

Let $a_0,\ldots,a_n\in\mathbb{R}^n$ be the points such that $a_1-a_0,\ldots,a_n-a_0$ are linearly independent. The existence and uniqueness of the representation of point $x\in\mathbb{R}^n$ by the affine combination of points $a_0,\ldots,a_n$ can be demonstrated in two simple steps. We first represent the difference $x-a_0$ by the unique linear combination

$$x-a_0=\sum_{i=1}^{n}\lambda_i(a_i-a_0),$$

and then express the point x, which yields the affine combination

$$x = \sum_{i=1}^{n} \lambda_i a_i + \left(1 - \sum_{i=1}^{n} \lambda_i\right) a_0.$$

We want to verify the coefficients in formula (7.11). According to the above consideration, we can suppose that

$$x = \sum_{j=0}^{n} \alpha_j(x) a_j = \left(\sum_{j=0}^{n} \alpha_j(x) a_{j1}, \ldots, \sum_{j=0}^{n} \alpha_j(x) a_{jn}\right) \tag{7.13}$$

with $\sum_{j=0}^{n} \alpha_j(x) = 1$. We still need the determinants

$$A_i(x) = \begin{vmatrix} a_{01} & \ldots & a_{0n} & 1 \\ \vdots & \ddots & \vdots & \vdots \\ x_1 & \ldots & x_n & 1 \\ \vdots & \ddots & \vdots & \vdots \\ a_{n1} & \ldots & a_{nn} & 1 \end{vmatrix}, \quad A = \begin{vmatrix} a_{01} & \ldots & a_{0n} & 1 \\ \vdots & \ddots & \vdots & \vdots \\ a_{i1} & \ldots & a_{in} & 1 \\ \vdots & \ddots & \vdots & \vdots \\ a_{n1} & \ldots & a_{nn} & 1 \end{vmatrix}. \tag{7.14}$$

By exposing the ith row of $A_i(x)$ with coordinates of x in formula (7.13), and $\sum_{j=0}^{n} \alpha_j(x)$ instead of 1, we induce

$$A_i(x) = \begin{vmatrix} a_{01} & \ldots & a_{0n} & 1 \\ \vdots & \ddots & \vdots & \vdots \\ \sum_{j=0}^{n} \alpha_j(x) a_{j1} & \ldots & \sum_{j=0}^{n} \alpha_j(x) a_{jn} & \sum_{j=0}^{n} \alpha_j(x) \\ \vdots & \ddots & \vdots & \vdots \\ a_{n1} & \ldots & a_{nn} & 1 \end{vmatrix}.$$

After decomposition into $n+1$ summands by the ith row, it follows that

$$A_i(x) = \sum_{j=0}^{n} \alpha_j(x) \begin{vmatrix} a_{01} & \ldots & a_{0n} & 1 \\ \vdots & \ddots & \vdots & \vdots \\ a_{j1} & \ldots & a_{jn} & 1 \\ \vdots & \ddots & \vdots & \vdots \\ a_{n1} & \ldots & a_{nn} & 1 \end{vmatrix} = \alpha_i(x) \begin{vmatrix} a_{01} & \ldots & a_{0n} & 1 \\ \vdots & \ddots & \vdots & \vdots \\ a_{i1} & \ldots & a_{in} & 1 \\ \vdots & \ddots & \vdots & \vdots \\ a_{n1} & \ldots & a_{nn} & 1 \end{vmatrix} = \alpha_i(x) A,$$

and thus, we get the expression $\alpha_i(x) = A_i(x)/A$ corresponding to formula (7.11).

The coefficients $\alpha_i(x)$ have a geometric meaning. To present it as simple as possible, we consider the simplex $\Delta_{a_0 \ldots a_{i-1} x a_{i+1} \ldots a_n}$ as a fictive n-simplex. If x belongs to the facet $\Delta_{a_0 \ldots a_{i-1} a_{i+1} \ldots a_n}$, we assume that $\Delta_{a_0 \ldots a_{i-1} x a_{i+1} \ldots a_n} = \Delta_{a_0 \ldots a_{i-1} a_{i+1} \ldots a_n}$. In this case, the coefficient $\alpha_i(x)$ and the n-volume of the facet are equal to zero. As for the coefficients' geometric meaning, since

$$A_i(x) = \pm n! \operatorname{vol}_n(\Delta_{a_0 \ldots a_{i-1} x a_{i+1} \ldots a_n}), \quad A = \pm n! \operatorname{vol}_n(\Delta_{a_0 \ldots a_n}), \tag{7.15}$$

it follows that

$$\alpha_i(x) = \pm\frac{\mathrm{vol}_n(\Delta_{a_0\ldots a_{i-1}xa_{i+1}\ldots a_n})}{\mathrm{vol}_n(\Delta_{a_0\ldots a_n})}. \tag{7.16}$$

The sign in formula (7.16) depends on the orientation in the space $\mathbb{R}^n$, which refers to the ordered $(n+1)$-tuples

$$(a_0,\ldots,a_{i-1},x,a_{i+1},\ldots,a_n) \text{ and } (a_0,\ldots,a_n).$$

The plus sign is related to the same orientation, and the minus sign to the opposite. If we consider only the points x belonging to the n-simplex $\Delta_{a_0\ldots a_n}$, then the above ordered $(n+1)$-tuples have the same orientation and the coefficients $\alpha_i(x)$ are nonnegative. In that case, the plus sign obviously stands in formula (7.16).

The coefficients $\alpha_i(x)$ in formula (7.11) indicate functions $\alpha_i : \mathbb{R}^n \to \mathbb{R}$. After expanding the numerator determinant $A_i(x)$ using the ith row containing coordinates of x, and rearranging, we reach the concise representation

$$\alpha_i(x_1,\ldots,x_n) = \alpha_{i0} + \sum_{j=1}^{n} \alpha_{ij}x_j, \tag{7.17}$$

where α_{i0} and α_{ij} are the constants as the ratios of determinants without coordinates of x. Following the above formula, it is easy to show that functions α_i are affine.

Let us look at the representation of an affine function $h : \mathbb{R}^n \to \mathbb{R}$. By combining the affinity of h with representations in formulas (7.10) and (7.17), we get

$$\begin{aligned} h(x_1,\ldots,x_n) &= \sum_{i=0}^{n} \alpha_i(x_1,\ldots,x_n)h(a_i) \\ &= \sum_{i=0}^{n}\Big(\alpha_{i0} + \sum_{j=1}^{n}\alpha_{ij}x_j\Big)h(a_i) \\ &= \kappa_0 + \kappa_1x_1 + \cdots + \kappa_nx_n, \end{aligned} \tag{7.18}$$

where $\kappa_0,\ldots,\kappa_n$ are the constants. The above representation is usually called the standard form of h. It is not difficult to prove the uniqueness of this form.

7.1.4 SUPPORT AND SECANT HYPERPLANES

Let $C \subseteq \mathbb{R}^n$ be a convex set with the nonempty interior, let c be an interior point of C, and let $f : C \to \mathbb{R}$ be a convex function. Then, the function f admits an affine function $h_1 : \mathbb{R}^n \to \mathbb{R}$, which meets the equality $h_1(c) = f(c)$, and for all $x \in C$ satisfies the inequality

$$h_1(x) \le f(x). \tag{7.19}$$

In the space $\mathbb{R}^{n+1}$, we can imagine that the graph of h_1 supports the graph of f at the point $F = (c, f(c))$. Accordingly, h_1 is called the support hyperplane of f at c. The support hyperplane of f at c is not necessarily unique. Formula (7.19) can be called the support hyperplane inequality. The existence of support hyperplanes arises from the separating and supporting hyperplane theorems.

Let $\Delta_{a_0 \dots a_n} \subset \mathbb{R}^n$ be an n-simplex, and let $f : \Delta_{a_0 \dots a_n} \to \mathbb{R}$ be a convex function. Then, the function f goes with the affine function $h_2 : \mathbb{R}^n \to \mathbb{R}$ determined by the equalities $f(a_0) = h_2(a_0), \dots, f(a_n) = h_2(a_n)$. It follows that the function h_2 for all $x \in \Delta_{a_0 \dots a_n}$ satisfies the inequality:

$$f(x) \leq h_2(x). \tag{7.20}$$

We can visualize that the graph of h_2 is a roof of the graph of f fixed at the points $F_0 = (a_0, f(a_0)), \dots, F_n = (a_n, f(a_n))$. Suitably, h_2 is called the secant hyperplane of f at vertices, and formula (7.20) can be called the secant hyperplane inequality. If we include c as an interior point of $\Delta_{a_0 \dots a_n}$ and h_1 as a support hyperplane of f at c, then for all $x \in \Delta_{a_0 \dots a_n}$, we have the support-secant hyperplane inequality:

$$h_1(x) \leq f(x) \leq h_2(x). \tag{7.21}$$

The equation of the secant hyperplane can be easily determined. By applying the affinity of h_2 to the affine combination in formula (7.10), and using the coincidences with f, we get the secant hyperplane equation:

$$h_2(x) = \sum_{i=0}^{n} \alpha_i(x) f(a_i). \tag{7.22}$$

If $n = 1$ ($n = 2$), we use terms "support" and "secant lines" (planes).

7.2 THE JENSEN INEQUALITY

The Jensen inequality is the most important. With more or less effort, almost all important and influential inequalities can be derived from Jensen's.

7.2.1 DISCRETE AND INTEGRAL FORMS OF THE JENSEN INEQUALITY

Let $\mathbb{X}$ be a real vector space, and let $\sum_{i=1}^{n} \lambda_i x_i$ be a linear combination of points $x_i \in \mathbb{X}$ such that $\lambda = \sum_{i=1}^{n} \lambda_i \neq 0$. The center of the given combination is defined as the point $\overline{x} \in \mathbb{X}$ satisfying the discrete equation

$$\sum_{i=1}^{n} \lambda_i (x_i - \overline{x}) = 0. \tag{7.23}$$

It follows that

$$\overline{x} = \sum_{i=1}^{n} \frac{\lambda_i}{\lambda} x_i. \tag{7.24}$$

So the center $\overline{x}$ is the affine combination of points x_i and coefficients $\kappa_i = \lambda_i / \lambda$. Accordingly, if the given combination is affine, then the combination center coincides with the combination itself, respectively, $\overline{x} = \sum_{i=1}^{n} \lambda_i x_i$. This is especially true for convex combinations.

Discrete form of Jensen's inequality includes a real vector space, convex combination with its center, and convex function. Regarding the strict form, we present the following theorem.

Theorem 7.1. *Let* $\mathbb{X}$ *be a real vector space, let* $\sum_{i=1}^{n} \lambda_i x_i$ *be a convex combination of points* $x_i \in \mathbb{X}$, *let* $C \subseteq \mathbb{X}$ *be a convex set containing* $\{x_1, \ldots, x_n\}$, *and let* $f : C \to \mathbb{R}$ *be a convex function.*

Then, we have the discrete inequality

$$f\left(\sum_{i=1}^{n} \lambda_i x_i\right) \le \sum_{i=1}^{n} \lambda_i f(x_i). \tag{7.25}$$

Proof. If $n = 1$, the trivial inequality $f(x_1) \le f(x_1)$ represents formula (7.25) regardless of the convexity of f. If $n \ge 2$, the method of mathematical induction enables us to perform formula (7.25) as follows.

The case $n = 2$ is regarded as the induction base, wherein the convexity of f provides formula (7.25).

The case $n > 2$ is considered as the induction step, wherein $\lambda_1 < 1$ is assumed without a reduction of generality. The induction base and premise are included through the next procedure. The right side of the representation

$$\sum_{i=1}^{n} \lambda_i x_i = \lambda_1 x_1 + (1 - \lambda_1) \sum_{i=2}^{n} \frac{\lambda_i}{1 - \lambda_1} x_i$$

can be regarded as a binomial convex combination. The induction base can be applied to the right side, and so gain the inequality

$$f\left(\sum_{i=1}^{n} \lambda_i x_i\right) \le \lambda_1 f(x_1) + (1 - \lambda_1) f\left(\sum_{i=2}^{n} \frac{\lambda_i}{1 - \lambda_1} x_i\right).$$

The induction premise can be applied to the $(n-1)$-member convex combination under the function f, and so obtain the inequality

$$f\left(\sum_{i=2}^{n} \frac{\lambda_i}{1 - \lambda_1} x_i\right) \le \sum_{i=2}^{n} \frac{\lambda_i}{1 - \lambda_1} f(x_i).$$

The coupling of the above two inequalities gives formula (7.25). □

Let X be a set with a nonnegative measure μ such that $\mu(X) > 0$, and let $g : X \to \mathbb{R}$ be a μ-integrable function. The μ-integral arithmetic mean of the function g is defined as the number $\overline{g} \in \mathbb{R}$, which satisfies the integral equation

$$\int_X (g(x) - \overline{g})\, d\mu(x) = 0. \tag{7.26}$$

A simple calculation gives

$$\overline{g} = \frac{\int_X g(x)\, d\mu(x)}{\mu(X)}. \tag{7.27}$$

If we include an affine function $h : \mathbb{R} \to \mathbb{R}$, then we have the integral equality

$$h\left(\frac{\int_X g(x)\, d\mu(x)}{\mu(X)}\right) = \frac{\int_X h(g(x))\, d\mu(x)}{\mu(X)}. \tag{7.28}$$

This is easy to prove if we use the standard representation $h(x) = \kappa_0 + \kappa_1 x$. The question is what changes in formula (7.28) if we replace the affine function h with a convex function f. This leads to Jensen's inequality.

Integral form of Jensen's inequality includes a measurable set X, integrable function $g : X \to \mathbb{R}$ with its integral arithmetic mean $\overline{g}$, and convex function f such that the composition $f(g) : X \to \mathbb{R}$ is integrable. The next appropriate lemma and theorem provide the strict form.

Lemma 7.1. *Let X be a set with a nonnegative measure μ such that $\mu(X) > 0$, and let $g : X \to \mathbb{R}$ be a μ-integrable function.*

Then, the μ-integral arithmetic mean of g is in the convex hull of the image of g.

Proof. Let $\overline{g}$ be the μ-integral arithmetic mean of g, and let $I \subseteq \mathbb{R}$ be an interval containing the image of g. Relying on proof by contradiction, we will suppose that $\overline{g}$ does not belong to I. Then, either $g(x) - \overline{g} > 0$ or $g(x) - \overline{g} < 0$ for all $x \in X$. Consequently, it implies either $\int_X (g(x) - \overline{g})\,d\mu(x) > 0$ or $\int_X (g(x) - \overline{g})\,d\mu(x) < 0$. Thus, $\overline{g}$ does not represent the μ-integral arithmetic mean of g.

The conclusion is that $\overline{g}$ belongs to each interval containing the image of g. This must be true even for the smallest one, the convex hull of the image of g. □

Theorem 7.2. *Let X be a set with a nonnegative measure μ such that $\mu(X) > 0$, let $g : X \to \mathbb{R}$ be a μ-integrable function, let $I \subseteq \mathbb{R}$ be an interval containing the image of g, and let $f : I \to \mathbb{R}$ be a convex function such that $f(g)$ is μ-integrable.*

Then, we have the integral inequality

$$f\left(\frac{\int_X g(x)\,d\mu(x)}{\mu(X)}\right) \le \frac{\int_X f(g(x))\,d\mu(x)}{\mu(X)}. \tag{7.29}$$

Proof. The mean $\overline{g} = \int_X g(x)\,d\mu(x)/\mu(X)$ belongs to the interval I by Lemma 7.1. Thus, $\overline{g}$ is either an interior or boundary point of I.

If $\overline{g}$ is an interior point of I, then f as a convex function admits a support line h at $\overline{g}$. Thus, $h(\overline{g}) = f(\overline{g})$, and for all $x \in X$, we have the inequality

$$h(g(x)) \le f(g(x)). \tag{7.30}$$

By using the coincidence $f(\overline{g}) = h(\overline{g})$, affinity equality in formula (7.28), and support line inequality in formula (7.30), we obtain the multiple relation

$$\begin{aligned} f\left(\frac{\int_X g(x)\,d\mu(x)}{\mu(X)}\right) &= h\left(\frac{\int_X g(x)\,d\mu(x)}{\mu(X)}\right) \\ &= \frac{\int_X h(g(x))\,d\mu(x)}{\mu(X)} \\ &\le \frac{\int_X f(g(x))\,d\mu(x)}{\mu(X)}, \end{aligned}$$

covering the inequality in formula (7.29).

If $\overline{g}$ is a boundary point of I, then either $g(x)-\overline{g}\geq 0$ or $g(x)-\overline{g}\leq 0$ for all $x\in X$. Integral equality in formula (7.26) implies $g(x)-\overline{g}=0$, and so $g(x)=\overline{g}$ for μ-almost all $x\in X$. The trivial inequality $f(\overline{g})\leq f(\overline{g})$ represents formula (7.29). □

Versions of the discrete and integrals forms of the inequalities in formulas (7.25) and (7.29) were proven by Jensen, the discrete version in [6], and integral version in [7]. In the period 1905–1906, Jensen also defined a convex function.

7.2.2 GENERALIZATIONS OF THE JENSEN INEQUALITY

For the sake of formulae conciseness, we will sometimes omit the variable x in integral expressions. This is the case in Theorem 7.3.

Let $g_1,\dots,g_n : X\to\mathbb{R}$ be μ-integrable functions. The μ-integral arithmetic mean of the mapping $g=(g_1,\dots,g_n) : X\to\mathbb{R}^n$ can be defined as the point $\overline{g}=(\overline{g}_1,\dots,\overline{g}_n)\in\mathbb{R}^n$ whose coordinates satisfy the integral equations

$$\int_X (g_1(x)-\overline{g}_1)\,d\mu(x)=0,\dots,\int_X (g_n(x)-\overline{g}_n)\,d\mu(x)=0. \tag{7.31}$$

Then, it follows that

$$\overline{g}=\left(\frac{\int_X g_1(x)\,d\mu(x)}{\mu(X)},\dots,\frac{\int_X g_n(x)\,d\mu(x)}{\mu(X)}\right). \tag{7.32}$$

If we include an affine function $h:\mathbb{R}^n\to\mathbb{R}$, then we have the integral equality

$$h\left(\frac{\int_X g_1(x)d\mu(x)}{\mu(X)},\dots,\frac{\int_X g_n(x)d\mu(x)}{\mu(X)}\right)=\frac{\int_X h(g_1(x),\dots,g_n(x))\,d\mu(x)}{\mu(X)} \tag{7.33}$$

due to the representation $h(t_1,\dots,t_n)=\kappa_0+\kappa_1 t_1+\cdots+\kappa_n t_n$.

Lemma 7.2. *Let X be a set with a nonnegative measure μ such that $\mu(X)>0$, let $g_1,\dots,g_n : X\to\mathbb{R}$ be μ-integrable functions, and let $g=(g_1,\dots,g_n) : X\to\mathbb{R}^n$ be the corresponding $\mathbb{R}^n$-valued mapping.*

Then, the μ-integral arithmetic mean of the mapping g is in the Cartesian product of the convex hulls of the images of $g_1,\dots,g_n$.

Proof. Let $I_1,\dots,I_n$ be the convex hulls of images of $g_1,\dots,g_n$, respectively. Since each $\overline{g}_i\in I_i$ by Lemma 7.1, it follows that $\overline{g}=(\overline{g}_1,\dots,\overline{g}_n)\in I_1\times\cdots\times I_n$. □

Integral forms of Jensen's inequality for convex functions of several variables are certainly very important. We present the following expansion of Theorem 7.2.

Theorem 7.3. *Let X be a set with a nonnegative measure μ such that $\mu(X)>0$, let $g_1,\dots,g_n : X\to\mathbb{R}$ be μ-integrable functions, let $I_1,\dots,I_n\subseteq\mathbb{R}$ be intervals containing the images of $g_1,\dots,g_n$, respectively, and let $f: I_1\times\cdots\times I_n\to\mathbb{R}$ be a convex function such that $f(g_1,\dots,g_n)$ is μ-integrable.*

Then, we have the integral inequality

$$f\left(\frac{\int_X g_1 d\mu}{\mu(X)},\dots,\frac{\int_X g_n d\mu}{\mu(X)}\right)\leq\frac{\int_X f(g_1,\dots,g_n)d\mu}{\mu(X)}. \tag{7.34}$$

Proof. If $n = 1$, Theorem 7.3 is reduced to Theorem 7.2. Therefore, we will assume that $n \geq 2$ and consider three positions of the mean $\overline{g} = (\overline{g}_1, \ldots, \overline{g}_n)$ located in the cuboid $I = I_1 \times \cdots \times I_n$. The first position refers to the interior of I, and the other two positions refer to the boundary (relative interior and vertices) of I.

If $\overline{g}$ is an interior point of I (i.e., if $\overline{g}_1, \ldots, \overline{g}_n$ are the interior points of $I_1, \ldots, I_n$, respectively), then f as a convex function admits a support hyperplane h at $\overline{g}$. The support hyperplane inequality applied to the points $g(x) = (g_1(x), \ldots, g_n(x))$ holds for all $x \in X$ in the form

$$h\big(g_1(x), \ldots, g_n(x)\big) \leq f\big(g_1(x), \ldots, g_n(x)\big). \tag{7.35}$$

By relying on the coincidence $f(\overline{g}) = h(\overline{g})$, affinity equality in formula (7.33), and support hyperplane inequality in formula (7.35), we generate the connections

$$\begin{aligned} f\left(\frac{\int_X g_1\, d\mu}{\mu(X)}, \ldots, \frac{\int_X g_n\, d\mu}{\mu(X)}\right) &= h\left(\frac{\int_X g_1\, d\mu}{\mu(X)}, \ldots, \frac{\int_X g_n\, d\mu}{\mu(X)}\right) \\ &= \frac{\int_X h(g_1, \ldots, g_n)\, d\mu}{\mu(X)} \\ &\leq \frac{\int_X f(g_1, \ldots, g_n)\, d\mu}{\mu(X)}, \end{aligned} \tag{7.36}$$

including the inequality in formula (7.34).

If $\overline{g}$ is a boundary point of I so that $\overline{g}_1, \ldots, \overline{g}_k$ are the interior points of $I_1, \ldots, I_k$, respectively, and that $\overline{g}_{k+1}, \ldots, \overline{g}_n$ are the boundary points of $I_{k+1}, \ldots, I_n$, respectively, where $1 \leq k \leq n-1$, then $\overline{g}$ is an interior point of the k-cuboid

$$J = I_1 \times \cdots \times I_k \times \{\overline{g}_{k+1}\} \times \cdots \times \{\overline{g}_n\}$$

contained in the affine plane $\mathbb{A} = \mathbb{R}^k \times \{\overline{g}_{k+1}\} \times \cdots \times \{\overline{g}_n\}$ of dimension k. Therefore, we will use the restriction $f_J = f/J$. The convex function $f_J : J \to \mathbb{R}$ admits a support hyperplane $\tilde{h}_J : \mathbb{A} \to \mathbb{R}$ at $\overline{g}$ as an interior point of J. The support hyperplane inequality holds for all $x \in X$ in the form

$$\tilde{h}_J\big(g_1(x), \ldots, g_k(x), \overline{g}_{k+1}, \ldots, \overline{g}_n\big) \leq f_J\big(g_1(x), \ldots, g_k(x), \overline{g}_{k+1}, \ldots, \overline{g}_n\big).$$

Let $\tilde{h} : \mathbb{R}^n \to \mathbb{R}$ be any affine extension of $\tilde{h}_J$. By utilizing the connections $\tilde{h}/J = \tilde{h}_J$, $\tilde{h}_J(\overline{g}) = f_J(\overline{g})$, and $f_J = f/J$, we get

$$\tilde{h}(\overline{g}) = \tilde{h}_J(\overline{g}) = f_J(\overline{g}) = f(\overline{g}).$$

The numbers $\overline{g}_{k+1}, \ldots, \overline{g}_n$ are the boundary points of the intervals $I_{k+1}, \ldots, I_n$, respectively, which guarantees that μ-almost all $x \in X$ satisfy the equalities

$$g_{k+1}(x) = \overline{g}_{k+1}, \ldots, g_n(x) = \overline{g}_n.$$

All considered indicates that μ-almost all $x \in X$ satisfy the multiple relation

$$\begin{aligned}\tilde{h}(g_1(x),\dots,g_n(x)) &= \tilde{h}(g_1(x),\dots,g_k(x),\overline{g}_{k+1},\dots,\overline{g}_n)\\ &= \tilde{h}_J(g_1(x),\dots,g_k(x),\overline{g}_{k+1},\dots,\overline{g}_n)\\ &\le f_J(g_1(x),\dots,g_k(x),\overline{g}_{k+1},\dots,\overline{g}_n)\\ &= f(g_1(x),\dots,g_k(x),\overline{g}_{k+1},\dots,\overline{g}_n)\\ &= f(g_1(x),\dots,g_n(x)),\end{aligned}$$

and so the support hyperplane inequality

$$\tilde{h}(g_1(x),\dots,g_n(x)) \le f(g_1(x),\dots,g_n(x)).$$

The coincidence $\tilde{h}(\overline{g}) = f(\overline{g})$ and the above μ-almost everywhere inequality encourage us to put $\tilde{h}$ instead of h into formula (7.36), and thus reach formula (7.34). Similar arguments apply to any k-cuboid of I.

If $\overline{g}$ is a vertex of I (i.e., if $\overline{g}_1,\dots,\overline{g}_n$ are the boundary points of $I_1,\dots,I_n$, respectively), then $g_1(x) = \overline{g}_1,\dots,g_n(x) = \overline{g}_n$ for μ-almost all $x \in X$. The trivial inequality $f(\overline{g}_1,\dots,\overline{g}_n) \le f(\overline{g}_1,\dots,\overline{g}_n)$ represents formula (7.34). □

Some extensions and generalizations of different variants of Jensen's inequality can be found in papers [13,16,17].

7.3 THE HERMITE-HADAMARD INEQUALITY

The Hermite-Hadamard inequality is the double inequality with harmonic form consisting of the integral member between two discrete members.

7.3.1 THE CLASSIC FORM OF THE HERMITE-HADAMARD INEQUALITY

In this subsection, we use the Riemann integral and a bounded closed interval $[a,b]$ in the line $\mathbb{R}$ with $a < b$.

By using the expression for the integral arithmetic mean

$$\overline{g} = \frac{\int_a^b g(x)\,dx}{b-a}$$

of an integrable function $g : [a,b] \to \mathbb{R}$ with $g(x) = x$, we get the barycenter

$$\overline{c} = \frac{\int_a^b x\,dx}{b-a} = \frac{a+b}{2} \tag{7.37}$$

of the interval $[a,b]$, also called the midpoint.

The classic form of the Hermite-Hadamard inequality is as follows.

Theorem 7.4. *Let $[a,b]$ be a bounded closed interval in the line $\mathbb{R}$, and let $f : [a,b] \to \mathbb{R}$ be a convex function.*

Then, we have the double inequality

$$f\left(\frac{a+b}{2}\right) \le \frac{\int_a^b f(x)\,dx}{b-a} \le \frac{f(a)+f(b)}{2}. \tag{7.38}$$

First Proof. By applying the convexity of f to the right side of the convex combinations equality

$$\frac{a+b}{2}=\frac{1}{2}\big((1-t)a+tb\big)+\frac{1}{2}\big((1-t)b+ta\big),$$

as well as to convex combinations $(1-t)a+tb$ and $(1-t)b+ta$, we obtain the multiple relation

$$\begin{aligned} f\left(\frac{a+b}{2}\right) &\le \frac{1}{2}f\big((1-t)a+tb\big)+\frac{1}{2}f\big((1-t)b+ta\big) \\ &\le \frac{1}{2}\big((1-t)f(a)+tf(b)\big)+\frac{1}{2}\big((1-t)f(b)+tf(a)\big) \\ &= \frac{f(a)+f(b)}{2}. \end{aligned}$$

Now we single out the double inequality

$$f\left(\frac{a+b}{2}\right)\le\frac{1}{2}f\big((1-t)a+tb\big)+\frac{1}{2}f\big((1-t)b+ta\big)\le\frac{f(a)+f(b)}{2},$$

and then integrating over the unit interval $[0,1]$, and using equalities

$$\int_0^1 f\big((1-t)a+tb\big)\,dt=\int_0^1 f\big((1-t)b+ta\big)\,dt=\frac{\int_a^b f(x)\,dx}{b-a},$$

we achieve the inequality in formula (7.38). □

Second Proof. Let h_1 be a support line of f at the midpoint $\overline{c}=(a+b)/2$, and let h_2 be the secant line of f. By integrating the support-secant line inequality

$$h_1(x)\le f(x)\le h_2(x)$$

over the interval $[a,b]$, dividing with $b-a$, and applying the affinity of h_1 and h_2 through formula (7.28) with $g(x)=x$, we get

$$h_1\left(\frac{\int_a^b x\,dx}{b-a}\right)\le\frac{\int_a^b f(x)\,dx}{b-a}\le h_2\left(\frac{\int_a^b x\,dx}{b-a}\right).$$

By using the barycenter representation in formula (7.37), we obtain

$$h_1\left(\frac{a+b}{2}\right)\le\frac{\int_a^b f(x)\,dx}{b-a}\le h_2\left(\frac{a+b}{2}\right).$$

By utilizing relations

$$h_1\left(\frac{a+b}{2}\right)=f\left(\frac{a+b}{2}\right)\ \text{and}\ h_2\left(\frac{a+b}{2}\right)=\frac{h_2(a)+h_2(b)}{2}=\frac{f(a)+f(b)}{2},$$

we attain the inequality in formula (7.38). □

The inequality of the first (last) two members in formula (7.38) is usually called the left-hand (right-hand) side of the Hermite-Hadamard inequality. The double inequality in formula (7.38) was discovered by Hermite in 1883, see [4]. Ten years later, the left-hand side of this inequality was rediscovered by Hadamard, see [3]. At that time, the notion of a convex function was not yet introduced.

7.3.2 GENERALIZATIONS OF THE HERMITE-HADAMARD INEQUALITY

In the first part of this subsection, we use versions of the Riemann integral like multiple integrals in higher dimensions.

Let Δ_{abc} a triangle in the plane $\mathbb{R}^2$. By using the expression for the integral arithmetic mean

$$\overline{g} = \left(\frac{\iint_{\Delta_{abc}} g_1(x,y)\,dxdy}{\operatorname{ar}(\Delta_{abc})}, \frac{\iint_{\Delta_{abc}} g_2(x,y)\,dxdy}{\operatorname{ar}(\Delta_{abc})} \right)$$

of the mapping $g = (g_1, g_2) : \Delta_{abc} \to \mathbb{R}^2$ with $g_1(x,y) = x$ and $g_2(x,y) = y$, we get the barycenter

$$\overline{d} = \left(\frac{\iint_{\Delta_{abc}} x\,dxdy}{\operatorname{ar}(\Delta_{abc})}, \frac{\iint_{\Delta_{abc}} y\,dxdy}{\operatorname{ar}(\Delta_{abc})} \right) = \frac{a+b+c}{3} \tag{7.39}$$

of the triangle Δ_{abc}.

The next is the generalization of the Hermite-Hadamard inequality to triangles.

Lemma 7.3. *Let Δ_{abc} be a triangle in the plane $\mathbb{R}^2$, and let $f : \Delta_{abc} \to \mathbb{R}$ be a convex function.*

Then, we have the double inequality

$$f\left(\frac{a+b+c}{3}\right) \leq \frac{\iint_{\Delta_{abc}} f(x,y)\,dxdy}{\operatorname{ar}(\Delta_{abc})} \leq \frac{f(a)+f(b)+f(c)}{3}. \tag{7.40}$$

First Proof. We put into practice the unit triangle in the plane $\mathbb{R}^2$ as the set

$$\Delta_2 = \{(t,s) : 0 \leq t \leq 1, 0 \leq s \leq 1-t\},$$

and thus, we have the correlation

$$\Delta_{abc} = \{(1-t-s)a + tb + sc : (t,s) \in \Delta_2\}.$$

Now, we can employ the bijection $g_{abc} : \Delta_2 \to \Delta_{abc}$ determined by

$$g_{abc}(t,s) = (1-t-s)a + tb + sc.$$

Then, the convexity of f applied to the convex combination on the right side produces the inequality

$$f(g_{abc}(t,s)) \leq (1-t-s)f(a) + tf(b) + sf(c) = g_{f(a)f(b)f(c)}(t,s).$$

If we put the vertices coordinates $a = (a_1, a_2)$, $b = (b_1, b_2)$, and $c = (c_1, c_2)$, and if we take the substitution

$$(x,y) = \left(g_{a_1b_1c_1}(t,s),\, g_{a_2b_2c_2}(t,s)\right) = g_{abc}(t,s),$$

then the corresponding Jacobian outputs as

$$\begin{aligned} J &= \begin{vmatrix} \partial x/\partial t & \partial x/\partial s \\ \partial y/\partial t & \partial y/\partial s \end{vmatrix} = \begin{vmatrix} b_1-a_1 & c_1-a_1 \\ b_2-a_2 & c_2-a_2 \end{vmatrix} = \begin{vmatrix} a_1 & a_2 & 1 \\ b_1 & b_2 & 1 \\ c_1 & c_2 & 1 \end{vmatrix} \\ &= \pm 2\mathrm{ar}(\Delta_{abc}) = \pm\frac{\mathrm{ar}(\Delta_{abc})}{\mathrm{ar}(\Delta_2)} \end{aligned}$$

because $\mathrm{ar}(\Delta_2) = 1/2$. It follows that $|J| = \mathrm{ar}(\Delta_{abc})/\mathrm{ar}(\Delta_2)$, and we accomplish the integral relation

$$\frac{\iint_{\Delta_{abc}} f(x,y)\,dx\,dy}{\mathrm{ar}(\Delta_{abc})} = \frac{\iint_{\Delta_2} f(g_{abc}(t,s))\,dt\,ds}{\mathrm{ar}(\Delta_2)}. \tag{7.41}$$

By applying the convexity of f to the right side of the convex combinations equality

$$\frac{a+b+c}{3} = \frac{1}{3}g_{abc}(t,s) + \frac{1}{3}g_{bca}(t,s) + \frac{1}{3}g_{cab}(t,s),$$

as well as to the convex combinations of $g_{abc}(t,s)$, $g_{bca}(t,s)$, and $g_{cab}(t,s)$, we get the multiple relation

$$\begin{aligned} f\left(\frac{a+b+c}{3}\right) &\leq \frac{1}{3}f(g_{abc}(t,s)) + \frac{1}{3}f(g_{bca}(t,s)) + \frac{1}{3}f(g_{cab}(t,s)) \\ &\leq \frac{1}{3}g_{f(a)f(b)f(c)}(t,s) + \frac{1}{3}g_{f(b)f(c)f(a)}(t,s) + \frac{1}{3}g_{f(c)f(a)f(b)}(t,s) \\ &= \frac{f(a)+f(b)+f(c)}{3}, \end{aligned}$$

from which we extricate the double inequality

$$f\left(\frac{a+b+c}{3}\right) \leq \frac{f(g_{abc}(t,s))+f(g_{bca}(t,s))+f(g_{cab}(t,s))}{3} \leq \frac{f(a)+f(b)+f(c)}{3}.$$

By integrating over the unit triangle Δ_2, using equalities

$$\iint_{\Delta_2} f(g_{abc}(t,s))\,dt\,ds = \iint_{\Delta_2} f(g_{bca}(t,s))\,dt\,ds = \iint_{\Delta_2} f(g_{cab}(t,s))\,dt\,ds,$$

and dividing with $\mathrm{ar}(\Delta_2)$, we obtain

$$f\left(\frac{a+b+c}{3}\right) \leq \frac{\iint_{\Delta_2} f(g_{abc}(t,s))\,dt\,ds}{\mathrm{ar}(\Delta_2)} \leq \frac{f(a)+f(b)+f(c)}{3}.$$

Respecting the integral relation in formula (7.41), we have formula (7.40). □

Second Proof. Let h_1 be a support plane of f at the barycenter $\overline{d} = (a+b+c)/3$, and let h_2 be the secant plane of f. By integrating the support-secant plane inequality

$$h_1(x,y) \leq f(x,y) \leq h_2(x,y)$$

over the triangle Δ_{abc}, dividing with $\mathrm{ar}(\Delta_{abc})$, and applying the affinity of h_1 and h_2 through formula (7.33) with $g_1(x,y) = x$ and $g_2(x,y) = y$, we get

$$\begin{aligned} h_1\left(\frac{\iint_{\Delta_{abc}} x\,dx\,dy}{\mathrm{ar}(\Delta_{abc})}, \frac{\iint_{\Delta_{abc}} y\,dx\,dy}{\mathrm{ar}(\Delta_{abc})}\right) &= \frac{\iint_{\Delta_{abc}} h_1(x,y)\,dx\,dy}{\mathrm{ar}(\Delta_{abc})} \\ &\leq \frac{\iint_{\Delta_{abc}} f(x,y)\,dx\,dy}{\mathrm{ar}(\Delta_{abc})} \\ &\leq \frac{\iint_{\Delta_{abc}} h_2(x,y)\,dx\,dy}{\mathrm{ar}(\Delta_{abc})} \\ &= h_2\left(\frac{\iint_{\Delta_{abc}} x\,dx\,dy}{\mathrm{ar}(\Delta_{abc})}, \frac{\iint_{\Delta_{abc}} y\,dx\,dy}{\mathrm{ar}(\Delta_{abc})}\right). \end{aligned}$$

By taking the odd members, and referring to the barycenter representation in formula (7.39), we obtain

$$h_1\left(\frac{a+b+c}{3}\right) \leq \frac{\iint_{\Delta_{abc}} f(x,y)\,dx\,dy}{\mathrm{ar}(\Delta_{abc})} \leq h_2\left(\frac{a+b+c}{3}\right).$$

By utilizing coincidences

$$h_1\left(\frac{a+b+c}{3}\right) = f\left(\frac{a+b+c}{3}\right) \text{ and } h_2\left(\frac{a+b+c}{3}\right) = \frac{f(a)+f(b)+f(c)}{3},$$

we attain the inequality in formula (7.40). □

The barycenter of the n-simplex $\Delta_{a_0 \ldots a_n}$ in the space $\mathbb{R}^n$ is the point

$$\overline{a} = \left(\frac{\int \cdots \int_{\Delta_{a_0 \ldots a_n}} x_1\,dx_1 \ldots dx_n}{\mathrm{vol}_n(\Delta_{a_0 \ldots a_n})}, \ldots, \frac{\int \cdots \int_{\Delta_{a_0 \ldots a_n}} x_n\,dx_1 \ldots dx_n}{\mathrm{vol}_n(\Delta_{a_0 \ldots a_n})}\right) = \frac{\sum_{i=0}^n a_i}{n+1}. \tag{7.42}$$

We can realize the Hermite-Hadamard inequality for simplices.

Theorem 7.5. *Let $\Delta_{a_0 \ldots a_n}$ be an n-simplex in the space $\mathbb{R}^n$, and let $f : \Delta_{a_0 \ldots a_n} \to \mathbb{R}$ be a convex function.*

Then, we have the double inequality

$$f\left(\frac{\sum_{i=0}^n a_i}{n+1}\right) \leq \frac{\int \cdots \int_{\Delta_{a_0 \ldots a_n}} f(x_1, \ldots, x_n)\,dx_1 \ldots dx_n}{\mathrm{vol}_n(\Delta_{a_0 \ldots a_n})} \leq \frac{\sum_{i=0}^n f(a_i)}{n+1}. \tag{7.43}$$

To establish the relevant inequalities for convex functions defined on simplices, we will use the simplex representations presented in Section 7.1.

Lemma 7.4. *Let X be a set with a nonnegative measure μ such that $\mu(X) > 0$, let $g : X \to \mathbb{R}$ be a bounded μ-integrable function, let $[a,b]$ be an interval containing the image of g, let $\overline{\alpha}a + \overline{\beta}b$ be the convex combination such that*

$$\frac{\int_X g(x)\,d\mu(x)}{\mu(X)} = \overline{\alpha}a + \overline{\beta}b, \tag{7.44}$$

and let $f : [a,b] \to \mathbb{R}$ be a convex function.

Then, we have the double inequality

$$f(\overline{\alpha}a+\overline{\beta}b) \leq \frac{\int_X f(g(x))\,d\mu(x)}{\mu(X)} \leq \overline{\alpha}f(a)+\overline{\beta}f(b). \tag{7.45}$$

Proof. As regards the equality in formula (7.44), we will determine the convex combination $\overline{\alpha}a+\overline{\beta}b$, which represents the μ-integral arithmetic mean

$$\overline{g} = \frac{\int_X g(x)\,d\mu(x)}{\mu(X)}.$$

By applying formula (7.4) to the point $\overline{g}$, and utilizing the affinity of functions α and β, we obtain the affine combination

$$\begin{aligned}\overline{g} &= \alpha(\overline{g})a+\beta(\overline{g})b\\ &= \frac{\int_X \alpha(g(x))\,d\mu(x)}{\mu(X)}a+\frac{\int_X \beta(g(x))\,d\mu(x)}{\mu(X)}b\\ &= \overline{\alpha}a+\overline{\beta}b\end{aligned}$$

with coefficients

$$\overline{\alpha} = \frac{\int_X \alpha(g(x))\,d\mu(x)}{\mu(X)} \text{ and } \overline{\beta} = \frac{\int_X \beta(g(x))\,d\mu(x)}{\mu(X)}. \tag{7.46}$$

These coefficients are nonnegative because $g(x) \in [a,b]$ for all $x \in X$, and so the affine combination $\overline{g} = \overline{\alpha}a+\overline{\beta}b$ is convex.

To prove the inequality in formula (7.45), we will consider two cases of the combination $\overline{g} = \overline{\alpha}a+\overline{\beta}b$, referring to the number of positive coefficients. So the cases specify the interior (a,b) and boundary $\{a,b\}$.

If $\overline{g} = \overline{\alpha}a+\overline{\beta}b$ with $\overline{\alpha},\overline{\beta} > 0$, then $\overline{g} \in (a,b)$. By including a support line h_1 of f at $\overline{g}$, and the secant line h_2 of f at a and b, we get the multiple inequality

$$\begin{aligned}f(\overline{\alpha}a+\overline{\beta}b) &= f(\overline{g}) = h_1(\overline{g}) = \frac{\int_X h_1(g(x))\,d\mu(x)}{\mu(X)}\\ &\leq \frac{\int_X f(g(x))\,d\mu(x)}{\mu(X)} \leq \frac{\int_X h_2(g(x))\,d\mu(x)}{\mu(X)} = h_2(\overline{g})\\ &= \overline{\alpha}h_2(a)+\overline{\beta}h_2(b) = \overline{\alpha}f(a)+\overline{\beta}f(b),\end{aligned}$$

which contains formula (7.45). The composition $f(g)$ is μ-integrable because it is bounded on X, and continuous μ-almost everywhere on X.

If $\overline{g} = a$, then the equation $\overline{\alpha} = 1$ via formula (7.46) implies that $g(x) = a$ for μ-almost all $x \in X$ (the same arises from the equation $\overline{\beta} = 0$). It turns out that

$$\frac{\int_X f(g(x))\,d\mu(x)}{\mu(X)} = \frac{\int_X f(a)\,d\mu(x)}{\mu(X)} = f(a),$$

and so the trivial double inequality $f(a) \leq f(a) \leq f(a)$ represents formula (7.45). A similar reasoning applies for $\overline{g} = b$. □

Lemma 7.5. *Let X be a set with a nonnegative measure μ such that $\mu(X) > 0$, let $g_1, g_2 : X \to \mathbb{R}$ be the bounded μ-integrable functions, let Δ_{abc} be a triangle containing the image of the mapping $g = (g_1, g_2) : X \to \mathbb{R}^2$, let $\overline{\alpha}a + \overline{\beta}b + \overline{\gamma}c$ be the convex combination such that*

$$\left(\frac{\int_X g_1(x)\,d\mu(x)}{\mu(X)}, \frac{\int_X g_2(x)\,d\mu(x)}{\mu(X)} \right) = \overline{\alpha}a + \overline{\beta}b + \overline{\gamma}c, \tag{7.47}$$

and let $f : \Delta_{abc} \to \mathbb{R}$ be a convex function.

Then, we have the double inequality

$$f(\overline{\alpha}a + \overline{\beta}b + \overline{\gamma}c) \leq \frac{\int_X f(g_1(x), g_2(x))\,d\mu(x)}{\mu(X)} \leq \overline{\alpha}f(a) + \overline{\beta}f(b) + \overline{\gamma}f(c). \tag{7.48}$$

Theorem 7.6. *Let X be a set with a nonnegative measure μ such that $\mu(X) > 0$, let $g_1, \ldots, g_n : X \to \mathbb{R}$ be the bounded μ-integrable functions, let $\Delta_{a_0 \ldots a_n}$ be an n-simplex containing the image of the mapping $g = (g_1, \ldots, g_n) : X \to \mathbb{R}^n$, let $\sum_{i=0}^n \overline{\alpha}_i a_i$ be the convex combination such that*

$$\left(\frac{\int_X g_1(x)\,d\mu(x)}{\mu(X)}, \ldots, \frac{\int_X g_n(x)\,d\mu(x)}{\mu(X)} \right) = \sum_{i=0}^{n} \overline{\alpha}_i a_i, \tag{7.49}$$

and let $f : \Delta_{a_0 \ldots a_n} \to \mathbb{R}$ be a convex function.

Then, we have the double inequality

$$f\left(\sum_{i=0}^{n} \overline{\alpha}_i a_i \right) \leq \frac{\int_X f(g_1(x), \ldots, g_n(x))\,d\mu(x)}{\mu(X)} \leq \sum_{i=0}^{n} \overline{\alpha}_i f(a_i). \tag{7.50}$$

Proof. The representation in formula (7.49) can be realized by applying formula (7.10) to the μ-integral arithmetic mean

$$\overline{g} = \left(\frac{\int_X g_1(x)\,d\mu(x)}{\mu(X)}, \ldots, \frac{\int_X g_n(x)\,d\mu(x)}{\mu(X)} \right),$$

and employing the affinity of functions α_i. This yields nonnegative coefficients

$$\overline{\alpha}_i = \frac{\int_X \alpha_i(g(x))\,dx}{\mu(X)}. \tag{7.51}$$

To prove the inequality in formula (7.50), we will consider three cases of the convex combination $\overline{g} = \sum_{i=0}^n \overline{\alpha}_i a_i$, referring to the number of positive coefficients. The first case connotes the interior of $\Delta_{a_0 \ldots a_n}$, and the other two cases connote the boundary (relative interior and vertices) of $\Delta_{a_0 \ldots a_n}$.

If $\overline{g} = \sum_{i=0}^n \overline{\alpha}_i a_i$ with $\overline{\alpha}_0, \ldots, \overline{\alpha}_n > 0$, then $\overline{g}$ is an interior point of $\Delta_{a_0 \ldots a_n}$, and f as a convex function admits a support hyperplane h_1 at $\overline{g}$. By including the secant hyperplane h_2 at vertices, we have the coincidences

$$h_1(\overline{g}) = f(\overline{g}), \ f(a_0) = h_2(a_0), \ldots, f(a_n) = h_2(a_n),$$

and for all $x \in X$ the support-secant hyperplane inequality

$$h_1(g(x)) \leq f(g(x)) \leq h_2(g(x)),$$

which is applied to the points $g(x) = (g_1(x), \dots, g_n(x))$. By using the above coincidences and inequality, we can obtain the multiple relation

$$\begin{aligned} f\left(\sum_{i=0}^{n} \overline{\alpha}_i a_i\right) &= f(\overline{g}) = h_1(\overline{g}) = \frac{\int_X h_1(g(x))\,d\mu(x)}{\mu(X)} \\ &\leq \frac{\int_X f(g(x))\,d\mu(x)}{\mu(X)} \leq \frac{\int_X h_2(g(x))\,d\mu(x)}{\mu(X)} = h_2(\overline{g}) \qquad (7.52) \\ &= \sum_{i=0}^{n} \overline{\alpha}_i h_2(a_i) = \sum_{i=0}^{n} \overline{\alpha}_i f(a_i), \end{aligned}$$

covering the inequality in formula (7.50).

If $\overline{g} = \sum_{i=0}^{k} \overline{\alpha}_i a_i$ with $\overline{\alpha}_0, \dots, \overline{\alpha}_k > 0$, where $1 \leq k \leq n-1$, then $\overline{g}$ is an interior point of the k-face

$$S = \Delta_{a_0 \dots a_k}$$

contained in the affine plane $\mathbb{A} = \text{aff}(S)$ of dimension k, and we rely on the restriction $f_S = f/S$. The convex function $f_S : S \to \mathbb{R}$ admits a support hyperplane $\tilde{h}_S : \mathbb{A} \to \mathbb{R}$ at $\overline{g}$, which provides the equality $\tilde{h}_S(\overline{g}) = f_S(\overline{g})$, and for all $s = (s_1, \dots, s_n) \in S$, the support hyperplane inequality

$$\tilde{h}_S(s) \leq f_S(s). \qquad (7.53)$$

Let $\tilde{h}_1 : \mathbb{R}^n \to \mathbb{R}$ be any affine extension of $\tilde{h}_S$. Now, we have the coincidences

$$\tilde{h}_1(\overline{g}) = \tilde{h}_S(\overline{g}) = f_S(\overline{g}) = f(\overline{g}).$$

The system of equations $\overline{\alpha}_{k+1} = 0, \dots, \overline{\alpha}_n = 0$ via formula (7.51) implies that

$$g(x) \in \Delta_{a_0 \dots a_k a_{k+2} \dots a_n} \cap \dots \cap \Delta_{a_0 \dots a_{n-1}} = \Delta_{a_0 \dots a_k} = S$$

for μ-almost all $x \in X$. Therefore, the support hyperplane inequality in formula (7.53) applies to μ-almost all $x \in X$ in the form

$$\tilde{h}_S(g(x)) \leq f_S(g(x)).$$

Now, we include the secant hyperplane h_2 of f at vertices. Then, μ-almost all $x \in X$ satisfy the multiple relation

$$\begin{aligned} \tilde{h}_1(g(x)) = \tilde{h}_S(g(x)) &\leq f_S(g(x)) \\ = f(g(x)) &\leq h_2(g(x)), \end{aligned}$$

and consequently, the support-secant hyperplane inequality

$$\tilde{h}_1(g(x)) \leq f(g(x)) \leq h_2(g(x)).$$

The coincidence $\tilde{h}_1(\overline{g}) = f(\overline{g})$ and the above μ-almost everywhere inequality enable us to put $\tilde{h}_1$ instead of h_1 into formula (7.52), and thus reach formula (7.50). The same applies to other k-faces.

If $\overline{g} = a_0$, then the equation $\overline{\alpha}_0 = 1$ via formula (7.51) implies that $g(x) = a_0$ for μ-almost all $x \in X$. It further produces the trivial inequality $f(a_0) \leq f(a_0) \leq f(a_0)$, representing formula (7.50). The same applies to other vertices. □

The inequality in formula (7.50) is reduced to the Hermite-Hadamard inequality in formula (7.43) if we take $X = \Delta_{a_0 \dots a_n}$ and $g_1(x) = x_1, \dots, g_n(x) = x_n$. In this reduction, we have the correlations $\overline{\alpha}_0 = \dots = \overline{\alpha}_n = 1/(n+1)$, $d\mu(x) = d_1 \dots dx_n$, and $\mu(X) = \mathrm{vol}_n(\Delta_{a_0 \dots a_n})$.

Formula (7.12) can be considered as the simplex definition in any real vector space of sufficiently large dimension. We will include this feature in the statement of the discrete version of the Hermite-Hadamard inequality.

Theorem 7.7. *Let $\mathbb{X}$ be a real vector space of dimension of at least n, let $\Delta_{a_0 \dots a_n}$ be an n-simplex in $\mathbb{X}$, let $\sum_{j=1}^m \lambda_j x_j$ be a convex combination of points $x_j \in \Delta_{a_0 \dots a_n}$, let $\sum_{i=0}^n \overline{\alpha}_i a_i$ be the convex combination of vertices a_i such that*

$$\sum_{j=1}^m \lambda_j x_j = \sum_{i=0}^n \overline{\alpha}_i a_i, \tag{7.54}$$

and let $f : \Delta_{a_0 \dots a_n} \to \mathbb{R}$ be a convex function.

Then, we have the double discrete inequality

$$f\left(\sum_{i=0}^n \overline{\alpha}_i a_i\right) \leq \sum_{j=1}^m \lambda_j f(x_j) \leq \sum_{i=0}^n \overline{\alpha}_i f(a_i). \tag{7.55}$$

Proof. By using the simplex definition in formula (7.12), we can represent the points x_j as the convex combinations

$$x_j = \sum_{i=0}^n \overline{\alpha}_{ij} a_i. \tag{7.56}$$

Then, using the assumption in formula (7.54), we obtain

$$\sum_{i=0}^n \overline{\alpha}_i a_i = \sum_{j=1}^m \lambda_j x_j = \sum_{j=1}^m \lambda_j \left(\sum_{i=0}^n \overline{\alpha}_{ij} a_i\right) = \sum_{i=0}^n \left(\sum_{j=1}^m \lambda_j \overline{\alpha}_{ij}\right) a_i,$$

and since the convex combination of simplex vertices is unique, it follows that

$$\sum_{j=1}^m \lambda_j \overline{\alpha}_{ij} = \overline{\alpha}_i.$$

By applying Jensen's inequality to the convex combination $\sum_{j=1}^m \lambda_j x_j$ and subsequently through formula (7.56), and utilizing the above correlation coefficients, we produce the multiple relation

$$f\left(\sum_{i=0}^{n}\overline{\alpha}_i a_i\right) = f\left(\sum_{j=1}^{m}\lambda_j x_j\right) \leq \sum_{j=1}^{m}\lambda_j f(x_j) \leq \sum_{j=1}^{m}\lambda_j \sum_{i=0}^{n}\overline{\alpha}_{ij} f(a_i)$$
$$= \sum_{i=0}^{n}\left(\sum_{j=1}^{m}\lambda_j\overline{\alpha}_{ij}\right) f(a_i) = \sum_{i=0}^{n}\overline{\alpha}_i f(a_i),$$

including the double inequality in formula (7.55). □

Another Proof of the Case $\mathbb{X} = \mathbb{R}^n$. We will adapt Theorem 7.6 to the given discrete case. Let $x_j = (x_{j1},\ldots,x_{jn})$ be the coordinate representations for $j = 1,\ldots,m$. Let $X = \{x_1,\ldots,x_m\}$ be an m-member set with the probability measure μ given by $\mu(\{x_j\}) = \lambda_j$, and let g_i be the functions determined by $g_i(x_j) = x_{ji}$. Then, we have

$$\begin{aligned}
\overline{g} &= \left(\frac{\int_X g_1(x)\,d\mu(x)}{\mu(X)},\ldots,\frac{\int_X g_n(x)\,d\mu(x)}{\mu(X)}\right) \\
&= \left(\sum_{j=1}^{m} g_1(x_j)\,\mu(\{x_j\}),\ldots,\sum_{j=1}^{m} g_n(x_j)\,\mu(\{x_j\})\right) \\
&= \left(\sum_{j=1}^{m}\lambda_j x_{j1},\ldots,\sum_{j=1}^{m}\lambda_j x_{jn}\right) = \sum_{j=1}^{m}\lambda_j x_j
\end{aligned}$$

and

$$\begin{aligned}
\overline{f(g)} &= \frac{\int_X f(g_1(x),\ldots,g_n(x))\,d\mu(x)}{\mu(X)} \\
&= \sum_{j=1}^{m} f\big(g_1(x_j),\ldots,g_n(x_j)\big)\,\mu(\{x_j\}) \\
&= \sum_{j=1}^{m}\lambda_j f(x_{j1},\ldots,x_{jn}) = \sum_{j=1}^{m}\lambda_j f(x_j).
\end{aligned}$$

Thus, equality in formula (7.49) turns into equality in formula (7.54), and inequality in formula (7.50) turns into inequality in formula (7.55). □

The inequality in formula (7.55) discusses the nature of the behavior of convex functions. It shows that the convex function values taken in the form of convex combinations grow from the center, across the middle, to the vertices as ends.

Simple proof and refinement of the Hermite-Hadamard inequality was demonstrated in [2]. The Hermite-Hadamard inequality for the simplex, its refinements, and generalizations were considered in papers [1,10,11,14,15].

7.4 THE ROGERS-HÖLDER INEQUALITY

The Rogers-Hölder inequality occupies an important place in mathematical analysis, especially in the study and development of L^p spaces. Although it comes down to nonnegative real-valued functions, complex-valued functions are used due to their wider applications. This is usually done because of L^p spaces.

7.4.1 INTEGRAL AND DISCRETE FORMS OF THE ROGERS-HÖLDER INEQUALITY

Let X be a set with a nonnegative measure μ such that $\mu(X) > 0$. Let $p > 0$ be a real number, and let $L^p_\mu(X) = L^p(\mu)$ be the space of all μ-measurable functions $f : X \to \mathbb{C}$ such that f^p is μ-integrable (Lebesgue integrable with respect to the measure μ). Each function $f \in L^p(\mu)$ has the fictive p-norm

$$\|f\|_p = \left(\int_X |f(x)|^p d\mu(x) \right)^{1/p}. \tag{7.57}$$

If $p \geq 1$, the above definition gives the (proper) p-norm, the inequality $|f(x)| \leq |f(x)|^p$ holds for every $x \in X$, and therefore, $L^p_\mu(X) \subseteq L^1_\mu(X)$. The fundamental space $L^1_\mu(X)$ contains all complex functions on X that are μ-integrable.

Initial forms of the Rogers-Hölder inequality use a pair of positive real numbers p and q such that $p+q = pq$. It is equivalent to conditions $1/p + 1/q = 1$ and $p, q > 1$. The numbers p and q are called the conjugate exponents.

All that is needed is prepared for Rogers-Hölder inequalities. The first one that follows is the integral form of the Rogers-Hölder inequality. The integral inequality in formula (7.34) adjusted to concave functions will serve as a basis of the proof.

Theorem 7.8. *Let X be a set with a nonnegative measure μ such that $\mu(X) > 0$, let $p, q > 1$ be the real numbers such that $1/p + 1/q = 1$, let $f : X \to \mathbb{C}$ be a function in $L^p_\mu(X)$, and let $g : X \to \mathbb{C}$ be a function in $L^q_\mu(X)$.*

Then, we have the integral inequality

$$\int_X |f(x)g(x)|\, d\mu(x) \leq \left(\int_X |f(x)|^p d\mu(x) \right)^{1/p} \left(\int_X |g(x)|^q d\mu(x) \right)^{1/q}. \tag{7.58}$$

Proof. Since $p, q > 1$, the functions f and g are in $L^1_\mu(X)$, and therefore, their product fg is also in $L^1_\mu(X)$.

Let $I_1 \subseteq \mathbb{R}$ be an interval containing the image of $|f|^p$, let $I_2 \subseteq \mathbb{R}$ be an interval containing the image of $|g|^q$, and let $\phi : I_1 \times I_2 \to \mathbb{R}$ be a concave function. The generalized integral inequality in formula (7.34) for integer $n = 2$ adjusted to functions $g_1 = |f|^p$, $g_2 = |g|^q$ and $f = -\phi$ takes the form

$$\frac{\int_X \phi\left(|f(x)|^p, |g(x)|^q\right) d\mu(x)}{\mu(X)} \leq \phi\left(\frac{\int_X |f(x)|^p d\mu(x)}{\mu(X)}, \frac{\int_X |g(x)|^q d\mu(x)}{\mu(X)} \right).$$

If we implement the concave function $\phi : [0,\infty) \times [0,\infty) \to \mathbb{R}$ defined by

$$\phi(x,y) = x^{1/p} y^{1/q},$$

then we obtain

$$\frac{\int_X |f(x)g(x)|\, d\mu(x)}{\mu(X)} \leq \left(\frac{\int_X |f(x)|^p d\mu(x)}{\mu(X)} \right)^{1/p} \left(\frac{\int_X |g(x)|^q d\mu(x)}{\mu(X)} \right)^{1/q}.$$

After multiplying by $\mu(X)$, we achieve the inequality in formula (7.58). □

The integral inequality in formula (7.58) can be presented as the norm inequality

$$\|fg\|_1 \leq \|f\|_p \, \|g\|_q. \tag{7.59}$$

The integral form of the Rogers-Hölder inequality can be transformed into discrete form by taking advantage of the counting measure.

Theorem 7.9. *Let $x_i, y_i \in \mathbb{C}$ be the complex numbers for $i = 1, \ldots, n$, and let $p, q > 1$ be the real numbers such that $1/p + 1/q = 1$.*

Then, we have the discrete inequality

$$\sum_{i=1}^{n} |x_i y_i| \leq \left(\sum_{i=1}^{n} |x_i|^p \right)^{1/p} \left(\sum_{i=1}^{n} |y_i|^q \right)^{1/q}. \tag{7.60}$$

Proof. Let $X = \{x_1, \ldots, x_n\}$ be an n-member set with the counting measure μ as $\mu(\{x_i\}) = 1$, let f be the function as $f(x_i) = x_i$, and let g be the function as $g(x_i) = y_i$. Then, the transformation of the left side in formula (7.58) shows

$$\int_X |f(x)g(x)| \, d\mu(x) = \sum_{i=1}^{n} |f(x_i)g(x_i)| \, \mu(\{x_i\}) = \sum_{i=1}^{n} |x_i y_i|,$$

pointing the left side in formula (7.60). The same applies to the right side. □

By promoting n-tuples

$$x = (x_1, \ldots, x_n) \text{ and } y = (y_1, \ldots, y_n),$$

the product

$$x \cdot y = (x_1 y_1, \ldots, x_n y_n),$$

and norms

$$\|x\|_p = \left(\sum_{i=1}^{n} |x_i|^p \right)^{1/p}, \quad \|y\|_q = \left(\sum_{i=1}^{n} |y_i|^q \right)^{1/q}, \quad \|x \cdot y\|_1 = \sum_{i=1}^{n} |x_i y_i|,$$

the discrete inequality in formula (7.60) can be written as the norm inequality

$$\|x \cdot y\|_1 \leq \|x\|_p \, \|y\|_q. \tag{7.61}$$

In the books, formulas (7.58) and (7.60) were proved by relying on the discrete form of the Young inequality. It says that the inequality

$$ab \leq \frac{1}{p} a^p + \frac{1}{q} b^q \tag{7.62}$$

holds for every pair of nonnegative real numbers a and b.

Inequalities in formulas (7.58) and (7.60) are usually called the Hölder inequalities. Respecting the chronological publication order of paper [19] and paper [5], we preferred the name Rogers-Hölder. In the case $p = q = 2$, these inequalities are also called the Cauchy-Bunyakovsky-Schwarz inequalities. As regards the above special case, the inequality for sums was published by Cauchy (1821), the corresponding inequality for integrals was proved by Bunyakovsky in (1859), and the improved inequality for integrals was given by Schwarz (1888).

7.4.2 GENERALIZATIONS OF THE ROGERS-HÖLDER INEQUALITY

The main tool in the generalization of Rogers-Hölder inequalities are positive real numbers $p_1,\ldots,p_m,p$ such that $\sum_{j=1}^{m} 1/p_j = 1/p$. The consequent relation

$$\frac{p}{p_1}+\cdots+\frac{p}{p_m}=1$$

is effective, implying that $p/p_j < 1$ and thus $p < p_j$ for $j=1,\ldots,m$. Note that numbers p_j may be less than or equal to 1.

The generalized integral form of the Rogers-Hölder inequality is as follows.

Corollary 7.1. *Let X be a set with a nonnegative measure μ such that $\mu(X)>0$, let $p_1,\ldots,p_m,p>0$ be the real numbers such that $\sum_{j=1}^{m} 1/p_j = 1/p$, and let $f_j : X \to \mathbb{C}$ be the functions in $L_\mu^{p_j}(X)$ for $j=1,\ldots,m$.*

Then, we have the integral inequality

$$\left(\int_X |f_1\ldots f_m|^p \, d\mu\right)^{1/p} \le \left(\int_X |f_1|^{p_1} \, d\mu\right)^{1/p_1} \cdots \left(\int_X |f_m|^{p_m} \, d\mu\right)^{1/p_m}. \tag{7.63}$$

Proof. Since $p < p_1,\ldots,p < p_m$, it follows that

$$|f_1\ldots f_m|^p \le |f_1|^{p_1}\ldots|f_m|^{p_m},$$

and so the product $f_1\ldots f_m$ is in $L_\mu^p(X)$.

Let $I_j \subseteq \mathbb{R}$ be an interval containing the image of $|f_j|^{p_j}$ for $j=1,\ldots,m$, and let $\phi : I_1\times\cdots\times I_m \to \mathbb{R}$ be a concave function. The generalized integral inequality in formula (7.34) for integer $n=m$ with functions $g_1=|f_1|^{p_1},\ldots,g_m=|f_m|^{p_m}$ and $f=-\phi$ comes out as

$$\frac{\int_X \phi\left(|f_1|^{p_1},\ldots,|f_m|^{p_m}\right) d\mu}{\mu(X)} \le \phi\left(\frac{\int_X |f_1|^{p_1} \, d\mu}{\mu(X)},\ldots,\frac{\int_X |f_m|^{p_m} \, d\mu}{\mu(X)}\right).$$

If we include the concave function $\phi : [0,\infty)\times\cdots\times[0,\infty) \to \mathbb{R}$ determined by

$$\phi(x_1,\ldots,x_m) = x_1^{p/p_1}\ldots x_m^{p/p_m},$$

then we realize

$$\frac{\int_X |f_1\ldots f_m|^p \, d\mu}{\mu(X)} \le \left(\frac{\int_X |f_1|^{p_1} \, d\mu}{\mu(X)}\right)^{p/p_1} \cdots \left(\frac{\int_X |f_m|^{p_m} \, d\mu}{\mu(X)}\right)^{p/p_m}.$$

The multiplication by $\mu(X)$ and raising to the power $1/p$ gives formula (7.63). □

The integral inequality in formula (7.63) can be expressed as the fictive norm inequality

$$\|f_1\ldots f_m\|_p \le \|f_1\|_{p_1}\cdots\|f_m\|_{p_m}. \tag{7.64}$$

If $m=2$ and $p=1$, formula (7.63) is reduced to formula (7.58), and formula (7.64) is reduced to formula (7.59).

The following is the generalized discrete form of the Rogers-Hölder inequality.

Corollary 7.2. *Let $x_{ij} \in \mathbb{C}$ be the complex numbers for $i = 1,\dots,n$ and $j = 1,\dots,m$, and let $p_1,\dots,p_m, p > 0$ be the real numbers such that $\sum_{j=1}^{m} 1/p_j = 1/p$.*

Then, we have the discrete inequality

$$\left(\sum_{i=1}^{n} |x_{i1}\dots x_{im}|^p\right)^{1/p} \leq \left(\sum_{i=1}^{n} |x_{i1}|^{p_1}\right)^{1/p_1} \dots \left(\sum_{i=1}^{n} |x_{im}|^{p_m}\right)^{1/p_m}. \tag{7.65}$$

Proof. Let $X = \{x_1,\dots,x_n\}$ be an n-member set with the counting measure μ given by $\mu(\{x_i\}) = 1$, and let f_j be the functions determined by $f_j(x_i) = x_{ij}$. Then, the transformation of pth power of the left side in formula (7.63) produces

$$\int_X |f_1(x)\dots f_m(x)|^p \, d\mu(x) = \sum_{i=1}^{n} |f_1(x_i)\dots f_m(x_i)|^p \mu(\{x_i\}) = \sum_{i=1}^{n} |x_{i1}\dots x_{im}|^p$$

as pth power of the left side in formula (7.65). The similar applies to the right side. □

By exploiting n-tuples

$$x_1 = (x_{11},\dots,x_{n1}), \; \dots, \; x_m = (x_{1m},\dots,x_{nm}),$$

the product

$$x_1 \cdot \dots \cdot x_m = (x_{11}\dots x_{1m},\dots,x_{n1}\dots x_{nm}),$$

and using the corresponding fictive norms, the discrete inequality in formula (7.65) can be presented as the fictive norm inequality

$$\|x_1 \cdot \dots \cdot x_m\|_p \leq \|x_1\|_{p_1} \dots \|x_m\|_{p_m}. \tag{7.66}$$

Formulas (7.63) and (7.65) can be proved by applying the method of mathematical induction to integer m, wherein the case $m = 2$ serves as the induction base. These formulae can also be obtained by relying on the generalized discrete form of the Young inequality. It can be demonstrated that the inequality

$$a_1\dots a_m \leq \frac{p}{p_1} a_1^{p_1/p} + \dots + \frac{p}{p_m} a_m^{p_m/p} \tag{7.67}$$

holds for every m-tuple of nonnegative real numbers $a_1,\dots,a_m$.

More details and better insight into Young's original inequalities can be found in the proceedings [20].

7.5 THE MINKOWSKI INEQUALITY

The Minkowski inequality is the key inequality of functional analysis. It has the greatest influence in the normed vector spaces.

7.5.1 INTEGRAL AND DISCRETE FORMS OF THE MINKOWSKI INEQUALITY

Minkowski inequalities demonstrate that p-norms satisfy the triangle inequality. These inequalities are fundamental in the field of functional analysis.

Let $x, y \in \mathbb{C}$ be a pair of complex numbers. The prominent triangle inequality for absolute value as

$$|x+y| \leq |x| + |y| \tag{7.68}$$

can be obtained by applying Jensen's inequality in formula (7.25) to the convex combination $x/2 + y/2$, and convex function $\varphi : \mathbb{C} \to \mathbb{R}$ given by $\varphi(z) = 2|z|$.

The following is the integral form of the Minkowski inequality. The integral inequality in formula (7.34) will be the mainstay of the proof.

Theorem 7.10. *Let X be a set with a nonnegative measure μ such that $\mu(X) > 0$, let $p \geq 1$ be a real number, and let $f, g : X \to \mathbb{C}$ be the functions in $L^p_\mu(X)$.*

Then, we have the integral inequality

$$\left(\int_X |f(x)+g(x)|^p d\mu(x)\right)^{1/p} \leq \left(\int_X |f(x)|^p d\mu(x)\right)^{1/p} + \left(\int_X |g(x)|^p d\mu(x)\right)^{1/p}. \tag{7.69}$$

Proof. A slightly modified formula (7.69) with the integrand $\big(|f(x)| + |g(x)|\big)^p$ on the left side will be discussed first. The sum $|f| + |g|$ is in $L^p_\mu(X)$ because

$$\big(|f(x)| + |g(x)|\big)^p \leq 2^{p-1}\big(|f(x)|^p + |g(x)|^p\big)$$

for every $x \in X$. The above inequality can be obtained by applying Jensen's inequality in formula (7.25) to the convex combination $|f(x)|/2 + |g(x)|/2$, and convex function $\varphi : [0, \infty) \to \mathbb{R}$ given by $\varphi(t) = 2^p t^p$.

Let $I_1 \subseteq \mathbb{R}$ be an interval containing the image of $|f|^p$, let $I_2 \subseteq \mathbb{R}$ be an interval containing the image of $|g|^p$, and let $\phi : I_1 \times I_2 \to \mathbb{R}$ be a concave function. The generalized integral inequality in formula (7.34) for integer $n = 2$ with functions $g_1 = |f|^p$, $g_2 = |g|^p$, and $f = -\phi$ arises in the form

$$\frac{\int_X \phi\big(|f(x)|^p, |g(x)|^p\big)\, d\mu(x)}{\mu(X)} \leq \phi\left(\frac{\int_X |f(x)|^p d\mu(x)}{\mu(X)}, \frac{\int_X |g(x)|^p d\mu(x)}{\mu(X)}\right).$$

If we utilize the concave function $\phi : [0, \infty) \times [0, \infty) \to \mathbb{R}$ defined by

$$\phi(x, y) = (x^{1/p} + y^{1/p})^p,$$

then we gain

$$\frac{\int_X \big(|f(x)| + |g(x)|\big)^p d\mu(x)}{\mu(X)} \leq \left[\left(\frac{\int_X |f(x)|^p d\mu(x)}{\mu(X)}\right)^{1/p} + \left(\frac{\int_X |g(x)|^p d\mu(x)}{\mu(X)}\right)^{1/p}\right]^p.$$

After multiplying by $\mu(X)$, and raising to the power $1/p$, we have

$$\left(\int_X \big(|f(x)| + |g(x)|\big)^p d\mu(x)\right)^{1/p} \leq \left(\int_X |f(x)|^p d\mu(x)\right)^{1/p} + \left(\int_X |g(x)|^p d\mu(x)\right)^{1/p}.$$

This inequality can be extended to the left by using the integrand $|f(x)+g(x)|^p$ because $|f(x)+g(x)|^p \leq \big(|f(x)|+|g(x)|\big)^p$ for every $x \in X$. □

The collection $L^p_\mu(X)$ with $p \geq 1$ becomes the normed vector space if the integral inequality in formula (7.69) is expressed as the norm inequality

$$\|f+g\|_p \leq \|f\|_p + \|g\|_p. \tag{7.70}$$

The discrete form of the Minkowski inequality is as follows.

Theorem 7.11. *Let $x_i, y_i \in \mathbb{C}$ be the complex numbers for $i = 1,\dots,n$, and let $p \geq 1$ be a real number.*

Then, we have the discrete inequality

$$\left(\sum_{i=1}^{n} |x_i+y_i|^p\right)^{1/p} \leq \left(\sum_{i=1}^{n} |x_i|^p\right)^{1/p} + \left(\sum_{i=1}^{n} |y_i|^p\right)^{1/p}. \tag{7.71}$$

Proof. The integral inequality in formula (7.69) is transformed into discrete inequality in formula (7.71) by exploring the counting measure as in Theorem 7.9. □

By using n-tuples $x = (x_1,\dots,x_n)$ and $y = (y_1,\dots,y_n)$, and required p-norms, the discrete inequality in formula (7.71) can be presented as the norm inequality

$$\|x+y\|_p \leq \|x\|_p + \|y\|_p. \tag{7.72}$$

In the books, formulas (7.69) and (7.71) were proved by relying on the triangle inequality and corresponding Rogers-Hölder inequalities.

7.5.2 GENERALIZATIONS OF THE MINKOWSKI INEQUALITY

To generalize the integral form of the Minkowski inequality, we need a real number $p \geq 1$, and complex functions in the space $L^p_\mu(X)$. It is as follows.

Corollary 7.3. *Let X be a set with a nonnegative measure μ such that $\mu(X) > 0$, let $p \geq 1$ be a real number, and let $f_1,\dots,f_m : X \to \mathbb{C}$ be the functions in $L^p_\mu(X)$.*

Then, we have the integral inequality

$$\left(\int_X |f_1+\cdots+f_m|^p \, d\mu\right)^{1/p} \leq \left(\int_X |f_1|^p \, d\mu\right)^{1/p} + \cdots + \left(\int_X |f_m|^p \, d\mu\right)^{1/p}. \tag{7.73}$$

Proof. The modified formula (7.73) having the integrand $\big(|f_1|+\cdots+|f_m|\big)^p$ on the left side will be considered first. The sum $|f_1|+\cdots+|f_m|$ is in $L^p_\mu(X)$ because

$$\big(|f_1(x)|+\cdots+|f_m(x)|\big)^p \leq m^{p-1}\big(|f_1(x)|^p+\cdots+|f_m(x)|^p\big)$$

for every $x \in X$. The above inequality can be obtained by applying the discrete form of Jensen's inequality to the convex combination $|f_1(x)|/m+\cdots+|f_m(x)|/m$, and convex function $\varphi : [0,\infty) \to \mathbb{R}$ given by $\varphi(t) = m^p t^p$.

Let $I_j \subseteq \mathbb{R}$ be an interval containing the image of $|f_j|^p$ for $j = 1,\dots,m$, and let $\phi : I_1 \times \cdots \times I_m \to \mathbb{R}$ be a concave function. The generalized integral inequality in formula (7.34) for integer $n = m$ with functions $g_1 = |f_1|^p, \dots, g_m = |f_m|^p$, and $f = -\phi$ appears as

$$\frac{\int_X \phi\big(|f_1|^p,\dots,|f_m|^p\big)\,d\mu}{\mu(X)} \le \phi\left(\frac{\int_X |f_1|^p\,d\mu}{\mu(X)},\dots,\frac{\int_X |f_m|^p\,d\mu}{\mu(X)}\right).$$

If we employ the concave function $\phi : [0,\infty) \times \cdots \times [0,\infty) \to \mathbb{R}$ defined by

$$\phi(x_1,\dots,x_m) = (x_1^{1/p} + \cdots + x_m^{1/p})^p,$$

then we obtain

$$\frac{\int_X (|f_1| + \cdots + |f_m|)^p\,d\mu}{\mu(X)} \le \left(\frac{\int_X |f_1|^p\,d\mu}{\mu(X)}\right)^p + \cdots + \left[\left(\frac{\int_X |f_m|^p\,d\mu}{\mu(X)}\right)^p\right]^p.$$

After multiplying by $\mu(X)$, and raising to the power $1/p$, we attain

$$\left(\int_X (|f_1| + \cdots + |f_m|)^p\,d\mu\right)^{1/p} \le \left(\int_X |f_1|^p\,d\mu\right)^{1/p} + \cdots + \left(\int_X |f_m|^p\,d\mu\right)^{1/p}.$$

This inequality can be extended to the left by using the integrand $|f_1 + \cdots + f_m|^p$ because $|f_1(x) + \cdots + f_m(x)|^p \le \big(|f_1(x)| + \cdots + |f_m(x)|\big)^p$ for every $x \in X$. □

The integral inequality in formula (7.73) can be represented as the norm inequality

$$\|f_1 + \cdots + f_m\|_p \le \|f_1\|_p + \cdots + \|f_m\|_p. \tag{7.74}$$

The following is the generalized discrete form of the Minkowski inequality.

Corollary 7.4. *Let $x_{ij} \in \mathbb{C}$ be the complex numbers for $i = 1,\dots,n$ and $j = 1,\dots,m$, and let $p \ge 1$ be a real number.*

Then, we have the discrete inequality

$$\left(\sum_{i=1}^{n} |x_{i1} + \cdots + x_{im}|^p\right)^{1/p} \le \left(\sum_{i=1}^{n} |x_{i1}|^p\right)^{1/p} + \cdots + \left(\sum_{i=1}^{n} |x_{im}|^p\right)^{1/p}. \tag{7.75}$$

Proof. The integral inequality in formula (7.73) is transformed into discrete inequality in formula (7.75) by exploring the counting measure as in Corollary 7.2. □

By using n-tuples $x_j = (x_{1j},\dots,x_{nj})$ for $j = 1,\dots,m$, and required p-norms, the discrete inequality in formula (7.75) can be presented as the norm inequality

$$\|x_1 + \cdots + x_m\|_p \le \|x_1\|_p + \cdots + \|x_m\|_p. \tag{7.76}$$

Formulas (7.74) and (7.76) can be easily achieved by applying the method of mathematical induction to integer m, wherein the case $m = 2$ is as the induction base. In doing so, formulas (7.73) and (7.76) would also be verified as equivalents of (7.74) and (7.76), respectively.

Original details concerning Minkowski's inequalities can be seen in his book [9].

BIBLIOGRAPHY

[1] M. Bessenyei, The Hermite-Hadamard inequality on simplices, *Amer. Math. Monthly* **115** (2008), 339–345.

[2] A. Farissi, Simple proof and refinement of Hermite-Hadamard inequality, *J. Math. Inequal.* **4** (2010), 365–369.

[3] J. Hadamard, Étude sur les propriétés des fonctions entières et en particulier d'une fonction considerée par Riemann, *J. Math. Pures Appl.*, **58** (1893), 171–215.

[4] Ch. Hermite, Sur deux limites d'une intégrale définie, *Mathesis*, **3** (1883), 82.

[5] O. Hölder, Uber einen Mittelwertsatz, *Nachr. Ges. Wiss. Goettingen* (1889), 38–47.

[6] J. L. W. V. Jensen, Om konvekse Funktioner og Uligheder mellem Middelværdier, *Nyt Tidsskr. Math. B* **16** (1905), 49–68.

[7] J. L. W. V. Jensen, Sur les fonctions convexes et les inégalités entre les valeurs moyennes, *Acta Math.* **30** (1906), 175–193.

[8] E. J. McShane, Jensen's inequality, *Bulletin of the American Mathematical Society*, **43** (1937), 521–527.

[9] H. Minkowski, Theorie der Konvexen Körper, *Insbesondere Begründung ihres Oberflächenbegriffs,* Gesammelte Abhandlungen II, Leipzig, (1911).

[10] F.-C. Mitroi and C. I. Spiridon, Refinements of Hermite-Hadamard inequality on simplices, *Math. Rep. (Bucur.)***15** (2013), arXiv:1105.5043.

[11] E. Neuman, Inequalities involving multivariate convex functions II, *Proc. Amer. Math. Soc.* **109** (1990), 965–974.

[12] C. P. Niculescu and L. E. Persson, *Convex Functions and Their Applications,* Canadian Mathematical Society, Springer, New York, (2006).

[13] Z. Pavić, Extension of Jensen's inequality to affine combinations, *J. Inequal. Appl.* **2014** (2014), Article 298.

[14] Z. Pavić, Geometric and analytic connections of the Jensen and Hermite-Hadamard inequality, *Math. Sci. Appl. E-Notes* **4** (2016), 69–76.

[15] Z. Pavić, Improvements of the Hermite-Hadamard inequality for the simplex, *J. Inequal. Appl.* **2017** (2017), Article 3.

[16] Z. Pavić, J. Pečarić and I. Perić, Integral, discrete and functional variants of Jensen's inequality, *J. Math. Inequal.* **5** (2011), 253–264.

[17] Z. Pavić, The Jensen and Hermite-Hadamard inequality on the triangle, *J. Math. Inequal.* **11** (2017), 1099–1112.

[18] A. W. Roberts and D. E. Varberg, *Convex Functions,* Academic Press, New York and London, (1973).

[19] L. J. Rogers, An extension of a certain theorem in inequalities, *Messenger of Math.* **17** (1888), 145–150.

[20] W. H. Young, On classes of summable functions and their Fourier series, *Proceedings of the Royal Society A* **87** (1912), 225–229.

8 Refinements of Young's Integral Inequality via Fundamental Inequalities and Mean Value Theorems for Derivatives[1]

Feng Qi
Henan Polytechnic University
Inner Mongolia University for Nationalities
Tianjin Polytechnic University

Wen-Hui Li
Zhengzhou University of Science and Technology

Guo-Sheng Wu
Sichuan Technology and Business University

Bai-Ni Guo
Henan Polytechnic University

CONTENTS

[1]Dedicated to people facing and fighting COVID-19.

8.1 YOUNG'S INTEGRAL INEQUALITY AND SEVERAL REFINEMENTS

In the first part of this paper, we mainly review several refinements of Young's integral inequality via several mean value theorems, such as Lagrange's and Taylor's mean value theorems of Lagrange's and Cauchy's type remainders, and via several fundamental inequalities, such as Čebyšev's integral inequality, Hermite-Hadamard's type integral inequalities, Hölder's integral inequality, and Jensen's discrete and integral inequalities, in terms of higher-order derivatives and their norms, and simply survey several applications of several refinements of Young's integral inequality.

8.1.1 YOUNG'S INTEGRAL INEQUALITY

One of the fundamental and general inequalities in mathematics is Young's integral inequality below.

Theorem 8.1 ([55]). *Let $h(x)$ be a real-valued, continuous, and strictly increasing function on $[0,c]$ with $c>0$. If $h(0)=0$, $a\in[0,c]$, and $b\in[0,h(c)]$, then*

$$\int_0^a h(x)\,\mathrm{d}x+\int_0^b h^{-1}(x)\,\mathrm{d}x\ge ab, \tag{8.1}$$

where h^{-1} denotes the inverse function of h. The equality in (8.1) *is valid if and only if $b=h(a)$.*

Proof. This proof is adapted from the proof of [25, Section 2.7, Theorem 1].

Set

$$f(a)=ab-\int_0^a h(x)\,\mathrm{d}x \tag{8.2}$$

and consider $b>0$ as a parameter. Since $f'(a)=b-h(a)$ and h is strictly increasing, one obtains

$$f'(a)\begin{cases}>0, & 0<a<h^{-1}(b);\\ =0, & a=h^{-1}(b);\\ <0, & a>h^{-1}(b).\end{cases}$$

This means that $f(a)$ has a maximum of f at $a=h^{-1}(b)$. Therefore, it follows that

$$f(a)\le\max\{f(x)\}=f\big(h^{-1}(b)\big). \tag{8.3}$$

Integrating by parts gives

$$f\big(h^{-1}(b)\big)=bh^{-1}(b)-\int_0^{h^{-1}(b)}h(x)\,\mathrm{d}x=\int_0^{h^{-1}(b)}xh'(x)\,\mathrm{d}x.$$

Substituting $y=h(x)$ into the above integral yields

$$f\big(h^{-1}(b)\big)=\int_0^b h^{-1}(y)\,\mathrm{d}y. \tag{8.4}$$

Putting (8.2) and (8.4) into (8.3) results in (8.1). The proof of Theorem 8.1 is complete. □

Remark 8.1. *The geometric interpretation of Young's integral inequality* (8.1) *can be demonstrated by Figures 8.1 and 8.2.*

In Figure 8.1, we have

$$A+C=\int_0^a h(x)\,\mathrm{d}x, \quad A+B=ab, \quad B=\int_0^b h^{-1}(x)\,\mathrm{d}x,$$
$$A+B+C=\int_0^a h(x)\,\mathrm{d}x+\int_0^b h^{-1}(x)\,\mathrm{d}x \geq ab = A+B.$$

Therefore, the inequality (8.1) *means that the area*

$$C=\int_{h^{-1}(b)}^a h(x)\,\mathrm{d}x - b\left[a-h^{-1}(b)\right] \geq 0. \tag{8.5}$$

In Figure 8.2, we have

$$A=\int_0^a h(x)\,\mathrm{d}x, \quad A+B=ab, \quad B+C=\int_0^b h^{-1}(x)\,\mathrm{d}x,$$
$$A+B+C=\int_0^a h(x)\,\mathrm{d}x+\int_0^b h^{-1}(x)\,\mathrm{d}x \geq ab = A+B.$$

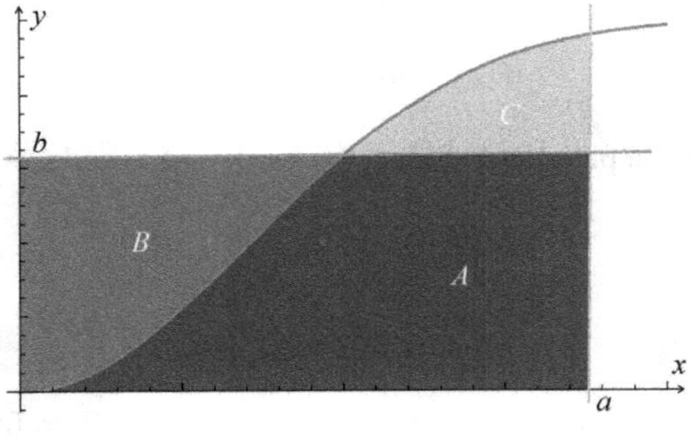

Figure 8.1 Geometric interpretation of the inequality (8.1).

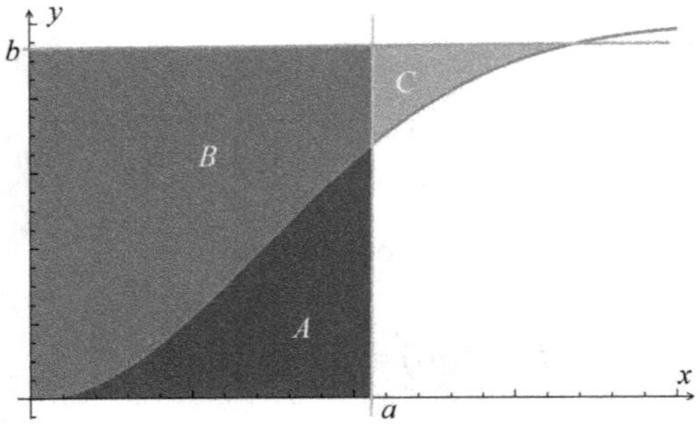

Figure 8.2 Geometric interpretation of the inequality (8.1).

Therefore, the inequality (8.1) *means that the area*

$$C = b\left[h^{-1}(b) - a\right] - \int_a^{h^{-1}(b)} h(x)\,\mathrm{d}x \geq 0. \tag{8.6}$$

Remark 8.2. *We notice that two expressions* (8.5) *and* (8.6) *are of the same form*

$$C = \int_{h^{-1}(b)}^{a} h(x)\,\mathrm{d}x - ab + bh^{-1}(b) \geq 0, \tag{8.7}$$

no matter which of a and $h^{-1}(b)$ is smaller or bigger.

Remark 8.3. *When $p > 1$, taking $h(x) = x^{p-1}$ in* (8.1) *derives*

$$\frac{1}{p}a^p + \frac{1}{q}b^q \geq ab$$

for $a,b \geq 0$ and $p,q > 1$ satisfying $\frac{1}{p} + \frac{1}{q} = 1$. Further replacing a^p and b^q by x and y, respectively, leads to

$$x^{1/p}y^{1/q} \leq \frac{x}{p} + \frac{y}{q} \tag{8.8}$$

for $x,y \geq 0$ and $a,q > 1$ satisfying $\frac{1}{p} + \frac{1}{q} = 1$. Perhaps this is why the weighted arithmetic-geometric inequality (8.8) *is also called Young's inequality in [13,14,21] and closely related references therein.*

Remark 8.4. *The inequality*

$$\sum_{k=1}^{n} \frac{\cos(k\theta)}{k} > -1, \quad n \geq 2, \quad \theta \in [0,\pi]$$

is also called Young's inequality in [2,3] and closely related references therein.

Remark 8.5. *In [25, Secton 2.7] and [26, Chapter XIV], a plenty of refinements, extensions, generalizations, and applications of Young's integral inequality* (8.1) *were collected, reviewed, and surveyed. For some new and recent development on this topic after the year 1990, please refer to the papers [4,43,47,51,56] and closely related references therein.*

8.1.2 REFINEMENTS OF YOUNG'S INTEGRAL INEQUALITY VIA LAGRANGE'S MEAN VALUE THEOREM

In 2008, Hoorfar and Qi refined Young's integral inequality (8.1) via Lagrange's mean value theorem for derivatives.

Theorem 8.2 ([18, Theorem 1]). *Let $h(x)$ be a differentiable and strictly increasing function on $[0,c]$ for $c > 0$, and let h^{-1} be the inverse function of h. If $h(0) = 0$, $a \in [0,c]$, $b \in [0,h(c)]$, and $h'(x)$ is strictly monotonic on $[0,c]$, then*

$$\frac{m}{2}\left[a - h^{-1}(b)\right]^2 \leq \int_0^a h(x)\,\mathrm{d}x + \int_0^b h^{-1}(x)\,\mathrm{d}x - ab \leq \frac{M}{2}\left[a - h^{-1}(b)\right]^2, \tag{8.9}$$

where

$$m = \min\{h'(a), h'(h^{-1}(b))\}$$

and

$$M = \max\{h'(a), h'(h^{-1}(b))\}.$$

The equalities in (8.9) *are valid if and only if* $b = h(a)$.

Proof. This is a modification of the proof of [18, Theorem 1] in [18, Section 2].

Changing the variable of integration by $x = h(y)$ and integrating by parts yield

$$\begin{aligned}\int_0^a h(x)\,\mathrm{d}x + \int_0^b h^{-1}(x)\,\mathrm{d}x &= \int_0^a h(x)\,\mathrm{d}x + \int_0^{h^{-1}(b)} yh'(y)\,\mathrm{d}y \\ &= \int_0^a h(x)\,\mathrm{d}x + bh^{-1}(b) - \int_0^{h^{-1}(b)} h(x)\,\mathrm{d}x \\ &= bh^{-1}(b) + \int_{h^{-1}(b)}^a h(x)\,\mathrm{d}x \\ &= ab + \int_{h^{-1}(b)}^a [h(x) - b]\,\mathrm{d}x. \end{aligned} \tag{8.10}$$

From the last line in (8.10), we can see that if $h^{-1}(b) = a$, then those equalities in (8.9) hold.

If $h^{-1}(b) < a$, since $h(x)$ is strictly increasing, then $h(x) - b > 0$ for $x \in (h^{-1}(b), a)$. By Lagrange's mean value theorem for derivatives, we can see that there exists $\xi = \xi(x)$, satisfying $h^{-1}(b) < \xi < x \le a$, such that

$$0 < h(x) - b = [x - h^{-1}(b)]h'(\xi).$$

By virtue of monotonicity of $h'(x)$ on $[0, c]$, we reveal that

$$0 < m = \min\{h'(a), h'(h^{-1}(b))\} < h'(\xi) < \max\{h'(a), h'(h^{-1}(b))\} = M.$$

Consequently, we have

$$0 < m[x - h^{-1}(b)] < h(x) - b < M[x - h^{-1}(b)].$$

As a result, we have

$$m\int_{h^{-1}(b)}^a [x - h^{-1}(b)]\,\mathrm{d}x < \int_{h^{-1}(b)}^a [h(x) - b]\,\mathrm{d}x < M\int_{h^{-1}(b)}^a [x - h^{-1}(b)]\,\mathrm{d}x$$

which is equivalent to

$$\frac{m}{2}[a - h^{-1}(b)]^2 < \int_{h^{-1}(b)}^a [h(x) - b]\,\mathrm{d}x < \frac{M}{2}[a - h^{-1}(b)]^2. \tag{8.11}$$

If $h^{-1}(b) > a$, we can derive inequalities in (8.11) by a similar argument as above.

Substituting the double inequality (8.11) into the equality (8.10) leads to the double inequality (8.9). The proof of Theorem 8.2 is complete. □

Remark 8.6. *The geometric interpretation of the double inequality* (8.9) *is that the areas C in Figures 8.3–8.6 satisfy*

$$\frac{m}{2}\left[a-h^{-1}(b)\right]^2 \le C \le \frac{M}{2}\left[a-h^{-1}(b)\right]^2. \tag{8.12}$$

When $h'(x)$ is strictly increasing, the double inequality (8.12) *can be equivalently written as*

$$\frac{h'(a)}{2}\left[a-h^{-1}(b)\right] \le \frac{\int_{h^{-1}(b)}^{a} h(x)\,\mathrm{d}x}{a-h^{-1}(b)} \le \frac{h'\left(h^{-1}(b)\right)}{2}\left[a-h^{-1}(b)\right]$$

and

$$\frac{h'(a)}{2}\left[h^{-1}(b)-a\right] \le \frac{\int_{a}^{h^{-1}(b)} h(x)\,\mathrm{d}x}{h^{-1}(b)-a} \le \frac{h'\left(h^{-1}(b)\right)}{2}\left[h^{-1}(b)-a\right]$$

corresponding to Figures 8.3 and 8.4, respectively.

Remark 8.7. *If Q is a convex function on J, then*

$$Q\left(\frac{\tau+\mu}{2}\right) \le \frac{1}{\mu-\tau}\int_{\tau}^{\mu} Q(x)\,\mathrm{d}x \le \frac{Q(\tau)+Q(\mu)}{2}; \tag{8.13}$$

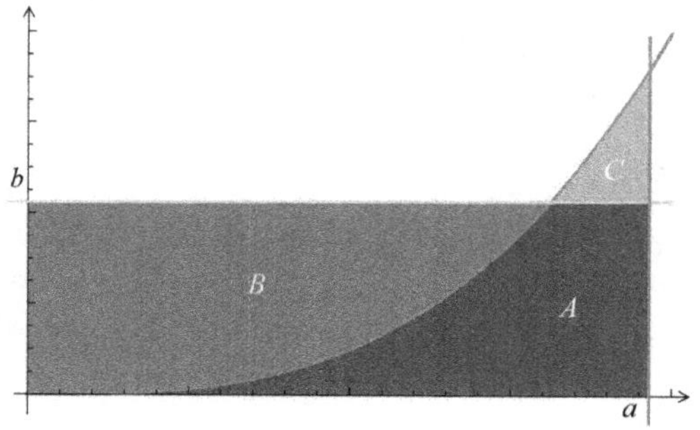

Figure 8.3 Geometric interpretation of the double inequality (8.9).

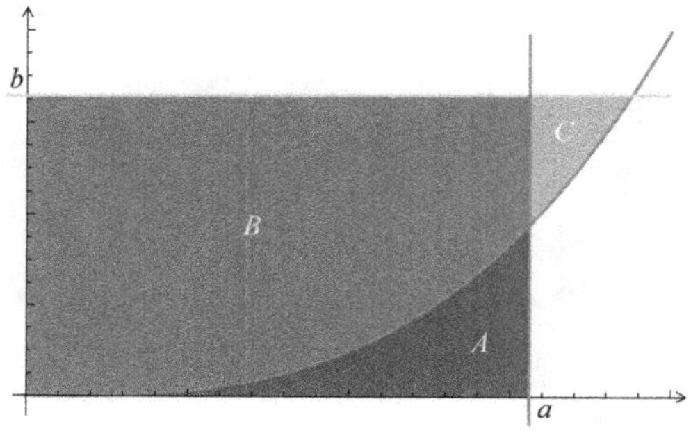

Figure 8.4 Geometric interpretation of the double inequality (8.9).

if Q is a concave function on J, then the double inequality (8.13) *is reversed, where $J \subseteq \mathbb{R}$ is a nonempty interval and $\tau, \mu \in J$ with $\tau < \mu$. The double inequality* (8.13) *is called Hermite-Hadamard's integral inequality for convex functions [7,38,45]. When $a > h^{-1}(b)$, as shown in Figures 8.3 and 8.5, and $h'(x)$ is strictly increasing, i.e., the function $h(x)$ is convex, as shown in Figures 8.3 and 8.4, applying the double inequality* (8.13) *yields*

$$h\left(\frac{a+h^{-1}(b)}{2}\right) \le \frac{\int_{h^{-1}(b)}^{a} h(x)\,\mathrm{d}x}{a-h^{-1}(b)} \le \frac{h(a)+b}{2}.$$

Substituting this into the third line in (8.10) *gives*

$$\int_0^a h(x)\,\mathrm{d}x + \int_0^b h^{-1}(x)\,\mathrm{d}x \le ab + \frac{h(a)-b}{2}\left[a-h^{-1}(b)\right]$$

and

$$\begin{aligned}\int_0^a h(x)\,\mathrm{d}x + \int_0^b h^{-1}(x)\,\mathrm{d}x &\ge bh^{-1}(b) + h\left(\frac{a+h^{-1}(b)}{2}\right)\left[a-h^{-1}(b)\right] \\ &= ab + \left[h\left(\frac{a+h^{-1}(b)}{2}\right) - b\right]\left[a-h^{-1}(b)\right].\end{aligned}$$

Equivalently speaking, it follows that the area C satisfies

$$\left[h\left(\frac{a+h^{-1}(b)}{2}\right) - b\right]\left[a-h^{-1}(b)\right] \le C \le \frac{h(a)-b}{2}\left[a-h^{-1}(b)\right].$$

Similarly, we can discuss other cases, corresponding to Figures 8.5 and 8.6, that the derivative $h'(x)$ is strictly decreasing.

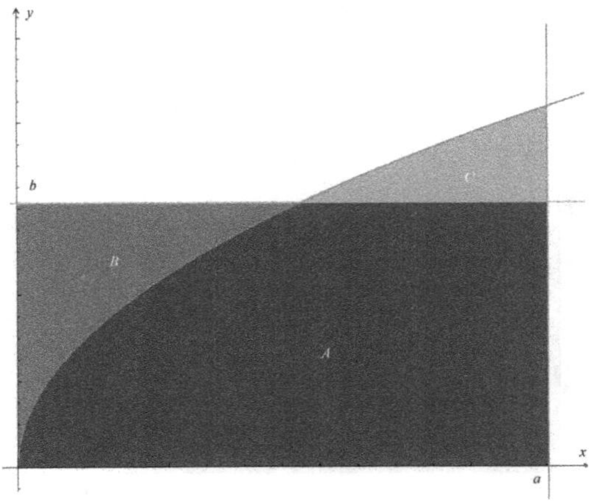

Figure 8.5 Geometric interpretation of the double inequality (8.9).

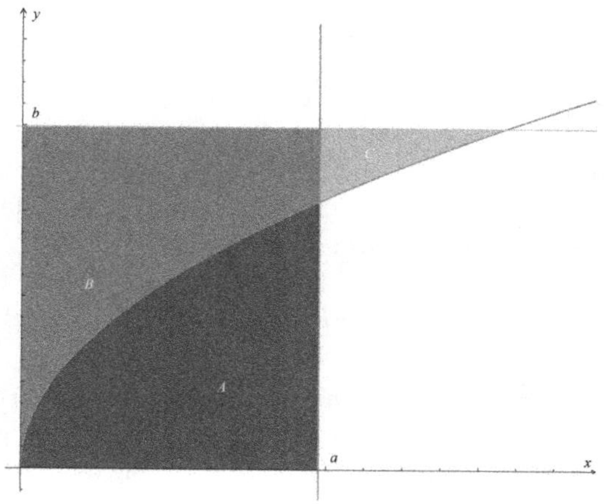

Figure 8.6 Geometric interpretation of the double inequality (8.9).

Remark 8.8. *Mercer has applied and employed the double inequality* (8.9) *in the paper [23] and in the Undergraduate Texts in Mathematics [24].*

8.1.3 REFINEMENTS OF YOUNG'S INTEGRAL INEQUALITY VIA HERMITE-HADAMARD'S AND ČEBYŠEV'S INTEGRAL INEQUALITIES

In 2009 and 2010, among other things, Jakšetić and Pečarić refined Young's integral inequality (8.1) and Hoorfar–Qi's double inequality (8.9) in [19,20].

Theorem 8.3 ([19, Theorem 2.1] and [20, Theorem 2.3]). *Let $h(x)$ be a differentiable and strictly increasing function on $[0,c]$ for $c>0$, $h(0)=0$, $a\in[0,c]$, $b\in[0,h(c)]$, and h^{-1} be the inverse function of h. Denote*

$$\alpha=\min\{a,h^{-1}(b)\} \quad \textit{and} \quad \beta=\max\{a,h^{-1}(b)\}. \tag{8.14}$$

1. *If $h'(x)$ is increasing on $[\alpha,\beta]$ and $b<h(a)$, or if $h'(x)$ is decreasing on $[\alpha,\beta]$ and $b>h(a)$, then*

$$\begin{aligned}[a-h^{-1}(b)]\left[h\left(\frac{a+h^{-1}(b)}{2}\right)-b\right] &\le \int_0^a h(x)\,\mathrm{d}x+\int_0^b h^{-1}(x)\,\mathrm{d}x-ab\\ &\le \frac{1}{2}[a-h^{-1}(b)][h(a)-b].\end{aligned} \tag{8.15}$$

2. *If $h'(x)$ is increasing on $[\alpha,\beta]$ and $b>h(a)$, or if $h'(x)$ is decreasing on $[\alpha,\beta]$ and $b<h(a)$, then the inequality* (8.15) *is reversed.*
3. *The equality in* (8.15) *is valid if and only if $h(x)=\lambda x$ for $\lambda>0$ or $b=h(a)$.*

Proof. This is the outline of proofs of [19, Theorem 2.1] and [20, Theorem 2.3].

From the third line in (8.10), it follows that

$$\int_0^a h(x)\,\mathrm{d}x + \int_0^b h^{-1}(x)\,\mathrm{d}x - ab = b\left[h^{-1}(b) - a\right] + \int_{h^{-1}(b)}^a h(x)\,\mathrm{d}x. \tag{8.16}$$

Considering monotonicity of $h'(x)$ and applying the double inequality (8.13) to the integrand in the last term of (8.16), we can derive the double inequality (8.15).

The last term in (8.10) can be rewritten as

$$\begin{aligned}\int_{h^{-1}(b)}^a [h(x) - b]\,\mathrm{d}x &= \int_{h^{-1}(b)}^a \left[h(x) - h(h^{-1}(b))\right]\mathrm{d}x \\ &= \int_{h^{-1}(b)}^a \int_{h^{-1}(b)}^x h'(u)\,\mathrm{d}u\,\mathrm{d}x = \int_{h^{-1}(b)}^a (a-u)h'(u)\,\mathrm{d}u. \end{aligned} \tag{8.17}$$

Let $f, g : [\mu, \nu] \to \mathbb{R}$ be the integrable functions satisfying that they are both increasing or both decreasing. Then

$$\int_\mu^\nu f(x)\,\mathrm{d}x \int_\mu^\nu g(x)\,\mathrm{d}x \le (\nu - \mu) \int_\mu^\nu f(x)g(x)\,\mathrm{d}x. \tag{8.18}$$

If one of the function f or g is nonincreasing and the other nondecreasing, then the inequality in (8.18) is reversed. The inequality (8.18) is called Čebyšev's integral inequality in the literature [26, Chapter IX] and [35,39]. Applying (8.18) to the last term in (8.17) leads to the right-hand side of the inequality (8.15). The proof of Theorem 8.3 is complete. □

Remark 8.9. *The double inequality* (8.15) *can be geometrically interpreted as*

$$\left[a - h^{-1}(b)\right]\left[h\left(\frac{a + h^{-1}(b)}{2}\right) - b\right] \le C \le \frac{1}{2}\left[a - h^{-1}(b)\right][h(a) - b],$$

where C denotes the area shown in Figures 8.1–8.6.

8.1.4 REFINEMENTS OF YOUNG'S INTEGRAL INEQUALITY VIA JENSEN'S DISCRETE AND INTEGRAL INEQUALITIES

In [20, Theorem 2.6], Jensen's discrete and integral inequalities were employed to establish the following inequalities which refine Young's integral inequality (8.1) and Hoorfar–Qi's double inequality (8.9).

Theorem 8.4 ([20, Theorem 2.6]). *Let $h(x)$ be a differentiable and strictly increasing function on $[0, c]$ for $c > 0$, and let h^{-1} be the inverse function of h. If $h(0) = 0$, $a \in [0, c]$, $b \in [0, h(c)]$, and $h'(x)$ is convex on $[\alpha, \beta]$, then*

$$\begin{aligned}\frac{\left[a - h^{-1}(b)\right]^2}{2} h'\left(\frac{a + 2h^{-1}(b)}{3}\right) &\le \int_0^a h(x)\,\mathrm{d}x + \int_0^b h^{-1}(x)\,\mathrm{d}x - ab \\ &\le \frac{\left[a - h^{-1}(b)\right]^2}{3}\left[\frac{h'(a)}{2} + h'\left(h^{-1}(b)\right)\right]. \end{aligned} \tag{8.19}$$

If $h'(x)$ is concave, then the double inequality (8.19) *is reversed.*

Proof. This is the outline of the proof of [20, Theorem 2.6].

Changing the variable of the last term in (8.17) results in

$$\int_{h^{-1}(b)}^{a}(a-u)h'(u)\,\mathrm{d}u=\int_{0}^{1}\left[a-h^{-1}(b)\right]^{2}(1-x)h'\left(xa+(1-x)h^{-1}(b)\right)\mathrm{d}x. \quad (8.20)$$

If f is a convex function on an interval $I\subseteq\mathbb{R}$ and if $n\ge 2$ and $x_k\in I$ for $1\le k\le n$, then

$$f\left(\frac{1}{\sum_{k=1}^{n}p_k}\sum_{k=1}^{n}p_kx_k\right)\le\frac{1}{\sum_{k=1}^{n}p_k}\sum_{k=1}^{n}p_kf(x_k), \quad (8.21)$$

where $p_k>0$ for $1\le k\le n$. If f is concave, the inequality (8.21) is reversed. The inequality (8.21) is called Jensen's discrete inequality for convex functions in the literature [25, Section 1.4] and [26, Chapter I]. Applying (8.21) to the third factor in the integrand of the right-hand side in (8.20) arrives at the right inequality in (8.19).

Let ϕ be a convex function on $[\mu,\nu]$, $f\in L_1(\mu,\nu)$, and σ be a nonnegative measure. Then,

$$\phi\left(\frac{\int_{\mu}^{\nu}f(x)\,\mathrm{d}\sigma}{\int_{\mu}^{\nu}\mathrm{d}\sigma}\right)\le\frac{\int_{\mu}^{\nu}\phi(f(x))\,\mathrm{d}\sigma}{\int_{\mu}^{\nu}\mathrm{d}\sigma}. \quad (8.22)$$

If ϕ is a concave function, then the inequality (8.22) is reversed. The inequality (8.22) is called Jensen's integral inequality for convex functions in the literature [26, p. 10, (7.15)]. Applying (8.22) yields

$$\begin{aligned}\int_{h^{-1}(b)}^{a}(a-x)h'(x)\,\mathrm{d}x&\ge\frac{[1-h^{-1}(b)]^{2}}{2}h'\left(\frac{\int_{h^{-1}(b)}^{a}(a-x)x\,\mathrm{d}x}{\int_{h^{-1}(b)}^{a}(a-x)\,\mathrm{d}x}\right)\\&=\frac{[1-h^{-1}(b)]^{2}}{2}h'\left(\frac{a+2h^{-1}(b)}{3}\right).\end{aligned}$$

The proof of Theorem 8.4 is complete. □

Remark 8.10. *The double inequality* (8.19) *can be geometrically interpreted as*

$$\frac{\left[a-h^{-1}(b)\right]^{2}}{2}h'\left(\frac{a+2h^{-1}(b)}{3}\right)\le C\le\frac{\left[a-h^{-1}(b)\right]^{2}}{3}\left[\frac{h'(a)}{2}+h'\left(h^{-1}(b)\right)\right],$$

where C denotes the area shown in Figures 8.1–8.6.

8.1.5 REFINEMENTS OF YOUNG'S INTEGRAL INEQUALITY VIA HÖLDER'S INTEGRAL INEQUALITY

In [20, Theorem 2.1], Hölder's integral inequality was utilized to present the following inequalities, which refine Young's integral inequality (8.1) and Hoorfar–Qi's double inequality (8.9), for the normed spaces.

Theorem 8.5 ([20, Theorem 2.1]). *Let $h(x)$ be a differentiable and strictly increasing function on $[0,c]$ for $c>0$, and let h^{-1} be the inverse function of h. If $h(0)=0$, $a\in[0,c]$, $b\in[0,h(c)]$, and $h'(x)$ is almost everywhere continuous with respect to Lebesgue measure on $[\alpha,\beta]$, then the double inequality*

$$C_u\|h'\|_v \le \int_0^a h(x)\,\mathrm{d}x+\int_0^b h^{-1}(x)\,\mathrm{d}x-ab\le C_p\|h'\|_q \tag{8.23}$$

is valid for all u,v and p,q satisfying

1. $\frac{1}{u}+\frac{1}{v}=1$ *for* $u,v\in(-\infty,0)\cup(0,1)$, *or* $(u,v)=(1,-\infty)$, *or* $(u,v)=(-\infty,1)$;
2. $\frac{1}{p}+\frac{1}{q}=1$ *for* $1<p,q<\infty$, *or* $(p,q)=(+\infty,1)$, *or* $(p,q)=(1,+\infty)$;

where

$$C_r=\begin{cases}\left[\frac{\left|a-h^{-1}(b)\right|^{r+1}}{r+1}\right]^{1/r}, & r\neq 0,\pm\infty;\\ \left|a-h^{-1}(b)\right|, & r=+\infty;\\ 0, & r=-\infty\end{cases}$$

and

$$\|h'\|_r=\begin{cases}\left[\int_\alpha^\beta [h'(t)]^r\,\mathrm{d}t\right]^{1/r}, & r\neq 0,\pm\infty;\\ \sup\{h'(t),t\in[\alpha,\beta]\}, & r=+\infty;\\ \inf\{h'(t),t\in[\alpha,\beta]\}, & r=-\infty.\end{cases}$$

Proof. This is the outline of the proof of [20, Theorem 2.1].

Let $\frac{1}{p}+\frac{1}{q}=1$ with $p>0$ and $p\neq 1$, let f and g be the real functions on $[\mu,\nu]$, and let $|f|^p$ and $|g|^q$ be integrable on $[\mu,\nu]$.

1. If $p>1$, then

$$\int_\mu^\nu |f(x)g(x)|\,\mathrm{d}x\le\left[\int_\mu^\nu |f(x)|^p\,\mathrm{d}x\right]^{1/p}\left[\int_\mu^\nu |g(x)|^q\,\mathrm{d}x\right]^{1/q}. \tag{8.24}$$

   The equality in (8.24) holds if and only if $A|f(x)|^p=B|g(x)|^q$ almost everywhere for two constants A and B.
2. If $0<p<1$, then the inequality (8.24) is reversed.

The inequality (8.24) is called Hölder's integral inequality in the lierature [26, Chapter V] and [44,48,49].

From (8.17), it follows that,

1. by a property of definite integrals, we have

$$\begin{aligned}\int_{h^{-1}(b)}^a (a-u)h'(u)\,\mathrm{d}u&=\int_\beta^\alpha |a-u|h'(u)\,\mathrm{d}u\\ &\le\left|h^{-1}(b)-a\right|\int_\beta^\alpha h'(u)\,\mathrm{d}u=C_\infty\|h'\|_1;\end{aligned}$$

2. by a property of definite integrals, we have

$$\int_{h^{-1}(b)}^{a}(a-u)h'(u)\,\mathrm{d}u=\int_{\beta}^{\alpha}|a-u|h'(u)\,\mathrm{d}u\le C_1\|h'\|_\infty;$$

3. by Hölder's integral inequality (8.24), we have

$$\begin{aligned}\int_{h^{-1}(b)}^{a}(a-u)h'(u)\,\mathrm{d}u&=\int_{\beta}^{\alpha}|a-u|h'(u)\,\mathrm{d}u\\&\le\left(\int_{\beta}^{\alpha}|a-u|^q\,\mathrm{d}u\right)^{1/q}\left(\int_{\beta}^{\alpha}[h'(u)]^p\,\mathrm{d}u\right)^{1/p}=C_q\|h'\|_p.\end{aligned}$$

The rest proofs are straightforward. The proofs of the double inequality (8.23) and Theorem 8.5 are thus complete. □

Remark 8.11. *The double inequality* (8.23) *can be geometrically interpreted as*

$$C_u\|h'\|_v\le C\le C_p\|h'\|_q,$$

where C denotes the area shown in Figures 8.1–8.6.

8.1.6 REFINEMENTS OF YOUNG'S INTEGRAL INEQUALITY VIA TAYLOR'S MEAN VALUE THEOREM OF LAGRANGE'S TYPE REMAINDER

In [50, Theorem 3.1], making use of Taylor's mean value theorem of Lagrange's type remainder, Wang, Guo, and Qi refined the above inequalities of Young's type via higher-order derivatives.

Theorem 8.6 ([50, Theorem 3.1])**.** *Let $h(0)=0$ and $h(x)$ be strictly increasing on $[0,c]$ for $c>0$, let $h^{(n)}(x)$ for $n\ge0$ be continuous on $[0,c]$, let $h^{(n+1)}(x)$ be finite and strictly monotonic on $(0,c)$, and let h^{-1} be the inverse function of h. For $a\in[0,c]$ and $b\in[0,h(c)]$,*

1. *if $b<h(a)$, then*

$$\begin{aligned}&\sum_{k=1}^{n}h^{(k)}\big(h^{-1}(b)\big)\frac{\big[a-h^{-1}(b)\big]^{k+1}}{(k+1)!}+m_n(a,b)\frac{\big[a-h^{-1}(b)\big]^{n+2}}{(n+2)!}\\&\le\int_0^a h(x)\,\mathrm{d}x+\int_0^b h^{-1}(x)\,\mathrm{d}x-ab\qquad(8.25)\\&\le\sum_{k=1}^{n}h^{(k)}\big(h^{-1}(b)\big)\frac{\big[a-h^{-1}(b)\big]^{k+1}}{(k+1)!}+M_n(a,b)\frac{\big[a-h^{-1}(b)\big]^{n+2}}{(n+2)!},\end{aligned}$$

where

$$m_n(a,b)=\min\big\{h^{(n+1)}\big(h^{-1}(b)\big),h^{(n+1)}(a)\big\}$$

and

$$M_n(a,b)=\max\big\{h^{(n+1)}\big(h^{-1}(b)\big),h^{(n+1)}(a)\big\};$$

2. *if* $b > h(a)$, *then*
   a. *when* $n = 2\ell$ *for* $\ell \ge 0$, *the double inequality* (8.25) *is valid;*
   b. *when* $n = 2\ell + 1$ *for* $\ell \ge 0$, *we have*

$$\begin{aligned}
&\sum_{k=1}^{n} h^{(k)}\big(h^{-1}(b)\big)\frac{\big[a-h^{-1}(b)\big]^{k+1}}{(k+1)!} - M_n(a,b)\frac{\big[a-h^{-1}(b)\big]^{n+2}}{(n+2)!}\\
&\le \int_0^a h(x)\,\mathrm{d}x + \int_0^b h^{-1}(x)\,\mathrm{d}x - ab\\
&\le \sum_{k=1}^{n} h^{(k)}\big(h^{-1}(b)\big)\frac{\big[a-h^{-1}(b)\big]^{k+1}}{(k+1)!} - m_n(a,b)\frac{\big[a-h^{-1}(b)\big]^{n+2}}{(n+2)!};
\end{aligned} \tag{8.26}$$

3. *if, and only if,* $b = h(a)$, *those equalities in* (8.25) *and* (8.26) *hold.*

Proof. This is the outline of the proof of [50, Theorem 3.1].

Let $f(x)$ be a function having finite nth derivative $f^{(n)}(x)$ everywhere in an open interval (μ, ν) and assume that $f^{(n-1)}(x)$ is continuous on the closed interval $[\mu, \nu]$. Then, for a fixed point $x_0 \in [\mu, \nu]$ and every $x \in [\mu, \nu]$ with $x \ne x_0$, there exists a point x_1 interior to the interval joining x and x_0 such that

$$f(x) = f(x_0) + \sum_{k=1}^{n-1} \frac{f^{(k)}(x_0)}{k!}(x-x_0)^k + \frac{f^{(n)}(x_1)}{n!}(x-x_0)^n. \tag{8.27}$$

Formula (8.27) is called Taylor's mean value theorem of Lagrange's type remainder in the literature [6, p. 113, Theorem 5.19]. Applying (8.27) in the last term of (8.10) reveals

$$\begin{aligned}
\int_{h^{-1}(b)}^{a} [h(x)-b]\,\mathrm{d}x &= \int_{h^{-1}(b)}^{a} \big[h(x) - h\big(h^{-1}(b)\big)\big]\,\mathrm{d}x\\
&= \sum_{k=1}^{n} \frac{h^{(k)}\big(h^{-1}(b)\big)}{k!} \int_{h^{-1}(b)}^{a} \big[x-h^{-1}(b)\big]^k\,\mathrm{d}x\\
&\quad + \frac{1}{(n+1)!}\int_{h^{-1}(b)}^{a} h^{(n+1)}(\xi)\big[x-h^{-1}(b)\big]^{n+1}\,\mathrm{d}x\\
&= \sum_{k=1}^{n} h^{(k)}\big(h^{-1}(b)\big)\frac{\big[a-h^{-1}(b)\big]^{k+1}}{(k+1)!}\\
&\quad + \int_{h^{-1}(b)}^{a} h^{(n+1)}(\xi)\frac{\big[x-h^{-1}(b)\big]^{n+1}}{(n+1)!}\,\mathrm{d}x,
\end{aligned}$$

where ξ is a point interior to the interval joining x and $h^{-1}(b)$. The rest proofs are straightforward discussions on various cases of the factor $h^{(n+1)}(\xi)$. The proof of Theorem 8.6 is complete. □

Remark 8.12. *The double inequalities* (8.25) *and* (8.26) *can be geometrically interpreted as*

$$\sum_{k=1}^{n} h^{(k)}\big(h^{-1}(b)\big)\frac{\big[a-h^{-1}(b)\big]^{k+1}}{(k+1)!}+m_n(a,b)\frac{\big[a-h^{-1}(b)\big]^{n+2}}{(n+2)!}\le C$$
$$\le \sum_{k=1}^{n} h^{(k)}\big(h^{-1}(b)\big)\frac{\big[a-h^{-1}(b)\big]^{k+1}}{(k+1)!}+M_n(a,b)\frac{\big[a-h^{-1}(b)\big]^{n+2}}{(n+2)!}$$

and

$$\sum_{k=1}^{n} h^{(k)}\big(h^{-1}(b)\big)\frac{\big[a-h^{-1}(b)\big]^{k+1}}{(k+1)!}-M_n(a,b)\frac{\big[a-h^{-1}(b)\big]^{n+2}}{(n+2)!}\le C$$
$$\le \sum_{k=1}^{n} h^{(k)}\big(h^{-1}(b)\big)\frac{\big[a-h^{-1}(b)\big]^{k+1}}{(k+1)!}-m_n(a,b)\frac{\big[a-h^{-1}(b)\big]^{n+2}}{(n+2)!},$$

where C denotes the area shown in Figures 8.1–8.6.

8.1.7 REFINEMENTS OF YOUNG'S INTEGRAL INEQUALITY VIA TAYLOR'S MEAN VALUE THEOREM OF CAUCHY'S TYPE REMAINDER AND HÖLDER'S INTEGRAL INEQUALITY

In [50, Theorem 3.2], employing Taylor's mean value theorem of Cauchy's type remainder and Hölder's integral inequality, Wang, Guo, and Qi refined the above inequalities of Young's type via norms of higher-order derivatives.

Theorem 8.7 ([50, Theorem 3.2]). *Let $n\ge 0$ and $h(x)\in C^{n+1}[0,c]$ such that $h(0)=0$, $h^{(n+1)}(x)\ge 0$ on $[\alpha,\beta]$, and $h(x)$ is strictly increasing on $[0,c]$ for $c>0$, let h^{-1} be the inverse function of h, and let $a\in[0,c]$ and $b\in[0,h(c)]$. Then*

1. *when $b>h(a)$ and $n=2\ell$ for $\ell\ge 0$ or when $b<h(a)$, we have*

$$\frac{C_{u,n}}{(n+1)!}\big\|h^{(n+1)}\big\|_v\le \int_0^a h(x)\,\mathrm{d}x+\int_0^b h^{-1}(x)\,\mathrm{d}x-ab$$
$$-\sum_{k=1}^{n} h^{(k)}\big(h^{-1}(b)\big)\frac{\big[a-h^{-1}(b)\big]^{k+1}}{(k+1)!}$$
$$\le\frac{C_{p,n}}{(n+1)!}\big\|h^{(n+1)}\big\|_q;$$

2. *when $b>h(a)$ and $n=2\ell+1$ for $\ell\ge 0$, we have*

$$-\frac{C_{p,n}}{(n+1)!}\big\|h^{(n+1)}\big\|_q\le \int_0^a h(x)\,\mathrm{d}x+\int_0^b h^{-1}(x)\,\mathrm{d}x-ab$$
$$-\sum_{k=1}^{n} h^{(k)}\big(h^{-1}(b)\big)\frac{\big[a-h^{-1}(b)\big]^{k+1}}{(k+1)!}$$
$$\le-\frac{C_{u,n}}{(n+1)!}\big\|h^{(n+1)}\big\|_v;$$

where α,β are defined as in (8.14),

$$C_{r,n}=\begin{cases}\left[\dfrac{\left|a-h^{-1}(b)\right|^{r(n+1)+1}}{r(n+1)+1}\right]^{1/r}, & r\neq 0,\pm\infty;\\ \left|a-h^{-1}(b)\right|^{n+1}, & r=+\infty;\\ 0, & r=-\infty,\end{cases}$$

$$\left\|h^{(n+1)}\right\|_r=\begin{cases}\left[\int_\alpha^\beta\left[h^{(n+1)}(t)\right]^r\mathrm{d}t\right]^{1/r}, & r\neq 0,\pm\infty;\\ \sup\{h^{(n+1)}(t),t\in[\alpha,\beta]\}, & r=+\infty;\\ \inf\{h^{(n+1)}(t),t\in[\alpha,\beta]\}, & r=-\infty,\end{cases}$$

and u,v,p,q satisfy

1. $u<1$ *and* $u\neq 0$ *with* $\frac{1}{u}+\frac{1}{v}=1$, *or* $(u,v)=(-\infty,1)$, *or* $(u,v)=(1,-\infty)$;
2. $1<p,q<\infty$ *with* $\frac{1}{p}+\frac{1}{q}=1$, *or* $(p,q)=(+\infty,1)$, *or* $(p,q)=(1,+\infty)$.

Proof. This is the outline of the proof of [50, Theorem 3.2].

If $f(x)\in C^{n+1}[\mu,\nu]$ and $x_0\in[\mu,\nu]$, then

$$f(x)=\sum_{k=0}^{n}\frac{f^{(k)}(x_0)}{k!}(x-x_0)^k+\frac{1}{n!}\int_{x_0}^{x}(x-t)^n f^{(n+1)}(t)\,\mathrm{d}t. \tag{8.28}$$

Formula (8.28) is called Taylor's mean value theorem of Cauchy's type remainder in the literature [5, p. 279, Theorem 7.6] and [27, p. 6, 1.4.37]. Applying formula (8.28) to the integrand in the last term of (8.10) yields

$$\begin{aligned}\int_{h^{-1}(b)}^{a}[h(x)-b]\,\mathrm{d}x&=\int_{h^{-1}(b)}^{a}\left[h(x)-h\left(h^{-1}(b)\right)\right]\mathrm{d}x\\ &=\sum_{k=1}^{n}h^{(k)}\left(h^{-1}(b)\right)\frac{\left[a-h^{-1}(b)\right]^{k+1}}{(k+1)!}\\ &\quad+\int_{h^{-1}(b)}^{a}\frac{1}{n!}\int_{h^{-1}(b)}^{x}(x-t)^n h^{(n+1)}(t)\,\mathrm{d}t\,\mathrm{d}x\\ &=\sum_{k=1}^{n}h^{(k)}\left(h^{-1}(b)\right)\frac{\left[a-h^{-1}(b)\right]^{k+1}}{(k+1)!}\\ &\quad+\int_{h^{-1}(b)}^{a}\frac{1}{n!}\int_{t}^{a}(x-t)^n h^{(n+1)}(t)\,\mathrm{d}x\,\mathrm{d}t\\ &=\sum_{k=1}^{n}h^{(k)}\left(h^{-1}(b)\right)\frac{\left[a-h^{-1}(b)\right]^{k+1}}{(k+1)!}\\ &\quad+\frac{1}{(n+1)!}\int_{h^{-1}(b)}^{a}(a-t)^{n+1}h^{(n+1)}(t)\,\mathrm{d}t\end{aligned}$$

$$= \sum_{k=1}^{n} h^{(k)}\left(h^{-1}(b)\right) \frac{\left[a-h^{-1}(b)\right]^{k+1}}{(k+1)!} + \begin{cases} \frac{1}{(n+1)!} \int_{\alpha}^{\beta} |a-t|^{n+1} h^{(n+1)}(t)\,\mathrm{d}t, & b < h(a); \\ \frac{(-1)^n}{(n+1)!} \int_{\alpha}^{\beta} |a-t|^{n+1} h^{(n+1)}(t)\,\mathrm{d}t, & b > h(a). \end{cases}$$

Discussing and making use of Hölder's integral inequality (8.24) as in the proof of Theorem 8.5, we can complete the proof of Theorem 8.7. □

Remark 8.13. *Two double inequalities in Theorem 8.7 can be geometrically interpreted as*

$$\frac{C_{u,n}}{(n+1)!} \left\| h^{(n+1)} \right\|_{v} \le C - \sum_{k=1}^{n} h^{(k)}\left(h^{-1}(b)\right) \frac{\left[a-h^{-1}(b)\right]^{k+1}}{(k+1)!} \le \frac{C_{p,n}}{(n+1)!} \left\| h^{(n+1)} \right\|_{q}$$

and

$$-\frac{C_{p,n}}{(n+1)!} \left\| h^{(n+1)} \right\|_{q} \le C - \sum_{k=1}^{n} h^{(k)}\left(h^{-1}(b)\right) \frac{\left[a-h^{-1}(b)\right]^{k+1}}{(k+1)!} \le -\frac{C_{u,n}}{(n+1)!} \left\| h^{(n+1)} \right\|_{v},$$

where C denotes the area shown in Figures 8.1–8.6.

8.1.8 REFINEMENTS OF YOUNG'S INTEGRAL INEQUALITY VIA TAYLOR'S MEAN VALUE THEOREM OF CAUCHY'S TYPE REMAINDER AND ČEBYŠEV'S INTEGRAL INEQUALITY

Theorem 8.8 ([50, Theorem 3.3]). *Let $n \ge 0$ and $h(x) \in C^{n+1}[0,c]$ such that $h(0) = 0$ and $h(x)$ is strictly increasing on $[0,c]$ for $c > 0$, let h^{-1} be the inverse function of h, let $a \in [0,c]$ and $b \in [0,h(c)]$, and let $\ell \ge 0$ be an integer. Then*

1. *when*
   a. *either $h(a) > b$ and $h^{(n+1)}(x)$ is increasing on $[\alpha,\beta]$;*
   b. *or $h(a) < b$, $h^{(n+1)}(x)$ is increasing on $[\alpha,\beta]$, and $n = 2\ell+1$;*
   c. *or $h(a) < b$, $h^{(n+1)}(x)$ is decreasing on $[\alpha,\beta]$, and $n = 2\ell$;*

   the inequality

$$\int_0^a h(x)\,\mathrm{d}x + \int_0^b h^{-1}(x)\,\mathrm{d}x - ab - \sum_{k=1}^{n} h^{(k)}\left(h^{-1}(b)\right) \frac{\left[a-h^{-1}(b)\right]^{k+1}}{(k+1)!} \le \frac{\left[a-h^{-1}(b)\right]^{n+1}}{(n+2)!} \left[h^{(n)}(a) - h^{(n)}\left(h^{-1}(b)\right)\right] \tag{8.29}$$

   is valid;

2. *when*
   a. *either* $h(a) > b$ *and* $h^{(n+1)}(x)$ *is decreasing on* $[\alpha, \beta]$*;*
   b. *or* $h(a) < b$, $h^{(n+1)}(x)$ *is increasing on* $[\alpha, \beta]$*, and* $n = 2\ell$*;*
   c. *or* $h(a) < b$, $h^{(n+1)}(x)$ *is decreasing on* $[\alpha, \beta]$*, and* $n = 2\ell + 1$*;*

   the inequality (8.29) *is reversed;*

where α, β *are defined as in* (8.14).

Proof. This is the outline of the proof of [50, Theorem 3.3].

This follows from applying formula (8.28) as in the proof of Theorem 8.7 and applying Čebyšev's integral inequality (8.18) to the integral

$$\int_{h^{-1}(b)}^{a} (a-t)^{n+1} h^{(n+1)}(t)\,\mathrm{d}t \tag{8.30}$$

in the proof of Theorem 8.7. The proof of Theorem 8.8 is complete. □

Remark 8.14. *The inequality* (8.29) *can be geometrically interpreted as*

$$C - \sum_{k=1}^{n} h^{(k)}\left(h^{-1}(b)\right) \frac{\left[a-h^{-1}(b)\right]^{k+1}}{(k+1)!} \le \frac{\left[a-h^{-1}(b)\right]^{n+1}}{(n+2)!}\left[h^{(n)}(a) - h^{(n)}\left(h^{-1}(b)\right)\right],$$

where C *denotes the area shown in Figures 8.1–8.6.*

8.1.9 REFINEMENTS OF YOUNG'S INTEGRAL INEQUALITY VIA TAYLOR'S MEAN VALUE THEOREM OF CAUCHY'S TYPE REMAINDER AND JENSEN'S INEQUALITIES

Theorem 8.9 ([50, Theorem 3.4]). *Let* $h(x) \in C^{n+1}[0,c]$ *such that* $h(0) = 0$ *and* $h(x)$ *is strictly increasing on* $[0,c]$ *for* $c > 0$*, let* h^{-1} *be the inverse function of* h*, and let* $a \in [0,c]$ *and* $b \in [0, h(c)]$*. If* $h^{(n+1)}(x)$ *is convex on* $[\alpha, \beta]$*, where* α, β *are defined as in* (8.14)*, then*

1. *when* $h(a) > b$ *or when* $h(a) < b$ *and* $n = 2\ell$*, we have*

$$\begin{aligned}
&\frac{\left[a-h^{-1}(b)\right]^{n+2}}{n+2} h^{(n+1)}\left(\frac{a+(n+2)h^{-1}(b)}{n+3}\right) \\
&\le \int_0^a h(x)\,\mathrm{d}x + \int_0^b h^{-1}(x)\,\mathrm{d}x - ab \quad - \sum_{k=1}^{n} h^{(k)}\left(h^{-1}(b)\right)\frac{\left[a-h^{-1}(b)\right]^{k+1}}{(k+1)!} \\
&\le \left[a-h^{-1}(b)\right]^{n+2} \frac{h^{(n+1)}(a) + (n+2)h^{(n+1)}\left(h^{-1}(b)\right)}{(n+3)!};
\end{aligned} \tag{8.31}$$

2. *when $h(a) < b$ and $n = 2\ell + 1$, the double inequality* (8.31) *is reversed,*

where $\ell \geq 0$ is an integer. If $h^{(n+1)}(x)$ is concave on $[\alpha, \beta]$, all the above inequalities are reversed for all corresponding cases.

Proof. This is the outline of the proof of [50, Theorem 3.4].

Considering the integral (8.30) and substituting integral variables give

$$\int_{h^{-1}(b)}^{a} (a-t)^{n+1} h^{(n+1)}(t)\,\mathrm{d}t = \left[a - h^{-1}(b)\right]^{n+2} \\ \times \int_0^1 (1-s)^{n+1} h^{(n+1)}\left(sa + (1-s)h^{-1}(b)\right)\mathrm{d}s.$$

Applying Jensen's inequalities (8.21) and (8.22) to $h^{(n+1)}\left(sa + (1-s)h^{-1}(b)\right)$ in the above equation yield the double inequality (8.31) and its reversed version. The proof of Theorem 8.9 is complete. □

Remark 8.15. *The double inequality* (8.31) *can be geometrically interpreted as*

$$\frac{\left[a - h^{-1}(b)\right]^{n+2}}{n+2} h^{(n+1)}\left(\frac{a + (n+2)h^{-1}(b)}{n+3}\right) \\ \leq C - \sum_{k=1}^{n} h^{(k)}\left(h^{-1}(b)\right) \frac{\left[a - h^{-1}(b)\right]^{k+1}}{(k+1)!} \\ \leq \left[a - h^{-1}(b)\right]^{n+2} \frac{h^{(n+1)}(a) + (n+2)h^{(n+1)}\left(h^{-1}(b)\right)}{(n+3)!},$$

where C denotes the area shown in Figures 8.1–8.6.

8.1.10 REFINEMENTS OF YOUNG'S INTEGRAL INEQUALITY VIA TAYLOR'S MEAN VALUE THEOREM OF CAUCHY'S TYPE REMAINDER AND INTEGRAL INEQUALITIES OF HERMITE-HADAMARD TYPE FOR THE PRODUCT OF TWO CONVEX FUNCTIONS

Theorem 8.10 ([50, Theorem 3.5]). *Let $n \geq 0$ and $h(x) \in C^{n+1}[0, c]$ such that $h(0) = 0$ and $h(x)$ is strictly increasing on $[0, c]$ for $c > 0$, let h^{-1} be the inverse function of h, let $a \in [0, c]$ and $b \in [0, h(c)]$, and let $h^{(n+1)}(x)$ be nonnegative and convex on $[\alpha, \beta]$, where α, β are defined as in* (8.14). *If $h(a) > b$, then*

$$\frac{\left[a - h^{-1}(b)\right]^{n+2}}{(n+1)!} \left[\frac{1}{2^n} h^{(n+1)}\left(\frac{a + h^{-1}(b)}{2}\right) - \frac{2h^{(n+1)}(a) + h^{(n+1)}\left(h^{-1}(b)\right)}{6}\right] \\ \leq \int_0^a h(x)\,\mathrm{d}x + \int_0^b h^{-1}(x)\,\mathrm{d}x - ab - \sum_{k=1}^{n} h^{(k)}\left(h^{-1}(b)\right) \frac{\left[a - h^{-1}(b)\right]^{k+1}}{(k+1)!} \\ \leq \frac{\left[a - h^{-1}(b)\right]^{n+2}}{(n+1)!} \frac{h^{(n+1)}(a) + 2h^{(n+1)}\left(h^{-1}(b)\right)}{6}. \tag{8.32}$$

If $h(a) < b$ and $n = 2\ell$ for $\ell \geq 0$, then

$$\begin{aligned}
&\frac{\left[h^{-1}(b)-a\right]^{n+2}}{(n+1)!}\left[\frac{1}{2^n}h^{(n+1)}\left(\frac{a+h^{-1}(b)}{2}\right)-\frac{2h^{(n+1)}(a)+h^{(n+1)}\left(h^{-1}(b)\right)}{6}\right]\\
&\leq \int_0^a h(x)\,\mathrm{d}x+\int_0^b h^{-1}(x)\,\mathrm{d}x-ab-\sum_{k=1}^{n} h^{(k)}\left(h^{-1}(b)\right)\frac{\left[a-h^{-1}(b)\right]^{k+1}}{(k+1)!}\\
&\leq \frac{\left[h^{-1}(b)-a\right]^{n+2}}{(n+1)!}\frac{h^{(n+1)}(a)+2h^{(n+1)}\left(h^{-1}(b)\right)}{6}. \qquad (8.33)
\end{aligned}$$

If $a < h^{-1}(b)$ and $n = 2\ell+1$ for $\ell \geq 0$, the double inequality (8.33) *is reversed.*

Proof. This is the outline of the proof of [50, Theorem 3.5].

Let $f(x)$ and $g(x)$ be nonnegative and convex functions on $[\mu,\nu]$. Then,

$$\begin{aligned}
&2f\left(\frac{\mu+\nu}{2}\right)g\left(\frac{\mu+\nu}{2}\right)-\frac{1}{6}M(\mu,\nu)-\frac{1}{3}N(\mu,\nu)\\
&\qquad\leq \frac{1}{\nu-\mu}\int_\mu^\nu f(x)g(x)\,\mathrm{d}x \leq \frac{1}{3}M(\mu,\nu)+\frac{1}{6}N(\mu,\nu), \qquad (8.34)
\end{aligned}$$

where

$$M(\mu,\nu)=f(\mu)g(\mu)+f(\nu)g(\nu) \quad \text{and} \quad N(\mu,\nu)=f(\mu)g(\nu)+f(\nu)g(\mu).$$

The double inequality (8.34) can be found in [28,52–54] and closely related references therein. Applying (8.34) in the integral (8.30) arrives at the double inequalities in (8.32) and (8.33). The proof of Theorem 8.10 is complete. □

Remark 8.16. *The double inequalities* (8.32) *and* (8.33) *can be geometrically interpreted as*

$$\begin{aligned}
&\frac{\left[a-h^{-1}(b)\right]^{n+2}}{(n+1)!}\left[\frac{1}{2^n}h^{(n+1)}\left(\frac{a+h^{-1}(b)}{2}\right)-\frac{2h^{(n+1)}(a)+h^{(n+1)}\left(h^{-1}(b)\right)}{6}\right]\\
&\leq C-\sum_{k=1}^{n} h^{(k)}\left(h^{-1}(b)\right)\frac{\left[a-h^{-1}(b)\right]^{k+1}}{(k+1)!} \leq \frac{\left[a-h^{-1}(b)\right]^{n+2}}{(n+1)!}\frac{h^{(n+1)}(a)+2h^{(n+1)}\left(h^{-1}(b)\right)}{6}
\end{aligned}$$

and

$$\begin{aligned}
&\frac{\left[h^{-1}(b)-a\right]^{n+2}}{(n+1)!}\left[\frac{1}{2^n}h^{(n+1)}\left(\frac{a+h^{-1}(b)}{2}\right)-\frac{2h^{(n+1)}(a)+h^{(n+1)}\left(h^{-1}(b)\right)}{6}\right]\\
&\leq C-\sum_{k=1}^{n} h^{(k)}\left(h^{-1}(b)\right)\frac{\left[a-h^{-1}(b)\right]^{k+1}}{(k+1)!} \leq \frac{\left[h^{-1}(b)-a\right]^{n+2}}{(n+1)!}\frac{h^{(n+1)}(a)+2h^{(n+1)}\left(h^{-1}(b)\right)}{6},
\end{aligned}$$

where C denotes the area shown in Figures 8.1–8.6.

8.1.11 THREE EXAMPLES SHOWING REFINEMENTS OF YOUNG'S INTEGRAL INEQUALITY

8.1.11.1 First Example

In [18, Section 3], the double inequality (8.9) was applied to obtain the estimate

$$\begin{aligned}9.000042866\ldots &= \frac{4\sqrt[4]{125}}{27}\left(3-2\sqrt[4]{5}\right)^2\\ &< \int_0^3 \sqrt[4]{x^4+1}\,\mathrm{d}x + \int_1^3 \sqrt[4]{x^4-1}\,\mathrm{d}x - 9\\ &< \frac{27}{2\sqrt[4]{82^3}}\left(3-2\sqrt[4]{5}\right)^2\\ &= 9.000042871\ldots\end{aligned}$$

whose gap between the upper and lower bounds is 0.000000005... and which refines a known result

$$9 < \int_0^3 \sqrt[4]{x^4+1}\,\mathrm{d}x + \int_1^3 \sqrt[4]{x^4-1}\,\mathrm{d}x < 9.0001$$

In [20, Example 2.5] and [20, Remark 2.7], it was obtained that

$$9.000042866 < \int_0^3 \sqrt[4]{x^4+1}\,\mathrm{d}x + \int_1^3 \sqrt[4]{x^4-1}\,\mathrm{d}x < 9.000042868880$$

and

$$9.000042868058 < \int_0^3 \sqrt[4]{x^4+1}\,\mathrm{d}x + \int_1^3 \sqrt[4]{x^4-1}\,\mathrm{d}x < 9.000042868066.$$

whose gaps between the upper and lower bounds are

$$0.0000000028\ldots \quad \text{and} \quad 0.000000000008\ldots$$

respectively.

In [19, Example 2.1], it was estimated that

$$9.00004286765564 < \int_0^3 \sqrt[4]{x^4+1}\,\mathrm{d}x + \int_1^3 \sqrt[4]{x^4-1}\,\mathrm{d}x < 9.00004286805781.$$

whose gap between the upper and lower bounds is 0.0000000004021....

In [50, Example 4.1], by virtue of the double inequality (8.31), the above double inequality was refined as

$$\begin{aligned}9.0000428983186013\ldots = \frac{\left(3-80^{1/4}\right)^3}{3}&\frac{3072\left(\sqrt[4]{95}\sqrt{2}+3\right)^2}{\left[\left(\sqrt[4]{95}\sqrt{2}+3\right)^4+256\right]^{7/4}}\\ &+9+\frac{8\times 5^{3/4}}{27}\frac{\left(3-80^{1/4}\right)^2}{2!}\end{aligned}$$

$$
\begin{aligned}
&\ge \int_0^3 \sqrt[4]{x^4+1}\,\mathrm{d}x + \int_1^3 \sqrt[4]{x^4-1}\,\mathrm{d}x \\
&\ge \frac{(3-80^{1/4})^3}{4!}\left(\frac{27}{82^{7/4}} + 3\times\frac{4\sqrt{5}}{729}\right) \\
&\quad +9+\frac{8\times 5^{3/4}}{27}\frac{(3-80^{1/4})^2}{2!} \\
&= 9.0000428680640760\ldots
\end{aligned}
$$

whose gap between the upper and lower bounds is 0.00000003025452....

8.1.11.2 Second Example

In [50, Example 4.2], by virtue of the double inequality (8.15), it follows that

$$
\begin{aligned}
0.364469045537996606\ldots &= \frac{1}{4}+\left(\frac{1}{2}-\frac{1}{\ln 2}\right)\left[\frac{1}{\exp[\frac{1}{2}(\frac{1}{2}+\frac{1}{\ln 2})]}-\frac{1}{2}\right] \\
&\le \int_0^{1/2}\frac{1}{e^{1/x}}\,\mathrm{d}x - \int_0^{1/2}\frac{1}{\ln x}\,\mathrm{d}x \\
&\le \frac{1}{4}+\frac{1}{2}\left(\frac{1}{2}-\frac{1}{\ln 2}\right)\left(\frac{1}{e^2}-\frac{1}{2}\right) \\
&= 0.421883810040011829\ldots.
\end{aligned}
\tag{8.35}
$$

The gap between the upper and lower bounds in the double inequality (8.35) is 0.057414764502015....

8.1.11.3 Third Example

In [50, Example 4.3], by virtue of the double inequality (8.32), we can obtain the estimate

$$
2.044751320\ldots \le \int_0^1 e^{t^2}\,\mathrm{d}t + \int_0^1 \sqrt{\ln(1+t)}\,\mathrm{d}t \le 2.060536019\ldots. \tag{8.36}
$$

The gap between the upper and lower bounds in the double inequality (8.36) is 0.01578469....

8.2 NEW REFINEMENTS OF YOUNG'S INTEGRAL INEQUALITY VIA PÓLYA'S TYPE INTEGRAL INEQUALITIES

In this section, by virtue of Pólya's type integral inequalities [33,46], we establish some new refinements in terms of higher-order derivatives.

8.2.1 REFINEMENTS OF YOUNG'S INTEGRAL INEQUALITY IN TERMS OF BOUNDS OF THE FIRST DERIVATIVE

Theorem 8.11. *Let $h(x)$ be a strictly increasing function on $[0,c]$ for $c>0$ and let h^{-1} be the inverse function of h. If $h(0)=0$, $a\in[0,c]$, $b\in[0,h(c)]$, L and U are the real constants, and $L\le h'(x)\le U$ on (α,β), then*

$$\frac{LU\left[a-h^{-1}(b)\right]^2-2\left[a-h^{-1}(b)\right][Lh(a)-Ub]+[h(a)-b]^2}{2(U-L)}$$
$$\le \int_0^a h(x)\,\mathrm{d}x+\int_0^b h^{-1}(x)\,\mathrm{d}x-bh^{-1}(b)$$
$$\le -\frac{LU\left[a-h^{-1}(b)\right]^2-2\left[a-h^{-1}(b)\right][Uh(a)-Lb]+[h(a)-b]^2}{2(U-L)}. \tag{8.37}$$

Proof. Let $f(x)$ be continuous on $[a,b]$ and differentiable on (a,b). If $f(x)$ is not identically a constant and $m \le f'(x) \le M$ in (a,b), then

$$\left|\frac{1}{b-a}\int_a^b f(x)\,\mathrm{d}x-\frac{f(a)+f(b)}{2}\right|$$
$$\le \frac{(M-m)(b-a)}{2}\left[\frac{1}{4}-\frac{\left(\frac{f(b)-f(a)}{b-a}-\frac{M+m}{2}\right)^2}{(M-m)^2}\right]. \tag{8.38}$$

The inequality (8.38) can be rearranged as a double inequality

$$\frac{mM(b-a)^2-2(b-a)[mf(b)-Mf(a)]+[f(b)-f(a)]^2}{2(M-m)} \le \int_a^b f(x)\,\mathrm{d}x$$
$$\le -\frac{mM(b-a)^2-2(b-a)[Mf(b)-mf(a)]+[f(b)-f(a)]^2}{2(M-m)}. \tag{8.39}$$

These inequalities can be found in [1, Theorem 2], the papers [8–12], [32, Proposition 2], [33, Section 5], and closely related references therein.

The area C can be computed by (8.7) in Remark 8.2, which can be estimated, by applying the double inequality (8.39), as

$$\frac{LU\left[a-h^{-1}(b)\right]^2-2\left[a-h^{-1}(b)\right][Lh(a)-Ub]+[h(a)-b]^2}{2(U-L)} \le \int_{h^{-1}(b)}^a h(x)\,\mathrm{d}x$$
$$\le -\frac{LU\left[a-h^{-1}(b)\right]^2-2\left[a-h^{-1}(b)\right][Uh(a)-Lb]+[h(a)-b]^2}{2(U-L)}.$$

Since

$$\int_{h^{-1}(b)}^a h(x)\,\mathrm{d}x-b\left[a-h^{-1}(b)\right]=\int_0^a h(x)\,\mathrm{d}x+\int_0^b h^{-1}(x)\,\mathrm{d}x-ab,$$

i.e.,

$$\int_{h^{-1}(b)}^a h(x)\,\mathrm{d}x=\int_0^a h(x)\,\mathrm{d}x+\int_0^b h^{-1}(x)\,\mathrm{d}x-bh^{-1}(b), \tag{8.40}$$

the double inequality (8.37) follows straightforwardly. The proof of Theorem 8.11 is complete. □

8.2.2 REFINEMENTS OF YOUNG'S INTEGRAL INEQUALITY IN TERMS OF BOUNDS OF THE SECOND DERIVATIVE

Theorem 8.12. *Let $h(x)$ be a strictly increasing function on $[0,c]$ for $c>0$, let h^{-1} be the inverse function of h, let $h(0)=0$, $a\in[0,c]$, and $b\in[0,h(c)]$, and let L and U be the real constants such that $L\le h''(x)\le U$ on (α,β). Then,*

$$\begin{aligned}&\frac{L\left[a^3-\left(h^{-1}(b)\right)^3\right]}{6}+\frac{\begin{pmatrix}b-h(a)+ah'(a)-h^{-1}(b)h'\left(h^{-1}(b)\right)\\ +L\left[\left(h^{-1}(b)\right)^2-a^2\right]/2\end{pmatrix}^2}{2\left[\left(h^{-1}(b)-a\right)L-h'\left(h^{-1}(b)\right)+h'(a)\right]}\\&\le\int_0^a h(x)\,\mathrm{d}x+\int_0^b h^{-1}(x)\,\mathrm{d}x-ah(a)+\frac{a^2h'(a)-\left[h^{-1}(b)\right]^2h'\left(h^{-1}(b)\right)}{2}\\&\le\frac{U\left[a^3-\left(h^{-1}(b)\right)^3\right]}{6}+\frac{\begin{pmatrix}b-h(a)+ah'(a)-h^{-1}(b)h'\left(h^{-1}(b)\right)\\ +U\left[\left(h^{-1}(b)\right)^2-a^2\right]/2\end{pmatrix}^2}{2\left[\left(h^{-1}(b)-a\right)U-h'\left(h^{-1}(b)\right)+h'(a)\right]}.\end{aligned}\tag{8.41}$$

Proof. In [29, Corollary] and [42, Corollary 1.2], it was acquired that if $f(x)\in C([a,b])$ satisfying $N\le f''(x)\le M$ on (a,b), then

$$\begin{aligned}&\frac{N(b^3-a^3)}{6}+\frac{\left[f(a)-f(b)+bf'(b)-af'(a)+N\left(a^2-b^2\right)/2\right]^2}{2[(a-b)N-f'(a)+f'(b)]}\\&\le\int_a^b f(x)\,\mathrm{d}x-bf(b)+af(a)+\frac{b^2f'(b)-a^2f'(a)}{2}\\&\le\frac{M(b^3-a^3)}{6}+\frac{\left[f(a)-f(b)+bf'(b)-af'(a)+M\left(a^2-b^2\right)/2\right]^2}{2[(a-b)M-f'(a)+f'(b)]}.\end{aligned}\tag{8.42}$$

Applying the double inequality (8.42) to the integral $\int_{h^{-1}(b)}^a h(x)\,\mathrm{d}x$ and considering Remark 8.2 yield

$$\begin{aligned}&\frac{L\left[a^3-\left(h^{-1}(b)\right)^3\right]}{6}+\frac{\begin{pmatrix}b-h(a)+ah'(a)-h^{-1}(b)h'\left(h^{-1}(b)\right)\\ +L\left[\left(h^{-1}(b)\right)^2-a^2\right]/2\end{pmatrix}^2}{2\left[\left(h^{-1}(b)-a\right)L-h'\left(h^{-1}(b)\right)+h'(a)\right]}\\&\le\int_{h^{-1}(b)}^a h(x)\,\mathrm{d}x-ah(a)+bh^{-1}(b)+\frac{a^2h'(a)-\left[h^{-1}(b)\right]^2h'\left(h^{-1}(b)\right)}{2}\\&\le\frac{U\left[a^3-\left(h^{-1}(b)\right)^3\right]}{6}+\frac{\begin{pmatrix}b-h(a)+ah'(a)-h^{-1}(b)h'\left(h^{-1}(b)\right)\\ +U\left[\left(h^{-1}(b)\right)^2-a^2\right]/2\end{pmatrix}^2}{2\left[\left(h^{-1}(b)-a\right)U-h'\left(h^{-1}(b)\right)+h'(a)\right]}.\end{aligned}\tag{8.43}$$

Substituting (8.40) into (8.43) results in the double inequality (8.41). The proof of Theorem 8.12 is complete. □

8.2.3 REFINEMENTS OF YOUNG'S INTEGRAL INEQUALITY IN TERMS OF BOUNDS OF HIGHER-ORDER DERIVATIVES

Theorem 8.13. *Let $h(x)$ be a strictly increasing function on $[0,c]$ for $c>0$, let h^{-1} be the inverse of h, let $h(0)=0$, $a\in[0,c]$, and $b\in[0,h(c)]$, and let $h(x)$ have the $(n+1)$-th derivative on $[0,c]$ such that $L\le h^{(n+1)}(x)\le U$ on (α,β). Then, for all t between a and $h^{-1}(b)$,*

1. *when n is a nonnegative odd integer,*

$$\sum_{i=0}^{n+2}\frac{(-1)^i}{i!}\Big[S_{n+2}^{(i)}(h;h^{-1}(b),h^{-1}(b),L)-S_{n+2}^{(i)}(h;a,a,L)\Big]t^i \le \int_0^a h(x)\,\mathrm{d}x+\int_0^b h^{-1}(x)\,\mathrm{d}x-bh^{-1}(b) \le \sum_{i=0}^{n+2}\frac{(-1)^i}{i!}\Big[S_{n+2}^{(i)}(h;h^{-1}(b),h^{-1}(b),U)-S_{n+2}^{(i)}(h;a,a,U)\Big]t^i; \quad (8.44)$$

2. *when n is a nonnegative even integer,*

$$\sum_{i=0}^{n+2}\frac{(-1)^i}{i!}\Big[S_{n+2}^{(i)}(h;h^{-1}(b),h^{-1}(b),L)-S_{n+2}^{(i)}(h;a,a,U)\Big]t^i \le \int_0^a h(x)\,\mathrm{d}x+\int_0^b h^{-1}(x)\,\mathrm{d}x-bh^{-1}(b) \le \sum_{i=0}^{n+2}\frac{(-1)^i}{i!}\Big[S_{n+2}^{(i)}(h;h^{-1}(b),h^{-1}(b),U)-S_{n+2}^{(i)}(h;a,a,L)\Big]t^i; \quad (8.45)$$

where L and U are the real constants,

$$S_n(h;u,v,w)=\sum_{k=1}^{n-1}\frac{(-1)^k}{k!}u^k h^{(k-1)}(v)+(-1)^n\frac{w}{n!}u^n,$$

and

$$S_n^{(k)}(h;u,v,w)=\frac{\partial^k S_n(h;u,v,w)}{\partial u^k}.$$

Proof. In [29, Theorem], it was discovered that if $f\in C^n([a,b])$ has derivative of $(n+1)$-th order satisfying $N\le f^{(n+1)}(x)\le M$ on (a,b), then, for all $t\in(a,b)$,

1. when n is a nonnegative odd integer,

$$\sum_{i=0}^{n+2}\frac{(-1)^i}{i!}\Big[S_{n+2}^{(i)}(f;a,a,N)-S_{n+2}^{(i)}(f;b,b,N)\Big]t^i \le \int_a^b f(x)\,\mathrm{d}x \le \sum_{i=0}^{n+2}\frac{(-1)^i}{i!}\Big[S_{n+2}^{(i)}(f;a,a,M)-S_{n+2}^{(i)}(f;b,b,M)\Big]t^i; \quad (8.46)$$

2. when n is a nonnegative even integer,

$$\sum_{i=0}^{n+2}\frac{(-1)^i}{i!}\left[S_{n+2}^{(i)}(f;a,a,N)-S_{n+2}^{(i)}(f;b,b,M)\right]t^i \le \int_a^b f(x)\,\mathrm{d}x$$
$$\le \sum_{i=0}^{n+2}\frac{(-1)^i}{i!}\left[S_{n+2}^{(i)}(f;a,a,M)-S_{n+2}^{(i)}(f;b,b,N)\right]t^i. \quad (8.47)$$

These inequalities can also be found in [30,31,34,41,46] and closely related references therein.

Applying (8.46) and (8.47) to the integral $\int_{h^{-1}(b)}^{a} h(x)\,\mathrm{d}x$ and considering Remark 8.2 yield

1. when n is a nonnegative odd integer,

$$\sum_{i=0}^{n+2}\frac{(-1)^i}{i!}\left[S_{n+2}^{(i)}\big(h;h^{-1}(b),h^{-1}(b),L\big)-S_{n+2}^{(i)}(h;a,a,L)\right]t^i \le \int_{h^{-1}(b)}^{a} h(x)\,\mathrm{d}x$$
$$\le \sum_{i=0}^{n+2}\frac{(-1)^i}{i!}\left[S_{n+2}^{(i)}\big(h;h^{-1}(b),h^{-1}(b),U\big)-S_{n+2}^{(i)}(h;a,a,U)\right]t^i; \quad (8.48)$$

2. when n is a nonnegative even integer,

$$\sum_{i=0}^{n+2}\frac{(-1)^i}{i!}\left[S_{n+2}^{(i)}\big(h;h^{-1}(b),h^{-1}(b),L\big)-S_{n+2}^{(i)}(h;a,a,U)\right]t^i \le \int_{h^{-1}(b)}^{a} h(x)\,\mathrm{d}x$$
$$\le \sum_{i=0}^{n+2}\frac{(-1)^i}{i!}\left[S_{n+2}^{(i)}\big(h;h^{-1}(b),h^{-1}(b),U\big)-S_{n+2}^{(i)}(h;a,a,L)\right]t^i. \quad (8.49)$$

Substituting (8.40) into (8.48) and (8.49) results in (8.44) and (8.45). The proof of Theorem 8.13 is complete. □

8.2.4 REFINEMENTS OF YOUNG'S INTEGRAL INEQUALITY IN TERMS OF L^p-NORMS

Theorem 8.14. *Let $h(x)$ be a strictly increasing function on $[0,c]$ for $c>0$, let h^{-1} be the inverse of h, let $h(0)=0$, $a\in[0,c]$, and $b\in[0,h(c)]$, and let $h(x)$ have the $(n+1)$-th derivative on $[\alpha,\beta]$ such that $h^{(n+1)}\in L^p([\alpha,\beta])$ for $p,q>0$ with $\frac{1}{p}+\frac{1}{q}=1$. Then, for all $t\in[\alpha,\beta]$,*

1. *when $p,q>1$, we have*

$$\left|\int_{h^{-1}(b)}^{a} h(x)\,\mathrm{d}x-\sum_{i=0}^{n}\frac{h^{(i)}\big(h^{-1}(b)\big)}{(i+1)!}\big(t-h^{-1}(b)\big)^{i+1}+\sum_{i=0}^{n}\frac{h^{(i)}(a)}{(i+1)!}(t-a)^{i+1}\right|$$

$$\leq \frac{\left(t-h^{-1}(b)\right)^{n+1+1/q}+(a-t)^{n+1+1/q}}{(n+1)!\sqrt[q]{nq+q+1}}\left\|h^{(n+1)}\right\|_{L^p([h^{-1}(b),a])}$$

$$\leq \frac{2\left(a-h^{-1}(b)\right)^{n+2}}{(n+1)!}\left\|h^{(n+1)}\right\|_{L^p([h^{-1}(b),a])}; \tag{8.50}$$

2. *when $p=\infty$, we have*

$$\left|\int_{h^{-1}(b)}^{a} h(x)\,\mathrm{d}x-\sum_{i=0}^{n}\frac{h^{(i)}\left(h^{-1}(b)\right)}{(i+1)!}\left(t-h^{-1}(b)\right)^{i+1}+\sum_{i=0}^{n}\frac{h^{(i)}(a)}{(i+1)!}(t-a)^{i+1}\right|$$

$$\leq \frac{\left(t-h^{-1}(b)\right)^{n+2}+(a-t)^{n+2}}{(n+2)!}\left\|h^{(n+1)}\right\|_{L^\infty([h^{-1}(b),a])}$$

$$\leq \frac{2\left(a-h^{-1}(b)\right)^{n+2}}{(n+2)!}\left\|h^{(n+1)}\right\|_{L^\infty([h^{-1}(b),a])}; \tag{8.51}$$

3. *when $p=1$, we have*

$$\left|\int_{h^{-1}(b)}^{a} h(x)\,\mathrm{d}x-\sum_{i=0}^{n}\frac{h^{(i)}\left(h^{-1}(b)\right)}{(i+1)!}\left(t-h^{-1}(b)\right)^{i+1}+\sum_{i=0}^{n}\frac{h^{(i)}(a)}{(i+1)!}(t-a)^{i+1}\right|$$

$$\leq \frac{\left(t-h^{-1}(b)\right)^{n+1}+(a-t)^{n+1}}{(n+1)!}\left\|h^{(n+1)}\right\|_{L([h^{-1}(b),a])}$$

$$\leq \frac{2\left(a-h^{-1}(b)\right)^{n+1}}{(n+1)!}\left\|h^{(n+1)}\right\|_{L([h^{-1}(b),a])}. \tag{8.52}$$

Proof. Let $f\in C^n([a,b])$ have derivative of $(n+1)$-th order on (a,b) and $f^{(n+1)}\in L^p([a,b])$ for positive numbers p and q satisfying $\frac{1}{p}+\frac{1}{q}=1$. In [16] and [17, Theorem 2], it was established that, for any $t\in(a,b)$,

1. when $p,q>1$, we have

$$\left|\int_a^b f(x)\,\mathrm{d}x-\sum_{i=0}^{n}\frac{f^{(i)}(a)}{(i+1)!}(t-a)^{i+1}+\sum_{i=0}^{n}\frac{f^{(i)}(b)}{(i+1)!}(t-b)^{i+1}\right|$$

$$\leq \frac{(t-a)^{n+1+1/q}+(b-t)^{n+1+1/q}}{(n+1)!\sqrt[q]{nq+q+1}}\left\|f^{(n+1)}\right\|_{L^p([a,b])}$$

$$\leq \frac{2(b-a)^{n+2}}{(n+1)!}\left\|f^{(n+1)}\right\|_{L^p([a,b])}; \tag{8.53}$$

2. when $p = \infty$, we have

$$\left| \int_a^b f(x)\,\mathrm{d}x - \sum_{i=0}^{n} \frac{f^{(i)}(a)}{(i+1)!}(t-a)^{i+1} + \sum_{i=0}^{n} \frac{f^{(i)}(b)}{(i+1)!}(t-b)^{i+1} \right|$$
$$\leq \frac{(t-a)^{n+2} + (b-t)^{n+2}}{(n+2)!} \|f^{(n+1)}\|_{L^\infty([a,b])}$$
$$\leq \frac{2(b-a)^{n+2}}{(n+2)!} \|f^{(n+1)}\|_{L^\infty([a,b])}; \tag{8.54}$$

3. when $p = 1$, we have

$$\left| \int_a^b f(x)\,\mathrm{d}x - \sum_{i=0}^{n} \frac{f^{(i)}(a)}{(i+1)!}(t-a)^{i+1} + \sum_{i=0}^{n} \frac{f^{(i)}(b)}{(i+1)!}(t-b)^{i+1} \right|$$
$$\leq \frac{(t-a)^{n+1} + (b-t)^{n+1}}{(n+1)!} \|f^{(n+1)}\|_{L([a,b])}$$
$$\leq \frac{2(b-a)^{n+1}}{(n+1)!} \|f^{(n+1)}\|_{L([a,b])}. \tag{8.55}$$

Applying three inequalities (8.53), (8.54), and (8.55) to the integral $\int_{h^{-1}(b)}^{a} h(x)\,\mathrm{d}x$ and considering Remark 8.2 lead to the following conclusions:

1. when $p, q > 1$, we have

$$\left| \int_{h^{-1}(b)}^{a} h(x)\,\mathrm{d}x - \sum_{i=0}^{n} \frac{h^{(i)}(h^{-1}(b))}{(i+1)!}\left(t - h^{-1}(b)\right)^{i+1} + \sum_{i=0}^{n} \frac{h^{(i)}(a)}{(i+1)!}(t-a)^{i+1} \right|$$
$$\leq \frac{\left(t - h^{-1}(b)\right)^{n+1+1/q} + (a-t)^{n+1+1/q}}{(n+1)!\sqrt[q]{nq+q+1}} \|h^{(n+1)}\|_{L^p([h^{-1}(b),a])}$$
$$\leq \frac{2\left(a - h^{-1}(b)\right)^{n+2}}{(n+1)!} \|h^{(n+1)}\|_{L^p([h^{-1}(b),a])}; \tag{8.56}$$

2. when $p = \infty$, we have

$$\left| \int_{h^{-1}(b)}^{a} h(x)\,\mathrm{d}x - \sum_{i=0}^{n} \frac{h^{(i)}(h^{-1}(b))}{(i+1)!}\left(t - h^{-1}(b)\right)^{i+1} + \sum_{i=0}^{n} \frac{h^{(i)}(a)}{(i+1)!}(t-a)^{i+1} \right|$$
$$\leq \frac{\left(t - h^{-1}(b)\right)^{n+2} + (a-t)^{n+2}}{(n+2)!} \|h^{(n+1)}\|_{L^\infty([h^{-1}(b),a])}$$
$$\leq \frac{2\left(a - h^{-1}(b)\right)^{n+2}}{(n+2)!} \|h^{(n+1)}\|_{L^\infty([h^{-1}(b),a])}; \tag{8.57}$$

3. when $p = 1$, we have

$$\begin{aligned}
&\left|\int_{h^{-1}(b)}^{a} h(x)\,\mathrm{d}x - \sum_{i=0}^{n} \frac{h^{(i)}\left(h^{-1}(b)\right)}{(i+1)!}\left(t-h^{-1}(b)\right)^{i+1} + \sum_{i=0}^{n} \frac{h^{(i)}(a)}{(i+1)!}(t-a)^{i+1}\right| \\
&\qquad \le \frac{\left(t-h^{-1}(b)\right)^{n+1} + (a-t)^{n+1}}{(n+1)!}\left\|h^{(n+1)}\right\|_{L([h^{-1}(b),a])} \\
&\qquad \le \frac{2\left(a-h^{-1}(b)\right)^{n+1}}{(n+1)!}\left\|h^{(n+1)}\right\|_{L([h^{-1}(b),a])}.
\end{aligned} \tag{8.58}$$

Substituting (8.40) into (8.56), (8.57), and (8.58) results in (8.50), (8.51), and (8.52). The proof of Theorem 8.14 is complete. □

8.2.5 THREE EXAMPLES FOR NEW REFINEMENTS OF YOUNG'S INTEGRAL INEQUALITIES

8.2.5.1 First Example

Let $h(x) = \sqrt[4]{x^4+1} - 1$ and let $a = 3$ and $b = 2$ in Theorem 8.11. Then,

$$\begin{aligned}
h'(x) &= \frac{x^3}{(x^4+1)^{3/4}}, \quad h''(x) = \frac{3x^2}{(x^4+1)^{7/4}} > 0, \\
h^{-1}(2) &= \left(3^4-1\right)^{1/4} = 2\sqrt[4]{5} = 2.990\ldots, \\
L = h'\left(2\sqrt[4]{5}\right) &= \frac{8\times 5^{3/4}}{27}, \quad U = h'(3) = \frac{27}{82^{3/4}}, \\
h(3) = \sqrt[4]{82} - 1, &\quad \int_0^3 h(x)\,\mathrm{d}x = \int_0^3 \sqrt[4]{x^4+1}\,\mathrm{d}x - 3, \\
\int_0^2 h^{-1}(x)\,\mathrm{d}x &= \int_0^2 \sqrt[4]{(x+1)^4-1} = \int_1^3 \sqrt[4]{x^4-1}\,\mathrm{d}x,
\end{aligned}$$

and

$$\begin{aligned}
&\frac{\begin{pmatrix} \frac{8\times 5^{3/4}}{27}\frac{27}{82^{3/4}}\left(3-2\sqrt[4]{5}\right)^2 + \left(\sqrt[4]{82}-3\right)^2 \\ -2\left(3-2\sqrt[4]{5}\right)\left[\frac{8\times 5^{3/4}}{27}\left(\sqrt[4]{82}-1\right) - \frac{2\times 27}{82^{3/4}}\right]\end{pmatrix}}{2\left(\frac{27}{82^{3/4}} - \frac{8\times 5^{3/4}}{27}\right)} \\
&\quad \le \int_0^3 \sqrt[4]{x^4+1}\,\mathrm{d}x + \int_1^3 \sqrt[4]{x^4-1}\,\mathrm{d}x - 3 - 4\sqrt[4]{5} \\
&\quad \le -\frac{\begin{pmatrix} \frac{8\times 5^{3/4}}{27}\frac{27}{82^{3/4}}\left(3-2\sqrt[4]{5}\right)^2 + \left(\sqrt[4]{82}-3\right)^2 \\ -2\left(3-2\sqrt[4]{5}\right)\left[\frac{27}{82^{3/4}}\left(\sqrt[4]{82}-1\right) - \frac{2\times 8\times 5^{3/4}}{27}\right]\end{pmatrix}}{2\left(\frac{27}{82^{3/4}} - \frac{8\times 5^{3/4}}{27}\right)}.
\end{aligned}$$

Consequently, we arrive at

$$9.00004286765564673\ldots < \int_0^3 \sqrt[4]{x^4+1}\,\mathrm{d}x + \int_1^3 \sqrt[4]{x^4-1}\,\mathrm{d}x < 9.00004287010602764\ldots \quad (8.59)$$

which is neither the best nor the weakest estimate among those in Section 8.1.11. The gap between the upper and lower bounds in the double inequality (8.59) is 0.0000000024506... which, comparing with those gaps in Section 8.1.11, is neither the smallest nor the biggest one.

8.2.5.2 Second Example

Let

$$h(x) = \begin{cases} e^{-1/x}, & x > 0; \\ 0, & x = 0. \end{cases}$$

Let $a = b = \frac{1}{2}$ in Theorem 8.11. Then,

$$h'(x) = \frac{e^{-1/x}}{x^2}, \quad h''(x) = \frac{e^{-1/x}(1-2x)}{x^4},$$

$$h^{-1}\left(\frac{1}{2}\right) = \frac{1}{\ln 2} = 1.44\ldots, \quad h\left(\frac{1}{2}\right) = \frac{1}{e^2},$$

$$U = h'\left(\frac{1}{2}\right) = \frac{4}{e^2} = 0.54134\ldots, \quad L = h'\left(\frac{1}{\ln 2}\right) = \frac{\ln^2 2}{2} = 0.24022\ldots,$$

and

$$\frac{\frac{\ln^2 2}{2}\frac{4}{e^2}\left(\frac{1}{2}-\frac{1}{\ln 2}\right)^2 - 2\left(\frac{1}{2}-\frac{1}{\ln 2}\right)\left(\frac{\ln^2 2}{2}\frac{1}{e^2}-\frac{4}{e^2}\frac{1}{2}\right) + \left(\frac{1}{e^2}-\frac{1}{2}\right)^2}{2\left(\frac{4}{e^2}-\frac{\ln^2 2}{2}\right)}$$
$$\le \int_0^{1/2} \frac{1}{e^{1/x}}\,\mathrm{d}x - \int_0^{1/2} \frac{1}{\ln x}\,\mathrm{d}x - \frac{1}{2\ln 2}$$
$$\le -\frac{\frac{\ln^2 2}{2}\frac{4}{e^2}\left(\frac{1}{2}-\frac{1}{\ln 2}\right)^2 - 2\left(\frac{1}{2}-\frac{1}{\ln 2}\right)\left(\frac{4}{e^2}\frac{1}{e^2}-\frac{\ln^2 2}{2}\frac{1}{2}\right) + \left(\frac{1}{e^2}-\frac{1}{2}\right)^2}{2\left(\frac{4}{e^2}-\frac{\ln^2 2}{2}\right)}.$$

Accordingly, it follows that

$$0.388457763460961578\ldots < \int_0^{1/2} \frac{1}{e^{1/x}}\,\mathrm{d}x - \int_0^{1/2} \frac{1}{\ln x}\,\mathrm{d}x < 0.455309856619062079\ldots \quad (8.60)$$

whose lower bound is better, but whose upper bound is worse, than the corresponding ones in (8.35). The gap between the upper and lower bounds in the double inequality (8.60) is 0.066852093209446... which is bigger than the gap 0.057414764502015... in the double inequality (8.35).

8.2.5.3 Third Example

Let $h(x) = e^{x^2} - 1$ for $x \ge 0$. Then, $h^{-1}(x) = \sqrt{\ln(1+x)}$ for $x \ge 0$. Let $a = b = 1$ in Theorem 8.11. Then,

$$h'(x) = 2xe^{x^2}, \quad h^{-1}(1) = \sqrt{\ln 2} = 0.83255\ldots, \quad h(1) = e - 1,$$
$$U = h'(1) = 2e = 5.4365\ldots, \quad L = h'\left(\sqrt{\ln 2}\right) = 4\sqrt{\ln 2} = 3.3302\ldots,$$

and

$$\frac{8e\sqrt{\ln 2}\left(1-\sqrt{\ln 2}\right)^2 - 2\left(1-\sqrt{\ln 2}\right)\left[4\sqrt{\ln 2}(e-1) - 2e\right] + (e-2)^2}{2\left(2e - 4\sqrt{\ln 2}\right)}$$
$$\le \int_0^1 \left(e^{x^2} - 1\right) \mathrm{d}x + \int_0^1 \sqrt{\ln(1+x)}\, \mathrm{d}x - \sqrt{\ln 2}$$
$$\le \frac{8e\sqrt{\ln 2}\left(1-\sqrt{\ln 2}\right)^2 - 2\left(1-\sqrt{\ln 2}\right)\left[2e(e-1) - 4\sqrt{\ln 2}\right] + (e-2)^2}{2\left(2e - 4\sqrt{\ln 2}\right)}.$$

As a result, we have

$$2.05281277502489567\ldots \le \int_0^1 e^{x^2}\, \mathrm{d}x + \int_0^1 \sqrt{\ln(1+x)}\, \mathrm{d}x$$
$$\le 2.06746020503978898\ldots \quad (8.61)$$

whose lower bound is better, but whose upper bound is worse, than the corresponding ones in (8.36). The gap between the upper and lower bounds in the double inequality (8.61) is 0.01464743001489... which is smaller than the corresponding gap 0.01578469... in the double inequality (8.36).

8.3 MORE REMARKS

Finally, we would like to list more remarks on our main results and possible developing directions.

Remark 8.17. *Theorems 8.11 and 8.12 are special cases of Theorem 8.13. In other words, Theorems 8.11 and 8.12 can be deduced from Theorem 8.13.*

Remark 8.18. *Some Taylor-like power expansions such as those in [15,22,37,40] and closely related references can be used to refine Young's integral inequality* (8.1).

Remark 8.19. *At the present position, we conclude that many estimates of definite integrals can be used to refine Young's integral inequality* (8.1).

Remark 8.20. *Essentially speaking, all refinements in this paper are estimates of the area C, which can be geometrically demonstrated in Figures 8.1–8.6 and analytically expressed by* (8.7) *in Remark 8.2.*

Remark 8.21. *This paper is a slightly revised version of the electronic preprint [36].*

ACKNOWLEDGMENTS

The authors are thankful to anonymous referees for their careful corrections, helpful suggestions, and valuable comments on the original version of this paper.

BIBLIOGRAPHY

[1] R. P. Agarwal and S. S. Dragomir, An application of Hayashi inequality for differentiable functions, *Computers Math. Appl.* **32** (1996), no. 6, 95–99; available online at https://doi.org/10.1016/0898-1221(96)00146-0.

[2] H. Alzer and M. K. Kwong, On Young's inequality, *J. Math. Anal. Appl.* **469** (2019), no. 2, 480–492; available online at https://doi.org/10.1016/j.jmaa.2018.06.061.

[3] H. Alzer and S. Koumandos, A new refinement of Young's inequality, *Proc. Edinb. Math. Soc.* (2) **50** (2007), no. 2, 255–262; available online at https://doi.org/10.1017/S0013091504000744.

[4] D. R. Anderson, Young's integral inequality on time scales revisited, *J. Inequal. Pure Appl. Math.* **8** (2007), no. 3, Art. 64; available online at http://www.emis.de/journals/JIPAM/article876.html.

[5] T. M. Apostol, *Calculus*, Vol. I: One-variable calculus, with an introduction to linear algebra; Second edition, Blaisdell Publishing Co. Ginn and Co., Waltham, Mass.-Toronto, Ont.-London, 1967.

[6] T. M. Apostol, *Mathematical Analysis*, Second edition, Addison-Wesley Publishing Co., Reading, Mass.-London-Don Mills, Ont., 1974.

[7] S.-P. Bai, S.-H. Wang, and F. Qi, On HT-convexity and Hadamard-type inequalities, *J. Inequal. Appl.* **2020**, Paper No. 3, 12 pages; available online at https://doi.org/10.1186/s13660-019-2276-3.

[8] P. Cerone, Generalised trapezoidal rules with error involving bounds of the n-th derivative, *Math. Inequal. Appl.* **5** (2002), no. 3, 451–462; available online at https://doi.org/10.7153/mia-05-44.

[9] P. Cerone, On Gini mean difference bounds via generalised Iyengar results, *Hacet. J. Math. Stat.* **44** (2015), no. 4, 789–799; available online at https://doi.org/10.15672/HJMS.2015449430.

[10] P. Cerone and S. S. Dragomir, Lobatto type quadrature rules for functions with bounded derivative, *Math. Inequal. Appl.* **3** (2000), no. 2, 197–209; available online at https://doi.org/10.7153/mia-03-23.

[11] P. Cerone and S. S. Dragomir, On a weighted generalization of Iyengar type inequalities involving bounded first derivative, *Math. Inequal. Appl.* **3** (2000), no. 1, 35–44; available online at https://doi.org/10.7153/mia-03-04.

[12] X.-L. Cheng, The Iyengar-type inequality, *Appl. Math. Lett.* **14** (2001), no. 8, 975–978; available online at https://doi.org/10.1016/S0893-9659(01)00074-X.

[13] S. S. Dragomir, A note on Young's inequality, *Rev. R. Acad. Cienc. Exactas Fís. Nat. Ser. A Mat. RACSAM* **111** (2017), no. 2, 349–354; available online at https://doi.org/10.1007/s13398-016-0300-8.

[14] S. S. Dragomir, Trace inequalities for positive operators via recent refinements and reverses of Young's inequality, *Spec. Matrices* **6** (2018), 180–192; available online at https://doi.org/10.1515/spma-2018-0015.

[15] S. S. Dragomir, F. Qi, G. Hanna, and P. Cerone, New Taylor-like expansions for functions of two variables and estimates of their remainders, *J. Korean Soc. Indust. Appl. Math.* **9** (2005), no. 2, 1–16.

[16] B.-N. Guo and F. Qi, Estimates for an integral in L^p norm of the $(n+1)$-th derivative of its integrand, *Inequality Theory and Applications,* Volume **3**, 127–131, Nova Science Publishers, Hauppauge, NY, 2003.

[17] B.-N. Guo and F. Qi, Some estimates of an integral in terms of the L^p-norm of the $(n+1)$st derivative of its integrand, *Anal. Math.* **29** (2003), no. 1, 1–6; available online at http://dx.doi.org/10.1023/A:1022894413541.

[18] A. Hoorfar and F. Qi, A new refinement of Young's inequality, *Math. Inequal. Appl.* **11** (2008), no. 4, 689–692; available online at https://doi.org/10.7153/mia-11-58.

[19] J. Jakšetić and J. Pečarić, An estimation of Young inequality, *Asian-Eur. J. Math.* **2** (2009), no. 4, 593–604; available online at https://doi.org/10.1142/S1793557109000509.

[20] J. Jakšetić and J. Pečarić, A note on Young inequality, *Math. Inequal. Appl.* **13** (2010), no. 1, 43–48; available online at https://doi.org/10.7153/mia-13-03.

[21] P. Kórus, A refinement of Young's inequality, *Acta Math. Hungar.* **153** (2017), no. 2, 430–435; available online at https://doi.org/10.1007/s10474-017-0735-1.

[22] Q.-M. Luo, F. Qi, and B.-N. Guo, K. Petr's formula of double integral and estimates of its remainder, *Int. J. Math. Sci.* **3** (2004), no. 1, 77–92.

[23] P. R. Mercer, Error terms for Steffensen's, Young's and Chebychev's inequalities, *J. Math. Inequal.* **2** (2008), no. 4, 479–486; available online at https://doi.org/10.7153/jmi-02-43.

[24] P. R. Mercer, Techniques of Integration, Chapter 11 in: *More Calculus of a Single Variable*, Undergraduate Texts in Mathematics, Springer, New York, 2014; available online at https://doi.org/10.1007/978-1-4939-1926-0_11.

[25] D. S. Mitrinović, *Analytic Inequalities*, In cooperation with P. M. Vasić, Die Grundlehren der mathematischen Wissenschaften, Band 165, Springer-Verlag, New York-Berlin, 1970.

[26] D. S. Mitrinović, J. E. Pečarić, and A. M. Fink, *Classical and New Inequalities in Analysis*, Kluwer Academic Publishers, 1993; available online at http://dx.doi.org/10.1007/978-94-017-1043-5.

[27] F. W. J. Olver, D. W. Lozier, R. F. Boisvert, and C. W. Clark (eds.), *NIST Handbook of Mathematical Functions*, Cambridge University Press, New York, 2010; available online at http://dlmf.nist.gov/.

[28] B. G. Pachpatte, On some inequalities for convex functions, *RGMIA Res. Rep. Coll.* **6** (2003), Suppl., Art. 1, 9 pages; available online at http://rgmia.org/v6(E).php.

[29] F. Qi, Further generalizations of inequalities for an integral, *Univ. Beograd. Publ. Elektrotehn. Fak. Ser. Mat.* **8** (1997), 79–83.

[30] F. Qi, Inequalities for a multiple integral, *Acta Math. Hungar.* **84** (1999), no. 1–2, 19–26; available online at https://doi.org/10.1023/A:1006642601341.

[31] F. Qi, Inequalities for a weighted multiple integral, *J. Math. Anal. Appl.* **253** (2001), no. 2, 381–388; available online at https://doi.org/10.1006/jmaa.2000.7138.

[32] F. Qi, Inequalities for an integral, *Math. Gaz.* **80** (1996), no. 488, 376–377; available online at https://doi.org/10.2307/3619581.

[33] F. Qi, Pólya type integral inequalities: origin, variants, proofs, refinements, generalizations, equivalences, and applications, *Math. Inequal. Appl.* **18** (2015), no. 1, 1–38; available online at http://dx.doi.org/10.7153/mia-18-01.

[34] F. Qi, P. Cerone, and S. S. Dragomir, Some new Iyengar type inequalities, *Rocky Mountain J. Math.* **35** (2005), no. 3, 997–1015; available online at https://doi.org/10.1216/rmjm/1181069718.

[35] F. Qi, L.-H. Cui, and S.-L. Xu, Some inequalities constructed by Tchebysheff's integral inequality, *Math. Inequal. Appl.* **2** (1999), no. 4, 517–528; available online at http://dx.doi.org/10.7153/mia-02-42.

[36] F. Qi, W.-H. Li, G.-S. Wu, and B.-N. Guo, Refinements of Young's integral inequality via fundamental inequalities and mean value theorems for derivatives, *arXiv preprint* (2020), available online at https://arxiv.org/abs/2002.04428.

[37] F. Qi, Q.-M. Luo, and B.-N. Guo, Darboux's formula with integral remainder of functions with two independent variables, *Appl. Math. Comput.* **199** (2008), no. 2, 691–703; available online at https://doi.org/10.1016/j.amc.2007.10.028.

[38] F. Qi, P. O. Mohammed, J.-C. Yao, and Y.-H. Yao, Generalized fractional integral inequalities of Hermite–Hadamard type for (α,m)-convex functions, *J. Inequal. Appl.* **2019**, Paper No. 135, 17 pages; available online at https://doi.org/10.1186/s13660-019-2079-6.

[39] F. Qi, G. Rahman, S. M. Hussain, W.-S. Du, and K. S. Nisar, Some inequalities of Čebyšev type for conformable k-fractional integral operators, *Symmetry* **10** (2018), no. 11, Article 614, 8 pages; available online at https://doi.org/10.3390/sym10110614.

[40] F. Qi and W. Ul-Haq, Some integral inequalities involving the expectation and variance via Darboux's expansion, *Adv. Appl. Math. Sci.* **18** (2019), no. 7, 545–552.

[41] F. Qi, Z.-L. Wei, and Q. Yang, Generalizations and refinements of Hermite–Hadamard's inequality, *Rocky Mountain J. Math.* **35** (2005), no. 1, 235–251; available online at http://dx.doi.org/10.1216/rmjm/1181069779.

[42] F. Qi and Y.-J. Zhang, Inequalities for a weighted integral, *Adv. Stud. Contemp. Math. (Kyungshang)* **4** (2002), no. 2, 93–101.

[43] D. Ruthing, On Young's inequality, *Internat. J. Math. Ed. Sci. Techn.* **25** (1994), no. 2, 161–164; available online at https://doi.org/10.1080/0020739940250201.

[44] J. Sándor and V. E. S. Szabó, On an inequality for the sum of infimums of functions, *J. Math. Anal. Appl.* **204** (1996), no. 3, 646–654; available online at https://doi.org/10.1006/jmaa.1996.0459.

[45] Y. Shuang and F. Qi, Integral inequalities of Hermite–Hadamard type for extended s-convex functions and applications, *Mathematics* **6** (2018), no. 11, Article 223, 12 pages; available online at https://doi.org/10.3390/math6110223.

[46] Y. Sun, H.-T. Yang, and F. Qi, Some inequalities for multiple integrals on the n-dimensional ellipsoid, spherical shell, and ball, *Abstr. Appl. Anal.* **2013** (2013), Article ID 904721, 8 pages; available online at https://doi.org/10.1155/2013/904721.

[47] T. Takahashi, Remarks on some inequalities, *Tôhoku Math. J.* **36** (1932), 99–106.

[48] J.-F. Tian, Extension of Hu Ke's inequality and its applications, *J. Inequal. Appl.* **2011**, 2011:77, 14 pages; available online at https://doi.org/10.1186/1029-242X-2011-77.

[49] J.-F. Tian and M.-H. Ha, Properties of generalized sharp Hölder's inequalities, *J. Math. Inequal.* **11** (2017), no. 2, 511–525; available online at https://doi.org/10.7153/jmi-11-42.

[50] J.-Q. Wang, B.-N. Guo, and F. Qi, Generalizations and applications of Young's integral inequality by higher order derivatives, *J. Inequal. Appl.* **2019**, Paper No. 243, 18 pages; available online at https://doi.org/10.1186/s13660-019-2196-2.

[51] A. Witkowski, On Young inequality, *J. Inequal. Pure Appl. Math.* **7** (2006), no. 5, Art. 164; available online at http://www.emis.de/journals/JIPAM/article782.html.

[52] Y. Wu, F. Qi, and D.-W. Niu, Integral inequalities of Hermite–Hadamard type for the product of strongly logarithmically convex and other convex functions, *Maejo Int. J. Sci. Technol.* **9** (2015), no. 3, 394–402.

[53] H.-P. Yin and F. Qi, Hermite–Hadamard type inequalities for the product of (α,m)-convex functions, *J. Nonlinear Sci. Appl.* **8** (2015), no. 3, 231–236; available online at https://doi.org/10.22436/jnsa.008.03.07.
[54] H.-P. Yin and F. Qi, Hermite-Hadamard type inequalities for the product of (α,m)-convex functions, *Missouri J. Math. Sci.* **27** (2015), no. 1, 71–79; available online at http://projecteuclid.org/euclid.mjms/1449161369.
[55] W. H. Young, On classes of summable functions and their Fourier series, *Proc. Roy. Soc. London Ser. A* **87** (1912), 225–229; available online at https://doi.org/10.1098/rspa.1912.0076.
[56] L. Zhu, On Young's inequality, *Internat. J. Math. Ed. Sci. Tech.* **35** (2004), no. 4, 601–603; available online at https://doi.org/10.1080/00207390410001686698.

9 On the Coefficient Estimates for New Subclasses of Bi-univalent Functions Associated with Subordination and Fibonacci Numbers

Şahsene Altınkaya
Bursa Uludag University

CONTENTS

9.1 THE DEFINITION AND ELEMENTARY PROPERTIES OF UNIVALENT FUNCTIONS

Let $\mathbb{R} = (-\infty, \infty)$ be the set of real numbers, $\mathbb{C}$ be the set of complex numbers, and

$$\mathbb{N} := \{1, 2, 3, \ldots\} = \mathbb{N}_0 \setminus \{0\}$$

be the set of positive integers. Let $\mathbb{D}(z_0,r)$ be the open unit disk of radius $r>0$ centered at $z_0 \in \mathbb{C}$,

$$\mathbb{D}(z_0,r) = \{z \in \mathbb{C} : |z-z_0| < r\}.$$

The open disk $\mathbb{D}(0,r)$ will be indicated by $\mathbb{D}_r$, and the unit disk $\mathbb{D}_1$ will be indicated by $\mathbb{D}$.

A complex-valued function $f : B \to \mathbb{C}$ of a complex variable is differentiable at a point $z_0 \in B \subset \mathbb{C}$ if it has a derivative at z_0; i.e.,

$$f'(z_0) = \lim_{z \to z_0} \frac{f(z)-f(z_0)}{z-z_0}.$$

A function f is said to be analytic at a point $z_0 \in B$ if it is differentiable in some neighborhood of z_0, and it is analytic in a domain B if it is analytic at all points in domain B. We can say that f is an entire function if it is analytic on the whole complex plane.

Definition 9.1. *An analytic function f is said to be univalent in a domain B if the conditions*

$$f(z_1) = f(z_2) \quad (z_1, z_2 \in B)$$

imply that $z_1 = z_2$.

Geometrically, this means that different points in the domain will be mapped into different points on the image domain ([25], p. 26). The theory of univalent functions is so vast and complicated that certain simplifying assumptions are necessary. The most obvious one is to replace the arbitrary domain B by one that is convenient, and the most attractive selection is the open unit disk $\mathbb{D}$.

For example, consider the function $f(z) = (1+z)^2$ in the open unit disk $\mathbb{D}$. The univalence of this function in $\mathbb{D}$ is easy to see on geometric grounds. Indeed, $1+z$ shifts the open unit disk to the right, and the effect of squaring $1+z$ is easy to visualize. The same type of argument shows that $w = (1+z)^3$ is not univalent in $\mathbb{D}$ (see [31]).

One of the most basic results in the theory of univalent functions in one variable is the Riemann mapping theorem. Its failure in several variables is one of the key differences between complex analysis in one variable and higher dimensions.

Theorem 9.1 (Riemann Mapping Theorem)**.** *Every simply connected domain B, which is a proper subset of $\mathbb{C}$, can be mapped conformally onto the unit disk. Moreover, if $z_0 \in B$, there is a unique conformal map of B onto $\mathbb{D}$ such that $f(z_0) = 0$ and $f'(z_0) > 0$.*

Let $\mathcal{A}$ represent the class of functions f of the form

$$f(z) = z + a_2 z^2 + a_3 z^3 + \cdots = z + \sum_{n=2}^{\infty} a_n z^n, \tag{9.1}$$

which are analytic in $\mathbb{D}$ and normalized by $f(0) = f'(0) - 1 = 0$.

The subclass of $\mathcal{A}$ consisting of univalent functions is indicated by $\mathcal{S}$. The leading example of a function of the class $\mathcal{S}$ is the *Koebe function*

$$k(z) = \frac{z}{(1-z)^2} = \sum_{n=1}^{\infty} nz^n \quad (z \in \mathbb{D}).$$

The Koebe function maps $\mathbb{D}$ onto the complex plane except for a slit along the half-line $\left(-\infty, -\frac{1}{4}\right]$ and is univalent. This is the best seen by writing

$$k(z) = \frac{1}{4}\left(\frac{1+z}{1-z}\right)^2 - \frac{1}{4}$$

and observing that the function

$$w(z) = \frac{1+z}{1-z}$$

maps $\mathbb{D}$ conformally onto the right half-plane $\Re(w) > 0$. The rotation of the Koebe function, i.e., the function given by

$$k_\theta(z) = \frac{z}{(1-e^{i\theta}z)^2} \quad (z \in \mathbb{D}),$$

belongs to the class $\mathcal{S}$ for each $\theta \in \mathbb{R}$. This image of the unit disk is in the complex plane except for a radial slit from ∞ to $-\frac{e^{i\theta}}{4}$. The Koebe function and its rotations are the only extremal functions for various problems in the class $\mathcal{S}$.

Other simple examples of functions in $\mathcal{S}$ are as follows:

i. $f(z) = z$, the identity mapping;
ii. $f(z) = \frac{z}{1-z}$, which maps $\mathbb{D}$ conformally onto the half-plane $\Re(w) > -\frac{1}{2}$;
iii. $f(z) = \frac{z}{1-z^2}$, which maps $\mathbb{D}$ onto the entire plane minus the two half-lines $\frac{1}{2} \le y < \infty$ and $-\infty < y \le -\frac{1}{2}$.

Among the first important papers that discuss topics from this area are Koebe [42], Alexander [5], and Bieberbach [17]. Koebe initiated in 1907 the study about univalent functions, while Bieberbach presented in 1916 would soon become a famous conjecture. In 1916, Bieberbach [17] conjectured that for $f \in \mathcal{S}$,

$$|a_n| \le n \ \ (n \ge 2).$$

He proved only for the case when $n = 2$. For many years, this conjecture has stood as a challenge to all mathematicians and has inspired the development of important new methods in complex analysis. For the cases $n = 3$ and $n = 4$, the conjecture was proved by Lowner [48], Garabedian, and Schiffer [30], respectively. Later, Pederson and Schiffer [58] proved the conjecture for $n = 5$; for $n = 6$, it was proved by Pederson [57] and Ozawa [56], independently. In 1985, Louis de Branges [24] proved the Bieberbach's conjecture for all the coefficients n. Although almost 70 years had passed until the Bieberbach conjecture was finally proved in this paper, bounds for the Taylor coefficients were obtained in the meantime for some subclasses of univalent functions. After the proof of the Bieberbach conjecture, the study of different subclasses of analytic and univalent functions has began to take shape, still remaining an interesting subject. Nowadays, there are many books dedicated to the study of univalent functions, of which we mention those of Conway [23], Duren [25], Goodman [31], Miller and Mocanu [51], and Pommerenke [60]. Further reading and some additional results on this subject can also be found in the wide variety of classical and modern textbooks devoted to the univalent function theory and related topics, including [2,16,21,22,37,44,52,62,63,71].

9.1.1 INTEGRAL OPERATORS

The study of operators plays an important role in geometric function theory in complex analysis and its related fields. Hence, in this part, we consider a few of the classic integral operators defined on analytic functions.

We recall the definition of the Komatu integral operator of a complex-valued function $f(z)$:

Definition 9.2. *(See [43]) The Komatu integral operator of $f \in \mathcal{A}$ is denoted by $\mathfrak{L}_t^{\mu} f(z)$ and defined by*

$$\mathfrak{L}_t^{\mu} f(z) = z + \sum_{n=2}^{\infty} \left(\frac{t}{t+n-1} \right)^{\mu} a_n z^n \quad (t > 0,\ \mu \geq 0,\ z \in \mathbb{D})$$

$$= \frac{t^{\mu}}{\Gamma(\mu)} \int_0^1 \xi^{t-2} \left(\log \frac{1}{\xi} \right)^{\mu-1} f(z\xi) d\xi.$$

It is easy to verify that

$$\mathfrak{L}_t^{\mu} \left(z f'(z) \right) = z \left(\mathfrak{L}_t^{\mu} f(z) \right)'.$$

Many subclasses of analytic univalent functions involving the Komatu integral operator have been studied by various authors (see, e.g., [14,15,20]). Further, by suitably specializing the parameters μ and t, we obtain the following operators studied by various authors:

i. $\mathfrak{L}_t^{0} f(z) = f(z)$;
ii. $\mathfrak{L}_2^{1} f(z) = A[f](z)$, called Libera operator [46];
iii. $\mathfrak{L}_1^{-k} f(z) = D^k f(z)$, $(k \in \mathbb{N}_0)$, called Sălăgean differential operator [61].

9.2 SUBCLASSES OF ANALYTIC AND UNIVALENT FUNCTIONS

In this part, we briefly discuss some of the well-known subclasses of the analytic univalent function class $\mathcal{S}$, including starlike and convex functions (see [31]).

Definition 9.3. *A set D in the plane is called convex if for every pair of points w_1 and w_2 in the interior of D, the line segment joining w_1 and w_2 is also in the interior of D. If a function $f(z)$ maps $\mathbb{D}$ onto a convex domain, then $f(z)$ is called a convex function.*

Definition 9.4. *A set D in the plane is said to be starlike with respect to w_0 an interior point of D if each ray with initial point w_0 intersects the interior of D in a set that is either a line segment or a ray. If a function $f(z)$ maps $\mathbb{D}$ onto a domain that is starlike with respect to w_0, then we say that $f(z)$ is starlike with respect to w_0. In the special case $w_0 = 0$, we say that $f(z)$ is a starlike function.*

Examples of convex and starlike domains are shown in Figures 9.1 and 9.2, respectively. The domain shown in Figure 9.2 is starlike with respect to w_0 but is not

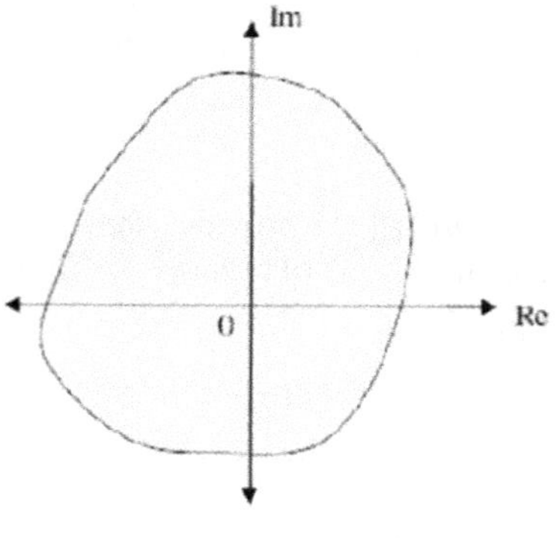

Figure 9.1 Convex domain.

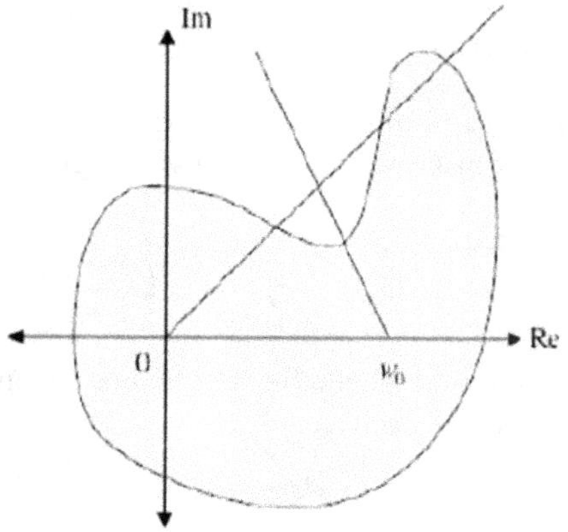

Figure 9.2 Starlike domain.

starlike with respect to the origin. The special function $w(z) = \frac{1+z}{1-z}$ is a convex function because it maps $\mathbb{D}$ onto a half-plane. The Koebe function is a starlike function. In fact, the domain $k(\mathbb{D})$ is starlike with respect to each $w_0 > -\frac{1}{4}$.

Further, two of the important and well-investigated subclasses of the analytic and univalent function class $\mathcal{S}$ are the class $\mathcal{S}^*(\beta)$ of starlike functions of order β in $\mathbb{D}$ and the class $\mathcal{K}(\beta)$ of convex functions of order β in $\mathbb{D}$, respectively. By definition, we have

$$\mathcal{S}^*(\alpha) = \left\{ f : f \in \mathcal{S} \text{ and } \Re\left(\frac{zf'(z)}{f(z)}\right) > \beta, \quad (z \in \mathbb{D}; 0 \leq \beta < 1) \right\} \tag{9.2}$$

and

$$\mathcal{K}(\alpha) = \left\{ f : f \in \mathcal{S} \text{ and } \Re\left(1 + \frac{zf''(z)}{f'(z)}\right) > \beta, \quad (z \in \mathbb{D}; 0 \leq \beta < 1) \right\}. \tag{9.3}$$

It is clear from definitions (9.2) and (9.3) that $\mathcal{K}(\beta) \subset \mathcal{S}^*(\beta)$. Also, we have

$$f(z) \in \mathcal{K}(\beta) \Leftrightarrow zf'(z) \in \mathcal{S}^*(\beta).$$

9.3 THE CLASS Σ

The *Koebe-One Quarter Theorem* [25] provides that the image of $\mathbb{D}$ under every univalent function $f \in \mathcal{A}$ contains a disk of radius $1/4$. Thus, every univalent function $f \in \mathcal{A}$ has an inverse f^{-1} satisfying

$$f^{-1}(f(z)) = z,$$

and

$$f\left(f^{-1}(w)\right) = w \quad \left(|w| < r_0(f),\ r_0(f) \geq \frac{1}{4}\right),$$

where

$$g(w) = f^{-1}(w) = w - a_2 w^2 + \left(2a_2^2 - a_3\right) w^3 - \left(5a_2^3 - 5a_2 a_3 + a_4\right) w^4 + \cdots. \tag{9.4}$$

A function $f \in \mathcal{A}$ is said to be bi-univalent in $\mathbb{D}$ if both f and f^{-1} are univalent in $\mathbb{D}$. Let Σ represent the class of bi-univalent functions in $\mathbb{D}$ given by (9.1). Examples of functions in the class Σ are

$$\frac{z}{1-z},\ -\log(1-z),\ \frac{1}{2}\log\left(\frac{1+z}{1-z}\right)$$

and so on. However, the familiar Koebe function is not a member of Σ. Other common examples of functions in $\mathcal{S}$ such as

$$z - \frac{z^2}{2},\ \frac{z}{1-z^2}$$

are also not the members of Σ.

Historically, Lewin [45] studied the class of bi-univalent functions, obtaining the bound 1.51 for the modulus of the second coefficient $|a_2|$. Subsequently, Brannan and Clunie [18] conjectured that $|a_2| \leqq \sqrt{2}$ for $f \in \Sigma$. Later on, Netanyahu [53] showed that $\max|a_2| = \frac{4}{3}$ if $f \in \Sigma$. Brannan and Taha [19] introduced certain subclasses of the bi-univalent function class Σ similar to the familiar subclasses $\mathcal{S}^{\star}(\beta)$ and $\mathcal{K}(\beta)$ of starlike and convex functions of order β $(0 \leqq \beta < 1)$ in $\mathbb{D}$, respectively (see [53]). The classes $\mathcal{S}_{\Sigma}^{\star}(\beta)$ and $\mathcal{K}_{\Sigma}(\beta)$ of bi-starlike functions of order β in $\mathbb{D}$ and bi-convex functions of order β in $\mathbb{D}$, corresponding to the function classes $\mathcal{S}^{\star}(\beta)$ and $\mathcal{K}(\beta)$, were also introduced analogously. For each of the function classes $\mathcal{S}_{\Sigma}^{\star}(\beta)$ and $\mathcal{K}_{\Sigma}(\beta)$, they found non-sharp estimates for the initial coefficients. Recently, motivated substantially by the aforementioned pioneering work on this subject by Srivastava *et al.* [69], many authors investigated the coefficient bounds for various subclasses of bi-univalent functions (see, for example, [3,9,12,13,38,50,68,74,75]). Not much is known about the bounds on the general coefficient $|a_n|$ for $n \geqq 4$. In the literature, there are only a few works determining the general coefficient bounds for $|a_n|$ for the analytic bi-univalent functions (see, for example, [7,10,35,36,70,72,73]). The coefficient estimate problem for each of the coefficients

$$|a_n| \qquad (n \in \mathbb{N} \setminus \{1,2\})$$

is presumably still an open problem for a number of subclasses of the bi-univalent function calss Σ.

Furthermore, the classical Fekete-Szegö inequality, presented by means of Loewner's method, for the coefficients of $f \in \mathcal{S}$ is

$$\left|a_3 - \rho a_2^2\right| \leq 1 + 2\exp(-2\rho/(1-\vartheta)) \;\; \text{for } \rho \in [0,1).$$

As $\rho \to 1^-$, we have the elementary inequality $\left|a_3 - a_2^2\right| \leq 1$. Moreover, the coefficient functional

$$F_\rho(f) = a_3 - \rho a_2^2$$

on the normalized analytic functions f in the unit disk $\mathbb{D}$ plays an important role in the function theory. The problem of maximizing the absolute value of the functional $F_\vartheta(f)$ is called the Fekete-Szegö problem (see [28]). The Fekete-Szegö problem has continued to receive attention of researches in geometric function theory (see, for example, [8,11,29,39,40,54,59,67,76]). Various interesting developments involving the Fekete-Szegö problem can be found in Abdel-Gawad and Thomas [1], Keogh and Merkes [41], and London [47].

9.4 FUNCTIONS WITH POSITIVE REAL PART

The well-known class of functions with positive real part, consisting of all functions p analytic in $\mathbb{D}$ satisfying $p(0) = 1$ and $\Re\, p(z) > 0$, is usually denoted by $\mathcal{P}$ and called the Carathéodory class. For example, the function

$$p(z) = \frac{1+z}{1-z} \qquad (z \in \mathbb{D})$$

belongs to $\mathcal{P}$. This function gives a conformal map of $\mathbb{D}$ onto the right-half-plane, and consequently, it plays a fundamental role in the class $\mathcal{P}$, similar to the Koebe function for the class $\mathcal{S}$. It should be noted that $p(z)$ is not required to be univalent. Thus, $p(z) = 1 + z^n$ is in $\mathcal{P}$ for any integer $n \geq 0$, but if $n \geq 2$, this function is not univalent.

Each $p \in \mathcal{P}$ has a Taylor series expansion

$$p(z) = 1 + x_1 z + x_2 z^2 + x_3 z^3 + \cdots \quad (x_1 > 0)$$

with coefficients satisfying $|x_n| \leq 2$ for $n \in \mathbb{N}$ (see [60]). More refinement coefficient bounds in the Carathéodory class were obtained by Grenander and Szegö [32].

9.4.1 SUBORDINATION

With a view to remaining the rule of subordination between analytic functions, let the functions f, g be analytic in $\mathbb{D}$, we say that the function f is subordinate to g, indicated as $f \prec g$ (or $f(z) \prec g(z)$) $(z \in \mathbb{D})$, if there exists a Schwarz function $\mathfrak{w} \in \Lambda$, where

$$\Lambda = \{\mathfrak{w} : \mathfrak{w}(0) = 0, \ |\mathfrak{w}(z)| < 1, \ z \in \mathbb{D}\},$$

such that

$$f(z) = g(\mathfrak{w}(z)) \qquad (z \in \mathbb{D}).$$

Furthermore, if the function g is univalent in $\mathbb{D}$, then we have the following equivalence that holds (see [51]):

$$f(z) \prec g(z) \Leftrightarrow f(0) = g(0) \text{ and } f(\mathbb{D}) \subset g(\mathbb{D}).$$

Ma and Minda [49] unified various subclasses of starlike and convex functions for which either of the quantities

$$\frac{zf'(z)}{f(z)} \quad \text{and} \quad 1 + \frac{zf''(z)}{f'(z)}$$

is subordinate to a more general superordinate function. For this purpose, they considered an analytic function φ with positive real part in the open unit disk $\mathbb{D}$, $\varphi(0) = 1, \varphi'(0) = 1$, and φ maps $\mathbb{D}$ onto a region starlike with respect to 1 and symmetric with respect to the real axis. The class of Ma-Minda starlike functions consists of functions $f \in \mathcal{A}$ satisfying the following subordination condition:

$$\frac{zf'(z)}{f(z)} \prec \varphi(z) \qquad (z \in \mathbb{D}).$$

Similarly, the class of Ma-Minda convex functions consists of functions $f \in \mathcal{A}$ satisfying the following subordination condition:

$$1 + \frac{zf''(z)}{f'(z)} \prec \varphi(z) \qquad (z \in \mathbb{D}).$$

A function f is said to be bi-starlike of Ma-Minda type or bi-convex of Ma-Minda type if both f and f^{-1} are, respectively Ma-Minda starlike or convex. These classes are denoted, respectively, by $\mathcal{S}^{\star}_{\Sigma}(\varphi)$ and $\mathcal{K}_{\Sigma}(\varphi)$ (see [6]).

9.5 BI-UNIVALENT FUNCTION CLASSES $\mathcal{S}_{t,\Sigma}^{\mu}(\widetilde{\mathfrak{P}})$ AND $\mathcal{K}_{t,\Sigma}^{\mu}(\widetilde{\mathfrak{P}})$

In this section, we can formulate the definition of the classes $\mathcal{S}_{t,\Sigma}^{\mu}(\widetilde{\mathfrak{p}})$ and $\mathcal{K}_{t,\Sigma}^{\mu}(\widetilde{\mathfrak{p}})$ by using subordination and the Komatu integral operator mentioned above.

Definition 9.5. *A function $f \in \Sigma$ is said to be in the class*

$$\mathcal{S}_{t,\Sigma}^{\mu}(\widetilde{\mathfrak{p}}) \quad (0 \leq \varepsilon \leq 1,\ t > 0,\ \mu \geq 0,\ z, w \in \mathbb{D})$$

if the following subordinations are satisfied:

$$\frac{z\left(\mathfrak{L}_t^{\mu} f(z)\right)'}{(1-\varepsilon)z + \varepsilon \mathfrak{L}_t^{\mu} f(z)} \prec \widetilde{\mathfrak{p}}(z) = \frac{1+\tau^2 z^2}{1-\tau z-\tau^2 z^2}$$

and

$$\frac{w\left(\mathfrak{L}_t^{\mu} g(w)\right)'}{(1-\varepsilon)w + \varepsilon \mathfrak{L}_t^{\mu} g(w)} \prec \widetilde{\mathfrak{p}}(w) = \frac{1+\tau^2 w^2}{1-\tau w-\tau^2 w^2},$$

where the function $g = f^{-1}$ and $\tau = \frac{1-\sqrt{5}}{2} \approx -0.618$.

Definition 9.6. *A function $f \in \Sigma$ is said to be in the class*

$$\mathcal{K}_{t,\Sigma}^{\mu}(\widetilde{\mathfrak{p}}) \quad (0 \leq \varepsilon \leq 1,\ t > 0,\ \mu \geq 0,\ z, w \in \mathbb{D})$$

if the following conditions are satisfied:

$$\frac{z\left(\mathfrak{L}_t^{\mu} f(z)\right)' + z^2\left(\mathfrak{L}_t^{\mu} f(z)\right)''}{(1-\varepsilon)z + \varepsilon z\left(\mathfrak{L}_t^{\mu} f(z)\right)'} \prec \widetilde{\mathfrak{p}}(z) = \frac{1+\tau^2 z^2}{1-\tau z-\tau^2 z^2}$$

and

$$\frac{w\left(\mathfrak{L}_t^{\mu} g(w)\right)' + w^2\left(\mathfrak{L}_t^{\mu} g(w)\right)''}{(1-\varepsilon)w + \varepsilon w\left(\mathfrak{L}_t^{\mu} g(w)\right)'} \prec \widetilde{\mathfrak{p}}(w) = \frac{1+\tau^2 w^2}{1-\tau w-\tau^2 w^2},$$

where the function $g = f^{-1}$ and $\tau = \frac{1-\sqrt{5}}{2} \approx -0.618$.

If we can take the parameters $\mu = 0$ and $\varepsilon = 1$ in the above definitions, we have the following function classes $\mathcal{SL}_{\Sigma}(\widetilde{\mathfrak{p}}(z))$ and $\mathcal{KL}_{\Sigma}(\widetilde{\mathfrak{p}}(z))$, respectively.

Definition 9.7. *(See [33]) A function $f \in \Sigma$ is said to be in the class $\mathcal{SL}_{\Sigma}(\widetilde{\mathfrak{p}}(z))$ if the following subordination holds:*

$$\frac{zf'(z)}{f(z)} \prec \widetilde{\mathfrak{p}}(z) = \frac{1+\tau^2 z^2}{1-\tau z-\tau^2 z^2}$$

and

$$\frac{wg'(w)}{g(w)} \prec \widetilde{\mathfrak{p}}(w) = \frac{1+\tau^2 w^2}{1-\tau w-\tau^2 w^2},$$

where the function $g = f^{-1}$ and $\tau = \frac{1-\sqrt{5}}{2} \approx -0.618$.

Definition 9.8. *(See [33]) A function $f \in \Sigma$ is said to be in the class $\mathcal{KL}_{\Sigma}(\widetilde{\mathfrak{p}}(z))$ if the following subordination holds:*

$$1+\frac{zf''(z)}{f'(z)} \prec \widetilde{\mathfrak{p}}(z) = \frac{1+\tau^2 z^2}{1-\tau z-\tau^2 z^2}$$

and

$$1+\frac{wg''(w)}{g'(w)} \prec \widetilde{\mathfrak{p}}(w) = \frac{1+\tau^2 w^2}{1-\tau w-\tau^2 w^2},$$

where the function $g = f^{-1}$ and $\tau = \frac{1-\sqrt{5}}{2} \approx -0.618$.

Remark 9.1. *The function $\widetilde{\mathfrak{p}}(z)$ is not univalent in $\mathbb{D}$, but it is univalent in the disk $|z| < \frac{3-\sqrt{5}}{2} \approx 0.38$. For example, $\widetilde{\mathfrak{p}}(0) = \widetilde{\mathfrak{p}}\left(-\frac{1}{2\tau}\right)$ and $\widetilde{\mathfrak{p}}\left(e^{\pm i \arccos(1/4)}\right) = \frac{\sqrt{5}}{5}$. Also, it can be written as*

$$\frac{1}{|\tau|} = \frac{|\tau|}{1-|\tau|}$$

which indicates that the number $|\tau|$ divides $[0,1]$ such that it fulfills the golden section (see for details Dziok et al. [26]).

Additionally, Dziok et al. [26] indicate a useful connection between the function $\widetilde{\mathfrak{p}}(z)$ and the Fibonacci numbers. Let $\{\Lambda_n\}$ be the sequence of Fibonacci numbers

$$\Lambda_0 = 0,\ \Lambda_1 = 1,\ \Lambda_{n+2} = \Lambda_n + \Lambda_{n+1}\ (n \in \mathbb{N}_0),$$

then

$$\Lambda_n = \frac{(1-\tau)^n - \tau^n}{\sqrt{5}}, \quad \tau = \frac{1-\sqrt{5}}{2}.$$

If we set

$$\widetilde{\mathfrak{p}}(z) = 1+\sum_{n=1}^{\infty} \widetilde{\mathfrak{p}}_n z^n = 1+(\Lambda_0+\Lambda_2)\tau z+(\Lambda_1+\Lambda_3)\tau^2 z^2$$

$$+\sum_{n=3}^{\infty}(\Lambda_{n-3}+\Lambda_{n-2}+\Lambda_{n-1}+\Lambda_n)\tau^n z^n,$$

then the coefficients $\widetilde{\mathfrak{p}}_n$ satisfy

$$\widetilde{\mathfrak{p}}_n = \begin{cases} \tau & (n=1) \\ 3\tau^2 & (n=2) \\ \tau\widetilde{\mathfrak{p}}_{n-1}+\tau^2\widetilde{\mathfrak{p}}_{n-2} & (n=3,4,\ldots) \end{cases}. \tag{9.5}$$

9.6 INEQUALITIES FOR THE TAYLOR-MACLAURIN COEFFICIENTS

In this part, we offer to get the estimates on the Taylor-Maclaurin coefficients and derive the Fekete-Szegö inequalities for functions in the classes $\mathcal{S}^{\mu}_{t,\Sigma}(\widetilde{\mathfrak{p}})$ and $\mathcal{K}^{\mu}_{t,\Sigma}(\widetilde{\mathfrak{p}})$.

Theorem 9.2. *Let the function f given by (9.1) be in the class $\mathcal{S}^{\mu}_{t,\Sigma}(\widetilde{\mathfrak{p}})$.Then,*

$$|a_2| \le \frac{|\tau|}{\sqrt{\left|(2-\varepsilon)^2\left(\frac{t}{t+1}\right)^{2\mu}+(3-\varepsilon)\left[\left(\frac{t}{t+2}\right)^{\mu}-2(2-\varepsilon)\left(\frac{t}{t+1}\right)^{2\mu}\right]\tau\right|}},$$

$$|a_3| \le \frac{|\tau|}{(3-\varepsilon)\left(\frac{t}{t+2}\right)^{\mu}}+\frac{\tau^2}{(2-\varepsilon)^2\left(\frac{t}{t+1}\right)^{2\mu}},$$

for any real number ρ,

$$|a_3-\rho a_2^2| \le$$

$$\begin{cases} \dfrac{|\tau|}{(3-\varepsilon)\left(\frac{t}{t+2}\right)^{\mu}}, & |\rho-1|\le T \\[2ex] \dfrac{|1-\rho|\tau^2}{\left|(2-\varepsilon)^2\left(\frac{t}{t+1}\right)^{2\mu}+(3-\varepsilon)\left[\left(\frac{t}{t+2}\right)^{\mu}-2(2-\varepsilon)\left(\frac{t}{t+1}\right)^{2\mu}\right]\tau\right|}, & |\rho-1|\ge T \end{cases},$$

where $T=\dfrac{\left|(2-\varepsilon)^2\left(\frac{t}{t+1}\right)^{2\mu}+(3-\varepsilon)\left[\left(\frac{t}{t+2}\right)^{\mu}-2(2-\varepsilon)\left(\frac{t}{t+1}\right)^{2\mu}\right]\tau\right|}{(3-\varepsilon)\left(\frac{t}{t+2}\right)^{\mu}|\tau|}.$

Proof. Suppose that $f\in\mathcal{S}^{\mu}_{t,\Sigma}(\widetilde{\mathfrak{p}})$. First, let $p\prec\widetilde{\mathfrak{p}}$. Then, by the definition of subordination, for two analytic functions $\mathfrak{u},\mathfrak{v}$ such that $\mathfrak{u}(0)=\mathfrak{v}(0)=0$, $|\mathfrak{u}(z)|<1, |\mathfrak{v}(w)|<1$ $(z,w\in\mathbb{D})$, we can write

$$\frac{z\left(\mathfrak{L}^{\mu}_{t}f(z)\right)'}{(1-\varepsilon)z+\varepsilon\mathfrak{L}^{\mu}_{t}f(z)}=\widetilde{\mathfrak{p}}(\mathfrak{u}(z)) \tag{9.6}$$

and

$$\frac{w\left(\mathfrak{L}^{\mu}_{t}g(w)\right)'}{(1-\varepsilon)w+\varepsilon\mathfrak{L}^{\mu}_{t}g(w)}=\widetilde{\mathfrak{p}}(\mathfrak{v}(w)). \tag{9.7}$$

Next, define the functions p_1 and p_2 by

$$p_1(z)=\frac{1+\mathfrak{u}(z)}{1-\mathfrak{u}(z)}=1+x_1z+x_2z^2+\cdots,$$

$$p_2(w) = \frac{1+\mathfrak{v}(w)}{1-\mathfrak{v}(w)} = 1 + y_1 w + y_2 w^2 + \cdots .$$

Since $\mathfrak{u}$ and $\mathfrak{v}$ are the Schwarz functions, p_1 and p_2 are the analytic functions in $\mathbb{D}$, with $p_1(0) = p_2(0) = 1$ and which have a positive real part in $\mathbb{D}$, we obtain the equations

$$\mathfrak{u}(z) = \frac{p_1(z)-1}{p_1(z)+1} = \frac{1}{2}\left[x_1 z + \left(x_2 - \frac{x_1^2}{2}\right) z^2\right] + \cdots ,$$

$$\mathfrak{v}(w) = \frac{p_2(w)-1}{p_2(w)+1} = \frac{1}{2}\left[y_1 w + \left(y_2 - \frac{y_1^2}{2}\right) w^2\right] + \cdots$$

leading to

$$\widetilde{\mathfrak{p}}(\mathfrak{u}(z)) = 1 + \frac{\widetilde{\mathfrak{p}}_1 x_1}{2} z + \left[\frac{1}{2}\left(x_2 - \frac{x_1^2}{2}\right)\widetilde{\mathfrak{p}}_1 + \frac{x_1^2}{4}\widetilde{\mathfrak{p}}_2\right] z^2 + \cdots ,$$

$$\widetilde{\mathfrak{p}}(\mathfrak{v}(w)) = 1 + \frac{\widetilde{\mathfrak{p}}_1 y_1}{2} w + \left[\frac{1}{2}\left(y_2 - \frac{y_1^2}{2}\right)\widetilde{\mathfrak{p}}_1 + \frac{y_1^2}{4}\widetilde{\mathfrak{p}}_2\right] w^2 + \cdots .$$

Now, upon comparing the corresponding coefficients in (9.6) and (9.7), we get

$$(2-\varepsilon)\left(\frac{t}{t+1}\right)^{\mu} a_2 = \frac{\widetilde{\mathfrak{p}}_1 x_1}{2}, \tag{9.8}$$

$$(3-\varepsilon)\left(\frac{t}{t+2}\right)^{\mu} a_3 - (2\varepsilon-\varepsilon^2)\left(\frac{t}{t+1}\right)^{2\mu} a_2^2 = \frac{1}{2}\left(x_2 - \frac{x_1^2}{2}\right)\widetilde{\mathfrak{p}}_1 + \frac{x_1^2}{4}\widetilde{\mathfrak{p}}_2, \tag{9.9}$$

$$-(2-\varepsilon)\left(\frac{t}{t+1}\right)^{\mu} a_2 = \frac{\widetilde{\mathfrak{p}}_1 y_1}{2}, \tag{9.10}$$

$$(3-\varepsilon)\left(\frac{t}{t+2}\right)^{\mu} (2a_2^2 - a_3) - (2\varepsilon-\varepsilon^2)\left(\frac{t}{t+1}\right)^{2\mu} a_2^2 = \frac{1}{2}\left(y_2 - \frac{y_1^2}{2}\right)\widetilde{\mathfrak{p}}_1 + \frac{y_1^2}{4}\widetilde{\mathfrak{p}}_2. \tag{9.11}$$

From equations (9.8) and (9.10), one can easily find that

$$x_1 = -y_1, \tag{9.12}$$

$$2(2-\varepsilon)^2\left(\frac{t}{t+1}\right)^{2\mu} a_2^2 = \frac{\widetilde{\mathfrak{p}}_1^2}{4}(x_1^2 + y_1^2). \tag{9.13}$$

If we add (9.9) to (9.11), we obtain

$$2\left[(3-\varepsilon)\left(\frac{t}{t+2}\right)^{\mu} - (2\varepsilon-\varepsilon^2)\left(\frac{t}{t+1}\right)^{2\mu}\right] a_2^2 = \frac{\widetilde{\mathfrak{p}}_1}{2}(x_2 + y_2) + \frac{(\widetilde{\mathfrak{p}}_2 - \widetilde{\mathfrak{p}}_1)}{4}(x_1^2 + y_1^2). \tag{9.14}$$

By making the use of (9.13) in (9.14), we have

$$a_2^2 = \frac{\widetilde{\mathfrak{p}}_1^3(x_2+y_2)}{4\left\{\left[(3-\varepsilon)\left(\frac{t}{t+2}\right)^{\mu}-(2\varepsilon-\varepsilon^2)\left(\frac{t}{t+1}\right)^{2\mu}\right]\widetilde{\mathfrak{p}}_1^2-(2-\varepsilon)^2\left(\frac{t}{t+1}\right)^{2\mu}(\widetilde{\mathfrak{p}}_2-\widetilde{\mathfrak{p}}_1)\right\}} \tag{9.15}$$

which yields

$$|a_2| \leq \frac{|\tau|}{\sqrt{\left|(2-\varepsilon)^2\left(\frac{t}{t+1}\right)^{2\mu}+(3-\varepsilon)\left[\left(\frac{t}{t+2}\right)^{\mu}-2(2-\varepsilon)\left(\frac{t}{t+1}\right)^{2\mu}\right]\tau\right|}}.$$

Next, if we subtract (9.11) from (9.9), we obtain

$$2(3-\varepsilon)\left(\frac{t}{t+2}\right)^{\mu}(a_3-a_2^2) = \frac{\widetilde{\mathfrak{p}}_1}{2}(x_2-y_2). \tag{9.16}$$

Then, in view of (9.13), the equation (9.16) becomes

$$a_3 = \frac{\widetilde{\mathfrak{p}}_1^2(x_1^2+y_1^2)}{8(2-\varepsilon)^2\left(\frac{t}{t+1}\right)^{2\mu}} + \frac{\widetilde{\mathfrak{p}}_1(x_2-y_2)}{4(3-\varepsilon)\left(\frac{t}{t+2}\right)^{\mu}}.$$

Notice that from (9.5), we get

$$|a_3| \leq \frac{|\tau|}{(3-\varepsilon)\left(\frac{t}{t+2}\right)^{\mu}} + \frac{\tau^2}{(2-\varepsilon)^2\left(\frac{t}{t+1}\right)^{2\mu}}.$$

From (9.15) and (9.16), we find that

$$a_3-\rho a_2^2 = \frac{(1-\rho)\widetilde{\mathfrak{p}}_1^3(x_2+y_2)}{4\left\{\left[(3-\varepsilon)\left(\frac{t}{t+2}\right)^{\mu}-(2\varepsilon-\varepsilon^2)\left(\frac{t}{t+1}\right)^{2\mu}\right]\widetilde{\mathfrak{p}}_1^2-(2-\varepsilon)^2\left(\frac{t}{t+1}\right)^{2\mu}(\widetilde{\mathfrak{p}}_2-\widetilde{\mathfrak{p}}_1)\right\}}$$
$$+\frac{\widetilde{\mathfrak{p}}_1(x_2-y_2)}{4(3-\varepsilon)\left(\frac{t}{t+2}\right)^{\mu}}$$
$$=\frac{\widetilde{\mathfrak{p}}_1}{4}\left[\left(h(\rho)+\frac{1}{(3-\varepsilon)\left(\frac{t}{t+2}\right)^{\mu}}\right)x_2+\left(h(\rho)-\frac{1}{(3-\varepsilon)\left(\frac{t}{t+2}\right)^{\mu}}\right)y_2\right],$$

where

$$h(\rho) = \frac{(1-\rho)\widetilde{\mathfrak{p}}_1^2}{\left[(3-\varepsilon)\left(\frac{t}{t+2}\right)^{\mu}-(2\varepsilon-\varepsilon^2)\left(\frac{t}{t+1}\right)^{2\mu}\right]\widetilde{\mathfrak{p}}_1^2-(2-\varepsilon)^2\left(\frac{t}{t+1}\right)^{2\mu}(\widetilde{\mathfrak{p}}_2-\widetilde{\mathfrak{p}}_1)}.$$

Thus, in view of (9.5), we get

$$\left|a_3-\rho a_2^2\right| \leq \begin{cases} \dfrac{|\widetilde{\mathfrak{p}}_1|}{(3-\varepsilon)\left(\dfrac{t}{t+2}\right)^{\mu}}, & 0 \leq |h(\rho)| \leq \dfrac{1}{(3-\varepsilon)\left(\frac{t}{t+2}\right)^{\mu}} \\ |h(\rho)|\,|\widetilde{\mathfrak{p}}_1|, & |h(\rho)| \geq \dfrac{1}{(3-\varepsilon)\left(\frac{t}{t+2}\right)^{\mu}} \end{cases},$$

which evidently completes the proof of Theorem 9.2. □

Analysis similar to that in the proof of the previous theorem shows that

Theorem 9.3. *Let the function f given by (9.1) be in the class $\mathcal{K}_{t,\Sigma}^{\mu}(\widetilde{\mathfrak{p}})$.Then,*

$$|a_2| \leq \frac{|\tau|}{\sqrt{\left|4(2-\varepsilon)^2\left(\dfrac{t}{t+1}\right)^{2\mu}+(3-\varepsilon)\left[\left(\dfrac{t}{t+2}\right)^{\mu}-8(2-\varepsilon)\left(\dfrac{t}{t+1}\right)^{2\mu}\right]\tau\right|}},$$

$$|a_3| \leq \frac{|\tau|}{3(3-\varepsilon)\left(\dfrac{t}{t+2}\right)^{\mu}}+\frac{\tau^2}{4(2-\varepsilon)^2\left(\dfrac{t}{t+1}\right)^{2\mu}},$$

for any real number ρ,

$$\left|a_3-\rho a_2^2\right| \leq$$

$$\begin{cases} \dfrac{|\tau|}{3(3-\varepsilon)\left(\dfrac{t}{t+2}\right)^{\mu}}, & |\rho-1| \leq Y \\ \dfrac{|1-\rho|\tau^2}{\left|4(2-\varepsilon)^2\left(\dfrac{t}{t+1}\right)^{2\mu}+(3-\varepsilon)\left[3\left(\dfrac{t}{t+2}\right)^{\mu}-8(2-\varepsilon)\left(\dfrac{t}{t+1}\right)^{2\mu}\right]\tau\right|}, & |\rho-1| \geq Y \end{cases},$$

where $Y = \dfrac{\left|4(2-\varepsilon)^2\left(\dfrac{t}{t+1}\right)^{2\mu}+(3-\varepsilon)\left[3\left(\dfrac{t}{t+2}\right)^{\mu}-8(2-\varepsilon)\left(\dfrac{t}{t+1}\right)^{2\mu}\right]\tau\right|}{3(3-\varepsilon)\left(\dfrac{t}{t+2}\right)^{\mu}|\tau|}.$

Taking $\mu = 0$ and $\varepsilon = 1$ in Theorems 9.2 and 9.3, we have the following corollaries.

Corollary 9.1. *(See [33]) Let the function f given by (9.1) be in the class $\mathcal{SL}_{\Sigma}(\widetilde{\mathfrak{p}}(z))$.Then*

$$|a_2| \leq \frac{|\tau|}{\sqrt{1-2\tau}},$$

$$|a_3| \leq \frac{|\tau|(1-4\tau)}{2(1-2\tau)},$$

for any real number ρ,

$$|a_3 - \rho a_2^2| \leq \begin{cases} \frac{|\tau|}{2}, & |\rho - 1| \leq \frac{1-2\tau}{2|\tau|} \\ \frac{|1-\rho|\tau^2}{1-2\tau}, & |\rho - 1| \geq \frac{1-2\tau}{2|\tau|} \end{cases}.$$

Corollary 9.2. *(See [33]) Let the function* f *given by (9.1) be in the class* $\mathcal{KL}_\Sigma(\widetilde{\mathfrak{p}}(z))$*.Then*

$$|a_2| \leq \frac{|\tau|}{\sqrt{4-10\tau}},$$

$$|a_3| \leq \frac{|\tau|(1-4\tau)}{3(2-5\tau)},$$

for any real number ρ,

$$|a_3 - \rho a_2^2| \leq \begin{cases} \frac{|\tau|}{6}, & |\rho - 1| \leq \frac{2-5\tau}{3|\tau|} \\ \frac{|1-\rho|\tau^2}{2(2-5\tau)}, & |\rho - 1| \geq \frac{2-5\tau}{3|\tau|} \end{cases}.$$

9.7 CONCLUDING REMARKS AND OBSERVATIONS

In our present investigation, we have introduced and studied the coefficient problems associated with the following new subclasses

$$\mathcal{S}_{t,\Sigma}^{\mu}(\widetilde{\mathfrak{p}}) \text{ and } \mathcal{K}_{t,\Sigma}^{\mu}(\widetilde{\mathfrak{p}}) \quad (t > 0,\ \mu \geq 0)$$

of the class of normalized bi-univalent functions in the open unit disk $\mathbb{D}$. For functions belonging to this bi-univalent function class, we have derived Taylor-Maclaurin coefficient inequalities and we have considered the celebrated Fekete-Szegö problem in Section 9.6. For the motivation and validity of our results, we have also pointed out relevant connections with those that were given in earlier works. Moreover, with a view to potentially motivate the interested reader, we choose to include a citation of a very recent survey-cum-expository article [64], which also provides a review of many other related recent works in geometric function theory of complex analysis.

The geometric properties of the function classes $\mathcal{S}_{t,\Sigma}^{\mu}(\widetilde{\mathfrak{p}})$, $\mathcal{K}_{t,\Sigma}^{\mu}(\widetilde{\mathfrak{p}})$ vary according to the values assigned to the parameters involved. Nevertheless, some results for the special cases of the parameters involved could be presented as illustrative examples.

If, on the other hand, we restrict our considerations for a given univalent function $\widetilde{\mathfrak{p}}(z)$ in $\mathbb{D}$, we can investigate the corresponding mapping problems for other regions of the complex z-plane. In this way, one can introduce many other subclasses of the function classes which we have studied in this paper.

Determination of extremal functions for bi-univalent functions (in general) and for bi-subordinate functions (in particular) remains a challenge.

We conclude our present investigation by observing that the interested reader will find several related recent developments concerning geometric function theory of complex analysis (see, for example, [4,27,34,55,65,66]) to be potentially useful for motivating further researches in this subject and on other related topics.

ACKNOWLEDGMENT

This work is supported by the Scientific and Technological Research Council of Turkey (TUBITAK 1002-Short Term R&D Funding Program) Project Number: 118F543.

BIBLIOGRAPHY

[1] H. R. Abdel-Gawad, D. K. Thomas, The Fekete-Szegö problem for strongly close-to-convex functions, *Proc. Amer. Math. Soc.,* 114 (1992) 345–349.

[2] L. V. Ahlfors, *Complex Analysis,* 2nd edn, McGraw-Hill, New York, 1966.

[3] A. Akgül, Ş. Altınkaya, Coefficient estimates associated with a new subclass of bi-univalent functions, *Acta Universitatis Apulensis,* 52 (2017) 121–128.

[4] O. Al-Refai, M. Ali, General coefficient estimates for bi-univalent functions: a new approach, *Turk. J. Math.,* 44 (2020) 240–251.

[5] J. W. Alexander, Functions which map the interior of the unit circle upon simple regions, *Ann. Math.,* 17 (1915) 12–22.

[6] R. M. Ali, S. K. Lee, V. Ravichandran, S. Supramanian, Coefficient estimates for bi-univalent Ma-Minda starlike and convex functions, *Appl. Math. Lett.,* 25 (2012) 344–351.

[7] Ş. Altınkaya, S. Yalçın, Coefficient bounds for a subclass of bi-univalent functions, *TWMS J. Pure Appl. Math.,* 6 (2015) 180–185.

[8] Ş. Altınkaya, S. Yalçın, Fekete-Szegö inequalities for certain classes of bi-univalent functions, Internat. *Scholar. Res. Notices,* 2014 (2014) Article ID 327962, 1–6.

[9] Ş. Altınkaya, S. Yalçın, Coefficient estimates for two new subclasses of bi-univalent functions with respect to symmetric points, *J. Funct. Spaces,* 2015 (2015) Article ID 145242, 1–5.

[10] Ş. Altınkaya, S. Yalçın, Faber polynomial coefficient bounds for a subclass of bi-univalent functions, *C. R. Acad. Sci. Paris Sér. I,* 353 (2015) 1075–1080.

[11] Ş. Altınkaya, S. Yalçın, The Fekete-Szegö problem for a general class of bi-univalent functions satisfying subordinate conditions, *Sahand Commun. Math. Anal.,* 5 (2017) 1–7.

[12] Ş. Altınkaya, S. Yalçın, Faber polynomial coefficient estimates for bi-univalent functions of complex order based on subordinate conditions involving of the Jackson (p,q)-derivative operator, *Miskolc Mathematical Notes,* 18 (2017) 555–572.

[13] Ş. Altınkaya, S. Yalçın, S. Çakmak, A subclass of bi-univalent functions based on the Faber polynomial expansions and the Fibonacci numbers, *Mathematics,* 160 (2019) 1–9.

[14] Ş. Altınkaya, S. Yalçın, Chebyshev polynomial bounds for a certain subclass of univalent functions defined by Komatu integral operator, *Afrika Matematika,* 30 (2019) 563–570.

[15] M. K. Aouf, Some inclusion relationships associated with the Komatu integral operator, *Mathe. Comput. Modell.* 50 (2009) 1360–1366.

[16] S. D. Bernardi, A survey of the development ofthe theory of schlicht functions, *Duke Math. J.,* 19 (1952) 263–287.

[17] L. Bieberbach, Uber die Koe zienten derjenigen Potenzreihen, welche eine schlichte Abbildung des Einheitskreises vernitteln, Sitzungsber. Preuss. *Akad. Wiss. Phys-Math.,* (1916) 940–955.

[18] D. A. Brannan, J. G. Clunie, Aspects of contemporary complex analysis, Proceedings of the NATO Advanced Study Instute Held at University of Durham, New York: Academic Press, 1979.

[19] D. A. Brannan, T. S. Taha, On some classes of bi-univalent functions, *Studia Universitatis Babeş-Bolyai Mathematica,* 31 (2) (1986) 70–77.

[20] S. Bulut, Fekete–Szegö problem for subclasses of analytic functions defined by Komatu integral operator, *Arab. J. Math.,* 2 (2013) 177–183.

[21] J. Clunie, On schlicht functions, *Ann. Math.,* 69 (1959) 511–519.

[22] J. Clunie, Ch. Pommerenke, On the coefficients of univalent functions, *Michigan Math. J.,* 14 (1967) 71–78.

[23] J. B. Conway, *Functions of one complex variable,* 2nd edition, Springer-Verlag, 1978.

[24] L. De Branges, A proof of the Bieberbach conjecture, *Acta Math.,* 154 (1985) 137–152.

[25] P. L. Duren, *Univalent Functions, Grundlehren der Mathematischen Wissenschaften,* Springer, New York, 259, 1983.

[26] J. Dziok, R. K. Raina, J. Sokół, On α-convex functions related to shell-like functions connected with Fibonacci numbers, *Applied Mathematics and Computation,* 218 (2011) 996–1002.

[27] S. M. El-Deeb, T. Bulboaca, B. M. El-Matary, Maclaurin coefficient estimates of bi-univalent functions connected with the q-derivative, *Mathematics,* 8 (2020) 1–14.

[28] M. Fekete, G. Szegö, Eine Bemerkung Über Ungerade Schlichte Funktionen, *J. London Math. Soc.,* (1933) [s1-8 (2)] 85–89.

[29] B. A. Frasin, M. Darus, On the Fekete-Szegö problem, *Int. J. Math. Math. Sci.,* 24 (2000) 577–581.

[30] P. R. Garabedian, M. Schiffer, A proof of the Bieberbach conjecture for the fourth coeffcient, *J. Rational Mech. Anal.,* 4 (1955) 427–465.

[31] A. W. Goodman, *Univalent Functions,* vols. I, II. Polygonal Publishing House, Washington, NJ, 1983.

[32] U. Grenander, G. Szegö, Toeplitz forms and their applications, California Monographs in Mathematical Sciences University, California Press, Berkeley, CA, 1958.

[33] H. Ö. Güney, G. Murugusundaramoorthy, J. Sokol, Subclasses of bi-univalent functions related to shell-like curves connected with Fibonacci numbers, *Acta Universitatis Sapientiae Mathematica,* 10 (2018) 70–84.

[34] H. Ö.Güney, G. Murugusundaramoorthy, New classes of pseudo-type bi-univalent functions, Revista de la Real Academia de Ciencias Exactas, *Físicas y Naturales. Serie A. Matemática,* 114 (2020) 1–8.

[35] S. G. Hamidi, J. M. Jahangiri, Faber polynomial coefficient estimates for analytic bi-close-to-convex functions, *C. R. Acad. Sci. Paris Sér. I,* 352 (2014) 17–20.

[36] S. G. Hamidi, J. M. Jahangiri, Faber polynomial coefficients of bi-subordinate functions, *C. R. Acad. Sci. Paris Sér. I,* 354 (2016) 365–370.

[37] W.K. Hayman, On successive coefficients of univalent functions, *J. London Math. Soc.*, 38 (1963) 228–243.

[38] P. Kamble, M. Shrigan, Ş. Altınkaya, On λ-pseudo q-bi-starlike functions, *Turkish J. Math.*, 43 (2019) 751–758.

[39] S. Kanas, An unified approach to the Fekete-Szegö problem, *Appl. Math, Comput.*, 218 (2012) 8453–8461.

[40] S. Kanas, H. E. Darwish, Fekete-Szegö problem for starlike and convex functions of complex order, *Appl. Math. Lett.* 23 (2010) 777–782.

[41] F. R. Keogh, E. P. Merkes, A coefficient inequality for certain classes of analytic functions, *Proc. Amer. Math. Soc.*, 20 (1969) 8–12.

[42] P. Koebe, Uber die uniformisierung beliebeger analytischer kurven, *Nach. Ges. Wiss. Gottingen,* (1907) 191–210.

[43] Y. Komatu, On analytic prolongation of a family of operators, *Math. (Cluj)* 32 (1990) 141–145.

[44] S. G. Krantz, *Goemtric Function Theory, Explorations in Complex Analysis,* Brikhäuser, Boston, MA, 2006.

[45] M. Lewin, On a coefficient problem for bi-univalent functions, *Proc. Amer. Math. Soc.*, 18 (1967) 63-68.

[46] R. J. Libera, Some classes of regular univalent functions, *Proc. Amer. Math. Soc.*, 16 (1965) 755–758.

[47] R. R. London, Fekete-Szegö inequalities for close-to convex functions, *Proc. Amer. Math. Soc.*, 117 (1993) 947–950.

[48] K. Lowner, Uniersuchungen uber schlichte konforme Abbildungen des Einheitskreises. *I., Math. Ann.*, 89 (1923) 103–121.

[49] W. C. Ma, D. Minda, A unified treatment of some special classes of univalent functions, in: *Proceedings of the Conference on Complex Analysis*, Tianjin, 1992, 157–169, Conference of Proceedings, Lecture Notes Analysis I, Int. Press, Cambridge, MA, 1994.

[50] N. Magesh, V. K. Balaji, Certain subclasses of bi-starlike and bi-convex functions of complex order, *Kyungpook Math. J.*, 55 (2015) 705–714.

[51] S. S. Miller, P. T. Mocanu, *Differential Subordinations: Theory and Applications,* Series on Monographs and Textbooks in Pure and Applied Mathematics, Vol. 225, Marcel Dekker, New York and Basel, 2000.

[52] Z. Nehari, *Conformal Mapping,* Mc. Grraw-Hill Book Comp., New York, 1952.

[53] E. Netanyahu, The minimal distance of the image boundary from the origin and the second coefficient of a univalent function in $|z| < 1$, *Arch. Ration. Mech. Anal.*, 32 (1969) 100–112.

[54] S. O. Olatunji, H. Dutta, Coefficient inequalities for pseudo subclasses of analytical functions related to Petal type domains defined by error function, *AIMS Math.*, 5 (2020) 2526–2538.

[55] H. Orhan, N. Magesh, V. K. Balaji, Fekete–Szegö problem for certain classes of Ma-Minda bi-univalent functions, *Afrika Matematika,* 27 (2016) 889–897.

[56] M. Ozawa, On the Bieberbach conjecture for the sixth coefficient, *Kodai Math. Sem. Rep.*, 21 (1969) 97–128.

[57] R. W. Pederson, A proof of the Bieberbach conjecture for the sixth coefficient, *Arch. Rational Mech. Anal.*, 31 (1968) 331–351.

[58] R. W. Pederson, M. Schiffer, A proof of the Bieberbach conjecture for the fifth coefficient, *Arch. Rational Mech. Anal.*, 45 (1972) 161–193.

[59] A. Pfluger, Fekete-Szegö inequality for complex parameters, *Complex Var. Theory Appl.*, 7 (1986) 149–160.

[60] C. Pommerenke, *Univalent Functions,* Vandenhoeck & Ruprecht, Göttingen. 1975.

[61] G. S. Sălăgean, *Subclasses of univalent functions,* in: Complex Analysis - Fifth Romanian Finish Seminar, Bucharest, 1 (1983) 362–372.

[62] A. C. Schaeffer, D. C. Spencer, The coefficients of schlicht functions, *Duke Math. J.,* 12 (1945) 107–125.

[63] G. Schober, *Univalent functions-selected topics,* Berlin-Heidelberg-New York: Springer, 1975.

[64] H. M. Srivastava, Operators of basic (or q-) calculus and fractional q-calculus and their applications in geometric function theory of complex analysis, *Iran. J. Sci. Technol. Trans. A Sci.,* 44 (2020) 327–344.

[65] H. M. Srivastava, Some general families of the Hurwitz-Lerch Zeta functions and their applications: Recent developments and directions for further researches, *Proc. Inst. Math. Mech. Nat. Acad. Sci Azerbaijan,* 45 (2019) 234–269.

[66] H. M. Srivastava, Ş. Altınkaya, S. Yalçın, Certain subclasses of bi-univalent functions associated with the Horadam polynomials, *Iran. J. Sci. Technol. Trans. A Sci.,* 43 (2019) 1873–1879.

[67] H. M. Srivastava, A. K. Mishra, M. K. Das, The Fekete-Szegö problem for a subclass of close-to-convex functions, *Complex Var. Theory Appl.,* 44 (2001) 145–163.

[68] H. M. Srivastava, S. B. Joshi, S. S. Joshi, H. Pawar, Coefficient estimates for certain subclasses of meromorphically bi-univalent functions, *Palest. J. Math.,* 5 (2016) 250–258.

[69] H. M. Srivastava, A. K. Mishra, P. Gochhayat, Certain subclasses of analytic and bi-univalent functions, *Appl. Math. Lett.,* 23 (2010) 1188–1192.

[70] H. M. Srivastava, F. M. Sakar, H. Ö. Güney, Some general coefficient estimates for a new class of analytic and bi-univalent functions defined by a linear combination, *Filomat,* 32 (2018) 1313–1322.

[71] H. M. Srivastava, S. Owa (Eds.), *Current Topics in Analytic Function Theory,* World Scientific Publishing Company, Singapore, New Jersey, London, and Hong Kong, 1992.

[72] H. M. Srivastava, S. Sümer Eker, S. G. Hamidi, J. M. Jahangiri, Faber polynomial coefficient estimates for bi-univalent functions defined by the Tremblay fractional derivative operator, *Bull. Iran. Math. Soc.,* 44 (2018) 149–157.

[73] H. M. Srivastava, S. Sümer Eker, R. M. Ali, Coefficient bounds for a certain class of analytic and bi-univalent functions, *Filomat,* 29 (2015) 1839–1845.

[74] Q.-H. Xu, Y.-C. Gui, H. M. Srivastava, Coefficient estimates for a certain subclass of analytic and bi-univalent functions, *Appl. Math. Lett.,* 25 (2012) 990–994.

[75] Q.-H. Xu, H.-G. Xiao, H. M. Srivastava, A certain general subclass of analytic and bi-univalent functions and associated coefficient estimate problems, *Appl. Math. Comput.,* 218 (2012) 11461–11465.

[76] Z. Zaprawa, On Fekete-Szegö problem for classes of bi-univalent functions, *Bull. Belg. Math. Soc. Simon Stevin,* 21 (2014) 169–178.

10 Fixed Point of Multivalued Cyclic Contractions

Talat Nazir
University of South Africa
COMSATS University Islamabad

Mujahid Abbas
Government College University Lahore

CONTENTS

10.1 MULTIVALUED MAPPINGS IN METRIC SPACES

The theory of multivalued mappings (also called set-valued mappings) has been extended by many researchers in different ways. The basic tool for generalization is Banach contraction principle [6] which has been used in different ways, either by considering generalized contractive conditions of mappings or extending the area of mappings. Following the Banach contraction principle, the hypothesis of multivalued maps has different applications in convex optimization, dynamical frameworks, commutative polynomial math, differential conditions, and economics. Nadler [14] gave the idea of set-valued contraction and established that a set-valued contraction has fixed point in complete metric space. Thus,the Nadler principle has been used by many researchers [1,2,4,5,9,11,13,19].

All through this work, $P(X)$, $P_{Cl}(X)$, $P_{CB}(X)$, $\mathbb{R}$, $\mathbb{C}$, $\mathbb{R}_+$, $\mathbb{R}^n_+$ and $\mathbb{N}$ signify the family of nonempty subsets of X, non-unfilled closed subsets of X, nonempty closed and bounded subsets of X, the set of real numbers, complex numbers, non-negative real numbers, n-tuples of non-negative real numbers, and natural numbers, respectively.

In this section, we examine some important definitions and major results from existing writing.

Definition 10.1. *For a multivalued mapping $T : X \to P(X)$, fixed point of T is defined as a point u in X if and only if $u \in Tu$. And if $Tu = \{u\}$, then it is called end point.*

Definition 10.2. *Let d be a metric space on X. The Hausdorff–Pompeiu metric on $P_{CB}(X)$ is represented by H, that is,*

$$H(A,B) = \max\{\sup_{u \in A}\{d(u,B)\}, \sup_{v \in B}\{d(v,A)\}\},$$

where A, $B \in P_{CB}(X)$ and

$$d(u,A) = \inf_{v \in B}\{d(u,v)\}.$$

If (X,d) is complete, then $(P_{CB}(X),H)$ is also complete metric space.

Lemma 10.1. *[3,14] Let d be a metric space on X, A, $B \in P_{CB}(X)$. For any $a \in A$, there exists $b = b(a) \in B$ such that*

$$d(a,b) \leq \lambda H(A,B),$$

where $\lambda \geq 1$.

10.2 MULTIVALUED CYCLIC *F*-CONTRACTIVE MAPPINGS

In 2003, Kirk et al. [10] extended the Banach contraction principle to cyclic contractive mappings, studied the cyclical contractive condition for self-mappings, and proved some fixed point results. They introduced cyclic representation and cyclic contractions. A mapping T such that $T : A_1 \cup A_2 \to A_1 \cup A_2$ is called cyclic if $T(A_1) \subset A_2$ and $T(A_2) \subset A_1$, while A_1 and A_2 are non-unfilled subsets of (X,d). And representation of T is cyclic if there exists $\alpha \in (0,1)$ such that $d(Tu,Tv) \leq \alpha d(u,v)$ where $u \in A$ and $v \in B$. It is important worth of cyclic contraction mappings that a map satisfying the Banach contraction is always continuous, but a map satisfying a cyclic contractive condition needs not to be continuous. Pacurar and Rus [16] obtained some fixed point results for maps satisfying cyclic weak contractive conditions. Petric [17] studied some results on cyclic Meir–Keeler contractions in metric spaces. Using fixed point result for weakly contractive map, Karapinar [7] gave some interesting fixed point results for cyclic $\phi-$weak contractive mappings.

Definition 10.3. *For a metric space (X,d), nonempty and closed subsets of X are $A_1, A_2, \ldots, A_p$, where p is a positive integer and $Y = \cup_{i=1}^{p} A_i$ and $T : Y \to P_{CB}(X)$. Then, cyclic representation of Y with respect to T is*

$$T(A_1) \subset A_2, T(A_2) \subset A_3, \ldots, T(A_{p-1}) \subset A_p, T(A_p) \subset A_1.$$

Wardowski [21] gave the following definition for class of mappings.

Definition 10.4. *Suppose $\mathcal{F}$ be the set of all continuous mappings $F : \mathbb{R}_+ \to \mathbb{R}$ satisfying the following conditions:*

$F_1)$ *F is strictly increasing, and for all $\gamma, \delta \in \mathbb{R}_+$ such that $\gamma < \delta$ implies that $F(\gamma) < F(\delta)$.*
$F_2)$ *For every sequence $\{\gamma_n\}$ of positive real numbers, $\lim_{n\to\infty} \gamma_n = 0$ and $\lim_{n\to\infty} F(\gamma_n) = -\infty$ are equivalent.*
$F_3)$ *There exists $\eta \in (0,1)$ such that $\lim_{\gamma\to 0^+} \gamma^\eta F(\gamma) = 0$.*

Definition 10.5. *Suppose (X,d) be a metric space, let $A_1, A_2, \ldots, A_p$ be nonempty closed subsets of X, where p is a positive integer, $Y = \cup_{i=1}^{p} A_i$. A multivalued mapping $T : Y \to P_{CB}(X)$ is called*

1. *cyclic $F-$contraction mapping if:*
   (a) Y has cyclic representation with respect to T.
   (b) For every $(u,v) \in A_i \times A_{i+1}$, $i = 1,2,\ldots,p$ and $A_{p+1} = A_1$, with $H(Tu,Tv) > 0$ implies

$$F(H(Tu,Tv)) + \tau(M_T(u,v)) \leq F(M_T(u,v)),$$

   where $F \in \mathcal{F}$, $\tau : \mathbb{R}_+ \to \mathbb{R}_+$ is a continuous function such that $\liminf_{s\to t^+} \tau(s) \geq 0$ for all $t \geq 0$, and

$$M_T(u,v) = \max\{d(u,v), d(u,Tu), d(v,Tv), \frac{d(u,Tv)+d(v,Tu)}{2}\}.$$

2. *Hardy Roger-type-cyclic contractive mapping if*
   (i) Y has cyclic representation with respect to T.
   (ii) For every $(u,v) \in A_i \times A_{i+1}$, $i = 1,2,\ldots,p$ and $A_{p+1} = A_1$,

$$H(Tu,Tv) \leq a_1 d(u,v) + a_2 d(u,Tu) + a_3 d(v,Tv) + a_4 d(u,Tv) + a_5 d(v,Tu)$$

   holds, where $a_1, \ldots, a_5 \geq 0$ and $a_1 + a_2 + a_3 + 2a_4 + a_5 < 1$.
3. *Reich-type-cyclic contractive mapping if*
   (i) Y has cyclic representation with respect to T.
   (ii) For every $(u,v) \in A_i \times A_{i+1}$, $i = 1,2,\ldots,p$ and $A_{p+1} = A_1$,

$$H(Tu,Tv) \leq ad(u,v) + bd(u,Tu) + cd(v,Tv)$$

   holds, where a, b, and $c \geq 0$ with $a + b + c < 1$.
4. *Kannan-type-cyclic contractive mapping if*
   (i) Y has cyclic representation with respect to T.
   (ii) For every $(u,v) \in A_i \times A_{i+1}$, $i = 1,2,\ldots,p$ and $A_{p+1} = A_1$,

$$H(Tu,Tv) \leq \alpha[d(u,Tu) + d(v,Tv)]$$

   holds, where $0 \leq \alpha < \frac{1}{2}$.

5. *Chatterjee-type-cyclic contractive mapping if*
   (i) Y *has cyclic representation with respect to* T.
   (ii) For every $(u,v) \in A_i \times A_{i+1}$, $i = 1,2,\ldots,p$ *and* $A_{p+1} = A_1$,

$$H(Tu,Tv) \leq \eta[d(u,Tv)+d(v,Tu)],$$

   holds, where $0 \leq \eta < \dfrac{1}{2}$.

Example 10.1. *Let* $X = \mathbb{R}$ *and* d *be a usual metric on* X. *Let* $A_1 = [-10,5]$ *and* $A_2 = [-5,10]$ *be any two closed subsets of* X, *such that* $Y = A_1 \cup A_2$ *and* $A_1 \cap A_2 = [-5,5]$. *Let* $T : Y \to P_{CB}(X)$ *be cyclic multivalued mapping such that*

$$T(x) = \begin{cases} [0, -\dfrac{x}{20}] & if\ x \leq 0, \\ [0, \dfrac{x}{20}] & if\ x > 0. \end{cases}$$

Note that the set Y *has cyclic representation with respect to* T, *that is,* $T(A_1) \subseteq A_2$ *and* $T(A_2) \subseteq A_1$.

1. *Now let us take* $F(x) = \ln x + x$ *for* $x > 0$ *and* $\tau : \mathbb{R}_+ \to \mathbb{R}_+$ *is a function defined as*

$$\tau(r) = \begin{cases} \frac{1}{30}\ if\ r = 0, \\ \frac{y}{30}\ if\ r > 0. \end{cases}$$

   Then for every $(u,v) \in A_i \times A_{i+1}$, $i = 1,2$ *with* $H(Tu,Tv) > 0$ *implies*

$$F(H(Tu,Tv)) + \tau(M_T(u,v)) \leq F(M_T(u,v)),$$

   where

$$M_T(u,v) = \max\{d(u,v), d(u,Tu), d(v,Tv), \frac{d(u,Tv)+d(v,Tu)}{2}\}.$$

   Thus, T *is the cyclic* $F-$*contraction mapping.*
2. *If we take* $a_1 = \frac{1}{4}$, $a_2 = a_3 = \frac{1}{8}, a_4 = \frac{1}{6}, a_5 = \frac{1}{4}$, *then for every* $(u,v) \in A_i \times A_{i+1}$, $i = 1,2$,

$$H(Tu,Tv) \leq a_1 d(u,v) + a_2 d(u,Tu) + a_3 d(v,Tv) + a_4 d(u,Tv) + a_5 d(v,Tu)$$

   holds with $a_1 + a_2 + a_3 + 2a_4 + a_5 < 1$. *Thus,* T *is the Hardy Roger-type-cyclic contractive mapping.*
3. *If we take* $a = \frac{1}{2}, b = \frac{1}{4}$, *and* $c = \frac{1}{5}$, *then for every* $(u,v) \in A_i \times A_{i+1}$, $i = 1,2$,

$$H(Tu,Tv) \leq ad(u,v) + bd(u,Tu) + cd(v,Tv)$$

   holds with $a + b + c < 1$. *Thus,* T *is the Reich-type-cyclic contractive mapping.*

4. *If we take* $\alpha = \frac{9}{20}$*, then for every* $(u,v) \in A_i \times A_{i+1}$ *for* $i = 1,2,$

$$H(Tu,Tv) \leq \alpha[d(u,Tu) + d(v,Tv)]$$

holds, where $0 \leq \alpha < \frac{1}{2}$*. Thus, T is the Kannan-type-cyclic contractive mapping.*

5. *If we take* $\eta = \frac{7}{15}$*, then for every* $(u,v) \in A_i \times A_{i+1}$ *for* $i = 1,2,$

$$H(Tu,Tv) \leq \eta[d(u,Tv) + d(v,Tu)],$$

holds, where $0 \leq \eta < \frac{1}{2}$*. Thus, T is the Chatterjee-type-cyclic contractive mapping.*

10.3 FIXED POINT RESULTS OF MULTIVALUED CYCLIC *F*-CONTRACTIVE MAPPINGS

In this section, we obtain the fixed point results for multivalued cyclic $F-$contractive mappings. We start with the following result:

Theorem 10.1. *Let d be a complete metric on X,* $p \in \mathbb{N}$*,* $A_1, A_2, \ldots, A_p$ *be nonempty closed subsets of X with* $\cap_{i=1}^{p} A_i \neq \emptyset$ *and* $Y = \cup_{i=1}^{p} A_i$*. If* $T : Y \to P_{CB}(X)$ *be a cyclic multivalued* $F-$*contraction mapping, then T has a fixed point in* $\cap_{i=1}^{p} A_i$*.*

Proof. Let $x_0 \in A_1$ and $x_1 \in T(x_0) \subset A_2$. Then, there exists $x_2 \in T(x_1) \subset A_3$, and then, by Lemma 10.1, we have

$$d(x_1,x_2) \leq \lambda H(Tx_0,Tx_1). \tag{10.1}$$

Similarly, there exists $x_3 \in T(x_2) \subset A_4$ such that

$$d(x_2,x_3) \leq \lambda H(Tx_1,Tx_2), \tag{10.2}$$

and there exists $x_4 \in T(x_3) \subset A_5$ such that

$$d(x_3,x_4) \leq \lambda H(Tx_2,Tx_3), \tag{10.3}$$

$$\vdots$$

and there exists $x_{p-2} \in T(x_{p-3}) \subset A_{p-1}$ and $x_{p-1} \in T(x_{p-2}) \subset A_p$ such that

$$d(x_{p-2},x_{p-1}) \leq \lambda H(Tx_{p-3},Tx_{p-2}). \tag{10.4}$$

Similarly, there exists $x_{n+1} \in T(x_n) \subset A_{n+2}$ such that

$$d(x_n,x_{n+1}) \leq \lambda H(Tx_{n-1},Tx_n). \tag{10.5}$$

Suppose that $d(x_n,x_{n+1}) > 0$, for all $n \in \mathbb{N}$. If not, then for some $k \in \mathbb{N}$, we have

$$d(x_k,x_{k+1}) = 0,$$

which implies $x_k = x_{k+1} \in Tx_k$. The proof is finished. Therefore, we assume that $d(x_n,x_{n+1}) > 0$, for all $n \in \mathbb{N}$. So we have

$$0 < d(x_n,x_{n+1}) \leq H(Tx_{n-1},Tx_n). \tag{10.6}$$

As $\lambda \geq 1$, we have $H(Tx_{n-1},Tx_n) > 0$.
This implies

$$F(H(Tx_{n-1},Tx_n)) + \tau(M_T(x_{n-1},x_n)) \leq F(M_T(x_{n-1},x_n)),$$

that is,

$$\begin{aligned} F(H(Tx_{n-1},Tx_n)) &\leq F(M_T(x_{n-1},x_n)) - \tau(M_T(x_{n-1},x_n)) \\ &< F(M_T(x_{n-1},x_n)). \end{aligned}$$

As F is strictly increasing, so we get

$$H(Tx_{n-1},Tx_n) < M_T(x_{n-1},x_n), \tag{10.7}$$

where

$$\begin{aligned} M_T(x_{n-1},x_n) &= \max\{d(x_{n-1},x_n), d(x_{n-1},Tx_{n-1}), d(x_n,Tx_n), \\ &\quad \frac{d(x_{n-1},Tx_n)+d(x_n,Tx_{n-1})}{2}\} \\ &\leq \max\{d(x_{n-1},x_n), d(x_{n-1},x_n), d(x_n,x_{n+1}), \\ &\quad \frac{d(x_{n-1},x_{n+1})+d(x_n,x_n)}{2}\} \\ &= \max\{d(x_{n-1},x_n), d(x_n,x_{n+1})\}. \end{aligned}$$

Now, if $M_T(x_{n-1},x_n) = d(x_n,x_{n+1})$, then from (10.6), we have

$$d(x_n,x_{n+1}) < d(x_n,x_{n+1})$$

a contradiction. Therefore, $M_T(x_{n-1},x_n) = d(x_{n-1},x_n)$ and

$$d(x_n,x_{n+1}) < d(x_{n-1},x_n)$$

for all $n = 0,1,2,\ldots$. Thus, the sequence $\{d(x_n,x_{n+1})\}$ is monotonically decreasing and consequently,

$$\tau(d(x_{n-1},x_n)) + F(d(x_n,x_{n+1})) \leq F(d(x_{n-1},x_n)) \text{ for all } n = 0,1,2,\ldots.$$

By the given assumption on τ, there exists $c > 0$ and $n \in \mathbb{N}$ such that $\tau(d(x_n,x_{n+1})) > c$ for all $n > n_0$. Thus, we obtain that

$$\begin{aligned} F(d(x_n,x_{n+1})) &\leq F(d(x_{n-1},x_n)) - \tau(d(x_{n-1},x_n)) \\ &\leq F(d(x_{n-2},x_{n-1})) - \tau(d(x_{n-2},x_{n-1})) - \tau(d(x_{n-1},x_n)) \end{aligned}$$

$$\leq \cdots \leq F(d(x_0,x_1)) - \{\tau(d(x_0,x_1)) + \cdots + \tau(d(x_{n_0-1},x_{n_0}))$$
$$-\{\tau(d(x_{n_0},x_{n_0+1}) + \cdots + \tau(d(x_{n-1},x_n)\}$$
$$= F(d(x_0,x_1)) - n_0.$$

On taking limit as $n \to \infty$, we obtain that

$$\lim_{n\to\infty} F(d(x_n,x_{n+1})) = -\infty.$$

Together with F_2, we have

$$\lim_{n\to\infty} d(x_n,x_{n+1}) = 0.$$

By F_3, there exists $r \in (0,1)$ such that

$$\lim_{n\to\infty} \{d(x_n,x_{n+1})\}^r F(d(x_n,x_{n+1})) = 0.$$

Hence, it follows that

$$[d(x_n,x_{n+1})]^r F(d(x_n,x_{n+1}) - [d(x_n,x_{n+1})]^r F(d(x_0,x_1))$$
$$\leq [d(x_n,x_{n+1})]^r \{F(d(x_0,x_1)) - n_0 - F(d(x_0,x_1))\}$$
$$\leq -n_0[d(x_n,x_{n+1})]^r \leq 0,$$

by applying limit as $n \to \infty$, we have

$$\lim_{n\to\infty} n[d(x_n,x_{n+1})]^r = 0,$$

that is

$$\lim_{n\to\infty} n^{\frac{1}{r}} d(x_n,x_{n+1}) = 0.$$

Now, let us consider that there exists $n_1 \in \mathbb{N}$ such that $n^{\frac{1}{r}} d(x_n,x_{n+1}) \leq 1$ for all $n \geq n_1$. Consequently, we have

$$d(x_n,x_{n+1}) \leq \frac{1}{n^{\frac{1}{r}}} \text{ for all } n \geq n_1.$$

In order to show that $\{x_n\}$ is a Cauchy sequence, consider $m,n \in \mathbb{N}$ such that $m > n$,

$$d(x_m,x_n) \leq d(x_m,x_{m-1}) + d(x_{m-1},x_{m-2}) + \cdots + d(x_{n+1},x_n)$$
$$< \sum_{i=n}^{\infty} d(x_{i+1},x_i) \leq \sum_{i=n}^{\infty} \frac{1}{i^{\frac{1}{r}}}.$$

If series $\sum_{i=1}^{\infty} \frac{1}{i^{\frac{1}{r}}}$ is convergent, then $\{x_n\}$ is a Cauchy sequence in X. Also d is complete metric on X, then there exists $x^* \in X$ such that $x_n \to x^*$ as $n \to \infty$. Thus, T has a fixed point in X. Since $A_1, A_2, \ldots, A_p$ are closed subsets in X, then $x^* \in A_i$ for $i = 1,2,\ldots,p$, which implies that $x^* \in \cap_{i=1}^{p} A_i$. □

Example 10.2. *Let $X = \mathbb{R}$ and d be a usual metric on X. Let $A_1 = (-\infty, 0]$ and $A_2 = [0, \infty)$ be any two closed subsets of X, such that $Y = A_1 \cup A_2$ and $A_1 \cap A_2 \neq \emptyset$. Let $T : Y \to P_{CB}(X)$ be cyclic multivalued mapping such that*

$$T(x) = \begin{cases} [0, \frac{-x}{40}] \text{ if } x \leq 0 \\ \\ [\frac{-x}{40}, 0] \text{ if } x > 0. \end{cases}$$

Clearly $T(A_1) \subseteq A_2$ and $T(A_2) \subseteq A_1$.
Consider a mapping $F : \mathbb{R}_+ \to \mathbb{R}$ defined as $F(x) = \ln x + x$ for $x > 0$ and $\tau : \mathbb{R}_+ \to \mathbb{R}_+$ is a function defined as

$$\tau(r) = \begin{cases} \frac{1}{20} \; if \; r = 0, \\ \\ \frac{y}{20} \; if \; r > 0. \end{cases}$$

To consider the contractive condition

$$H(Tx, Ty)e^{H(Tx,Ty) - M_T(x,y) + \tau(M_T(x,y))} \leq M_T(x, y)$$

for $H(Tx, Ty) > 0$, we have the following two cases:
Case (i): For $x \in A_1$, $y \in A_2$, $T(x) = [0, \frac{-x}{40}]$ and $T(y) = [\frac{-y}{40}, 0]$. Then

$$\begin{aligned} H(Tx, Ty) &= \max\left\{ \sup_{u \in [0, \frac{-x}{40}]} d\left(u, \left[\frac{-y}{40}, 0\right]\right), \sup_{v \in [\frac{-y}{40}, 0]} d\left(v, [0, \frac{-x}{40}]\right) \right\} \\ &= \max\left\{ \frac{-x}{40}, \frac{y}{40} \right\} = \frac{1}{40} \max\{-x, y\} \end{aligned}$$

and

$$\begin{aligned} M_T(x,y) &= \max\left\{ d(x,y), d(x,Tx), d(y,Ty), \frac{d(x,Ty) + d(y,Tx)}{2} \right\} \\ &= \max\left\{ 0, |x|, y, \frac{1}{80}(|x + 40y| + |40x + y|) \right\} \\ &= \max\{|x|, y\}. \end{aligned}$$

So we get

$$\begin{aligned} H(Tx, Ty)e^{H(Tx,Ty) - M_T(x,y) + \tau(M_T(x,y))} &= \frac{1}{40} \max\{-x, y\} e^{\frac{1}{40} \max\{-x,y\} - \frac{19}{20} \max\{|x|,y\}} \\ &= \frac{1}{40} \max\{-x, y\} e^{\frac{-37}{40} \max\{-x,y\}} \\ &\leq \max\{|x|, y\} \\ &= M_T(x, y). \end{aligned}$$

Case (ii): When $x \in A_2$ and $y \in A_1$, $T(x) = [\frac{-x}{40}, 0]$ and $T(y) = [0, \frac{-y}{40}]$. Then,

$$H(Tx,Ty) = \max\left\{ \sup_{u \in [0,\frac{-y}{40}]} d\left(u, \left[\frac{-x}{40}, 0\right]\right), \sup_{v \in [\frac{-x}{40},0]} d\left(v, \left[0, \frac{-y}{40}\right]\right) \right\}$$
$$= \max\left\{ \frac{x}{40}, \frac{-y}{40} \right\} = \frac{1}{40}\max\{x, -y\}$$

and

$$M_T(x,y) = \max\left\{ d(x,y), d(x,Tx), d(y,Ty), \frac{d(x,Ty)+d(y,Tx)}{2} \right\}$$
$$= \max\left\{ 0, x, \frac{1}{80}(|40x+y| + |x+40y|) \right\}$$
$$= \max\{x, |y|\}.$$

So we get

$$H(Tx,Ty)e^{H(Tx,Ty)-M_T(x,y)+\tau(M_T(x,y))} = \frac{1}{40}\max\{x,-y\}e^{\frac{1}{40}\max\{x,-y\}-\frac{19}{20}\max\{|x|,y\}}$$
$$= \frac{1}{40}\max\{x,-y\}e^{\frac{-37}{40}\max\{x,-y\}}$$
$$\leq \max\{x,|y|\}$$
$$= M_T(x,y).$$

Therefore, T is cyclic $F-$contraction mapping. Thus, all the conditions of Theorem 10.1 are satisfied. Moreover, $\frac{1}{2}$ is the fixed point of T.

Theorem 10.2. *Let (X,d) be a complete metric space, $p \in \mathbb{N}$, $A_1, A_2, \ldots, A_p$ be nonempty closed subsets of X with $\cap_{i=1}^{p} A_i \neq \emptyset$ and $Y = \cup_{i=1}^{p} A_i$. If $T : Y \to P_{CB}(X)$ be a mapping such that*
(a) $Y = \cup_{i=1}^{p} A_i$ has cyclic representation with respect to T.
(b) For any $(x,y) \in A_i \times A_{i+1}$, $i = 1,2,\ldots,p$ and $A_{p+1} = A_1$ where $H(Tx,Ty) > 0$ implies

$$\tau\left(\overline{M}_T(x,y)\right) + F\left(H(Tx,Ty)\right) \leq F\left(\overline{M}_T(x,y)\right), \tag{10.8}$$

where $F \in \mathcal{F}$, $\tau : \mathbb{R}_+ \to \mathbb{R}_+$ is a continuous function such that $\liminf_{s\to t^+} \tau(s) \geq 0$ for all $t \geq 0$, and

$$\overline{M}_T(x,y) = \alpha d(x,y) + \beta d(x,Tx) + \gamma d(y,Ty) + \delta_1 d(x,Ty) + \delta_2 d(y,Tx),$$

where $\alpha, \beta, \gamma, \delta_1, \delta_2 \geq 0$, $\delta_1 \leq \delta_2$ with $\alpha+\beta+\gamma+\delta_1+\delta_2 \leq 1$.
Then, T has a fixed point in $\cap_{i=1}^{p} A_i$.

Proof. Let $x_0 \in A_1$ and $x_1 \in T(x_0) \subset A_2$. Then, there exists $x_2 \in T(x_1) \subset A_3$, and then, by Lemma 10.1, we have

$$d(x_1,x_2) \leq \lambda H(Tx_0,Tx_1). \tag{10.9}$$

Similarly, there exists $x_3 \in T(x_2) \subset A_4$ such that

$$d(x_2,x_3) \leq \lambda H(Tx_1,Tx_2), \tag{10.10}$$

and there exists $x_4 \in T(x_3) \subset A_5$ such that

$$d(x_3,x_4) \leq \lambda H(Tx_2,Tx_3), \tag{10.11}$$

$$\vdots$$

and there exists $x_{p-2} \in T(x_{p-3}) \subset A_{p-1}$ and $x_{p-1} \in T(x_{p-2}) \subset A_p$ such that

$$d(x_{p-2},x_{p-1}) \leq \lambda H(Tx_{p-3},Tx_{p-2}). \tag{10.12}$$

Similarly, there exists $x_{n+1} \in T(x_n) \subset A_{n+2}$ such that

$$d(x_n,x_{n+1}) \leq \lambda H(Tx_{n-1},Tx_n). \tag{10.13}$$

Since $d(x_n,x_{n+1}) > 0$, for all $n \in \mathbb{N}$.
If not, then for some $k \in \mathbb{N}$, we have

$$d(x_k,x_{k+1}) = 0,$$

which implies $x_k = x_{k+1} \in Tx_k$. Thus, proof is finished. Therefore, we assume that $d(x_n,x_{n+1}) > 0$, for all $n \in \mathbb{N}$. So, we have

$$0 < d(x_n,x_{n+1}) \leq \lambda H(Tx_{n-1},Tx_n),$$

As $\lambda \geq 1$, so $H(Tx_{n-1},Tx_n) > 0$. Therefore,

$$d(x_n,x_{n+1}) \leq H(Tx_{n-1},Tx_n).$$

This implies that

$$\begin{aligned} F(H(Tx_{n-1},Tx_n)) &\leq F(\overline{M}_T(x_{n-1},x_n)) - \tau(\overline{M}_T(x_{n-1},x_n)) \\ &< F(\overline{M}_T(x_{n-1},x_n)). \end{aligned}$$

As F is strictly increasing, this implies

$$H(Tx_{n-1},Tx_n) < \overline{M}_T(x_{n-1},x_n),$$

where

$$\begin{aligned} \overline{M}_T(x_{n-1},x_n) &= \alpha d(x_{n-1},x_n) + \beta d(x_{n-1},Tx_{n-1}) + \gamma d(x_n,Tx_n) \\ &\quad + \delta_1 d(x_{n-1},Tx_n) + \delta_2 d(x_n,Tx_{n-1}) \\ &\leq \alpha d(x_{n-1},x_n) + \beta d(x_{n-1},x_n) + \gamma d(x_n,x_{n+1}) \\ &\quad + \delta_1 d(x_{n-1},x_{n+1}) + \delta_2 d(x_n,x_n) \\ &\leq (\alpha+\beta+\delta_1) d(x_{n-1},x_n) + (\gamma+\delta_1) d(x_n,x_{n+1}). \end{aligned}$$

Now if $d(x_{n-1},x_n) < d(x_n,x_{n+1})$, then

$$\begin{aligned} d(x_n,x_{n+1}) &< (\alpha+\beta+\gamma+2\delta_1) d(x_n,x_{n+1}) \\ &< d(x_n,x_{n+1}), \end{aligned}$$

a contradiction. Therefore, $d(x_n,x_{n+1}) < d(x_{n-1},x_n)$, and this implies

$$\begin{aligned} d(x_n,x_{n+1}) &< (\alpha+\beta+\gamma+2\delta_1)d(x_{n-1},x_n) \\ &< d(x_{n-1},x_n). \end{aligned}$$

Thus, the sequence $\{d(x_n,x_{n+1})\}$ is monotonically decreasing. Consequently,

$$F(d(x_n,x_{n+1})) \leq F(d(x_{n-1},x_n)) - \tau(d(x_{n-1},x_n)) \text{ for all } n = 0,1,2,\ldots,$$

and by the given assumption on τ, there exists $c > 0$ such that $\tau(d(x_n,x_{n+1})) > c$ for all $n \geq n_0$. Now

$$\begin{aligned} F(d(x_n,x_{n+1})) &\leq F(d(x_{n-1},x_n)) - \tau(d(x_{n-1},x_n)) \\ &\leq F(d(x_{n-2},x_{n-1})) - \tau(d(x_{n-2},x_{n-1})) - \tau(d(x_{n-1},x_n)) \\ &\leq \cdots \leq F(d(x_0,x_1)) - \{\tau(d(x_0,x_1)) + \cdots + \tau(d(x_{n_0-1},x_{n_0})) \\ &+ \tau(d(x_{n_0},x_{n_0+1}) + \cdots + \tau(d(x_{n-1},x_n)\} \\ &\leq F(d(x_0,x_1)) - n_0. \end{aligned}$$

By taking limit as $n \to \infty$, we get

$$\lim_{n\to\infty} F(d(x_n,x_{n+1})) = -\infty,$$

and together with F_2, it gives

$$\lim_{n\to\infty} d(x_n,x_{n+1}) = 0.$$

Now by F_3, there exists $r \in (0,1)$ such that

$$\begin{aligned} \lim_{n\to\infty} [d(x_n,x_{n+1})]^r F(d(x_n,x_{n+1})) &= 0 \\ \lim_{n\to\infty} n[d(x_n,x_{n+1})]^r &= 0. \end{aligned}$$

This implies

$$\lim_{n\to\infty} n^{\frac{1}{r}}[d(x_n,x_{n+1})] = 0.$$

Now, let us consider that there exists $n_1 \in \mathbb{N}$ such that $n^{\frac{1}{r}} d(x_n,x_{n+1}) \leq 1$ for all $n \geq n_1$. Consequently, we have

$$d(x_n,x_{n+1}) \leq \frac{1}{n^{\frac{1}{r}}} \text{ for all } n \geq n_1.$$

In order to show that $\{x_n\}$ is a Cauchy sequence, consider $m,n \in \mathbb{N}$ such that $m > n$,

$$\begin{aligned} d(x_m,x_n) &\leq d(x_m,x_{m-1}) + d(x_{m-1},x_{m-2}) + \cdots + d(x_{n+1},x_n) \\ &< \sum_{i=n}^{\infty} d(x_{i+1},x_i) \leq \sum_{i=n}^{\infty} \frac{1}{i^{\frac{1}{r}}}. \end{aligned}$$

If series $\sum_{i=1}^{\infty} \frac{1}{i^{\frac{1}{r}}}$ is convergent, then $\{x_n\}$ is a Cauchy sequence in X. Since (X,d) is complete, there exists $x^* \in X$ such that $x_n \to x^*$ as $n \to \infty$. Thus, T has a fixed point in X. Since $A_1, A_2, \ldots, A_p$ are closed subsets in X, $x^* \in A_i$, $i = 1, 2, \ldots, p$ which implies that $x^* \in \cap_{i=1}^{p} A_i$. □

Remark 10.1. *Let (X,d) be a complete metric space, $p \in \mathbb{N}$, $A_1, A_2, \ldots, A_p$ be nonempty closed subsets of X with $\cap_{i=1}^{p} A_i \neq \emptyset$ and $Y = \cup_{i=1}^{p} A_i$. If a map $T : Y \to P_{CB}(X)$ is satisfying either one of the following:*

1. *Hardy Roger-type-cyclic contractive mapping.*
2. *Reich-type-cyclic contractive mapping.*
3. *Kannan-type-cyclic contractive mapping.*
4. *Chatterjee-type-cyclic contractive mapping.*
   Then T has a fixed point in $\cap_{i=1}^{p} A_i$.

10.4 STABILITY OF FIXED POINT SETS OF CYCLIC *F*-CONTRACTIONS

Stability is a concept associated with limiting behaviors of a system. It has been studied in the contexts of both discrete and continuous dynamical systems. The study of the relationship between the convergence of a sequence of mappings and their fixed points, known as the stability of fixed points, has also been widely studied in various settings. The fixed point sets of a sequence of mappings are said to be stable if they converge to the set of fixed points of the limit mapping in the Hausdorff metric. Multivalued mappings often have more fixed points than their single-valued mappings. Therefore, the set of fixed points of multivalued mappings becomes larger and hence more interesting for the study of stability.

Definition 10.6. *The stability of fixed point sets is defined as in the Hausdorff metric the fixed point sets of a sequence of mappings $\{T_n\}$ converges to the fixed point sets of the limit mapping T, that is, $\lim_{n\to\infty} H(F(T_n), F(T)) = 0$.*

We investigate the stability of fixed point sets of multivalued mappings that are satisfying cyclic $F-$contraction condition.

Theorem 10.3. *Let (X,d) be a complete metric space, p be a positive integer, $A_1, A_2, \ldots, A_p$ be nonempty closed subsets of X with $\cap_{i=1}^{p} A_i \neq \emptyset$ such that $Y = \cup_{i=1}^{p} A_i$ and $\{T_i\}_{i=1}^{m} : Y \to P_{CB}(X)$ be family of multivalued mappings satisfying cyclic $F-$contraction condition.*

(i) $Y = \cup_{i=1}^{p} A_i$ *has cyclic representation of with respect to T_i for $i \in \{1, 2, \ldots, m\}$.*
(ii) *For any $(x,y) \in A_i \times A_{i+1}$, $i = 1, 2, \ldots, p$ and $A_{p+1} = A_1$ such that $H(Tx, Ty) > 0$ implies*

$$F(H(T_i x, T_i y)) + \tau(M_{T_i}(x,y)) \leq F(M_{T_i}(x,y)), \tag{10.14}$$

where $F \in F$, $\tau : \mathbb{R}_+ \to \mathbb{R}_+$ *is a continuous function such that* $\liminf_{s \to t^+} \tau(s) \geq 0$ *for all* $t \geq 0$, *and*

$$M_{T_i}(x,y) = \max\{d(x,y), d(x,T_ix), d(y,T_iy), \frac{d(x,T_iy)+d(y,T_ix)}{2}\}.$$

Then,

$$H(F(T_i), F(T_{i+1})) \leq k,$$

where $k = \sup_{x \in X} H(T_ix, T_{i+1}x)$.

Proof. From Theorem 10.1, the set of fixed points of T_i $(i = 1, 2, \ldots, m)$ is nonempty, that is, $F(T_i) \neq \emptyset$, for $i = 1, 2, \ldots, m$. Let $y_0 \in A_1$, and $y_1 \in F(T_1)$ that is $y_1 \in T_1y_1 \subset A_2$. Then, by Lemma 10.1, there exists $y_2 \in T_2y_1 \subset A_3$ such that

$$d(y_1, y_2) \leq H(T_1y_1, T_2y_1). \tag{10.15}$$

Since $y_2 \in T_2y_1 \subset A_3$, again by Lemma 10.1, there exists $y_3 \in T_2y_2 \subset A_4$ such that

$$d(y_2, y_3) \leq H(T_2y_1, T_2y_2).$$

By continuing in this way, we obtain a sequence $y_{n+1} \in T_2y_n \subset A_{n+2}$ such that

$$d(y_n, y_{n+1}) \leq H(T_2y_{n-1}, T_2y_n). \tag{10.16}$$

Since $d(y_n, y_{n+1}) > 0$ for all $n \in \mathbb{N}$, $H(T_2y_{n-1}, T_2y_n) > 0$.
This implies that

$$\begin{aligned} F(H(T_2y_{n-1}, T_2y_n)) &\leq F(M_T(y_{n-1}, y_n)) - \tau(M_T(y_{n-1}, y_n)) \\ &< F(M_T(y_{n-1}, y_n)), \end{aligned}$$

since F is monotonically increasing, we get

$$\begin{aligned} H(T_2y_{n-1}, T_2y_n) &< M_T(y_{n-1}, y_n) \\ &= \max\{d(y_{n-1}, y_n), d(y_{n-1}, Ty_{n-1}), d(y_n, Ty_n), \\ &\quad \frac{d(y_{n-1}, Ty_n) + d(y_n, Ty_{n-1})}{2}\} \\ &\leq \max\{d(y_{n-1}, y_n), d(y_{n-1}, y_n), d(y_n, y_{n+1}), \\ &\quad \frac{d(y_{n-1}, y_{n+1}) + d(y_n, y_n)}{2}\} \\ &\leq \max\{d(y_{n-1}, y_n), d(y_n, y_{n+1})\}. \end{aligned}$$

Now, if $M_T(y_{n-1}, y_n) = d(y_n, y_{n+1})$, then it implies

$$H(T_2y_{n-1}, T_2y_n) < d(y_n, y_{n+1}), \tag{10.17}$$

and from (10.16) and (10.17), we obtain

$$d(y_n, y_{n+1}) < d(y_n, y_{n+1}),$$

a contradiction.
Also if $M_T(y_{n-1}, y_n) = d(y_{n-1}, y_n)$, then we obtain

$$H(T_2y_{n-1}, T_2y_n) < d(y_{n-1}, y_n). \tag{10.18}$$

From (10.17) and (10.18), we get

$$d(y_n, y_{n+1}) < d(y_{n-1}, y_n).$$

Thus, the sequence $\{d(y_n, y_{n+1})\}$ is monotonically decreasing. Consequently, we have

$$\begin{aligned} F(d(y_n, y_{n+1})) &\leq F(d(y_{n-1}, y_n)) - \tau(d(y_{n-1}, y_n)) \\ &< F(d(y_{n-1}, y_n)). \end{aligned}$$

By the given assumption on τ, there exists $c > 0$ and $n \in \mathbb{N}$ such that $\tau(d(y_n, y_{n+1})) > c$ for all $n > n_0$.

$$\tau(d(y_n, y_{n+1})) > c \text{ for all } n > n_0.$$

Thus, we obtain that

$$\begin{aligned} F(d(y_n, y_{n+1})) &\leq F(d(y_{n-1}, y_n)) - \tau(d(y_{n-1}, y_n)) \\ &\leq F(d(y_{n-2}, y_{n-1})) - \tau(d(y_{n-2}, y_{n-1})) - \tau(d(y_{n-1}, y_n)) \\ &\leq \cdots \\ &\leq F(d(y_0, y_1)) - \{\tau(d(y_0, y_1)) + \cdots + \tau(d(y_{n_0-1}, y_{n_0})) \\ &\tau(d(y_{n_0}, y_{n_0+1})) + \cdots + \tau(d(y_{n-1}, y_n))\} \\ &= F(d(y_0, y_1)) - n_0. \end{aligned}$$

On taking limit as $n \to \infty$, we obtain that

$$\lim_{n\to\infty} F(d(y_n, y_{n+1})) = -\infty.$$

And together with F_2, we have

$$\lim_{n\to\infty} d(y_n, y_{n+1}) = 0.$$

By F_3, there exists $r \in (0,1)$ such that

$$\lim_{n\to\infty} \{d(y_n, y_{n+1})\}^r F(d(y_n, y_{n+1})) = 0.$$

Hence, it follows that

$$\begin{aligned} &[d(y_n, y_{n+1})]^r F(d(y_n, y_{n+1})) - [d(y_n, y_{n+1})]^r F(d(y_0, y_1)) \\ &\leq -n_0[d(y_n, y_{n+1})]^r \leq 0, \end{aligned}$$

by taking limit $n \to \infty$, we get

$$\lim_{n\to\infty} n[d(y_n, y_{n+1})]^r = 0.$$

Now, let us consider that there exists $n_1 \in \mathbb{N}$, such that $n^{\frac{1}{r}} d(y_n, y_{n+1}) \leq 1$ for all $n \geq n_1$. Consequently, we have

$$d(y_n, y_{n+1}) \leq \frac{1}{n^{\frac{1}{r}}} \text{ for all } n \geq n_1.$$

In order to show that $\{y_n\}$ is a Cauchy sequence, consider $m, n \in \mathbb{N}$ such that $m > n > n_1$.

$$\begin{aligned} d(y_m, y_n) &\leq d(y_m, y_{m-1}) + d(y_{m-1}, y_{m-2}) + \cdots + d(y_{n+1}, y_n) \\ &\leq \sum_{i=n}^{\infty} d(y_{i+1}, y_i) \\ &\leq \sum_{i=n}^{\infty} \frac{1}{i^{\frac{1}{r}}}. \end{aligned}$$

Thus, from the convergence of the series $\sum_{i=1}^{\infty} \frac{1}{i^{\frac{1}{r}}}$, we obtain that $\{y_n\}$ is a Cauchy sequence in X. Since (X, d) is complete, there exists $u \in X$ such that $y_n \to u$ as $n \to \infty$. Also u is a fixed point of T_2 that is $u \in T_2(u)$. Then by definition of k and (10.15), we have

$$d(y_1, y_2) \leq H(T_1 y_1, T_2 y_1) \leq k = \sup_{x \in X} H(T_i x, T_{i+1} x).$$

Again by the triangle inequality, we have

$$\begin{aligned} d(y_1, u) &\leq \sum_{i=1}^{n} d(y_i, y_{i+1}) + d(y_{n+1}, u) \\ &\leq \sum_{i=1}^{n} \frac{1}{i^{\frac{1}{r}}} + d(y_{n+1}, u), \end{aligned}$$

Taking the limit as $n \to \infty$, we obtain that

$$d(y_1, u) \leq \sum_{i=1}^{\infty} \frac{1}{i^{\frac{1}{r}}} \leq k,$$

that is, for $y_1 \in F(T_1)$ and $u \in F(T_2)$, we have

$$d(y_1, u) \leq k.$$

Similarly, for an arbitrary $z_0 \in F(T_2)$, we can find $v \in F(T_1)$ such that

$$d(z_0, v) \leq k.$$

Thus, we conclude that

$$H(F(T_1), F(T_2)) \leq k.$$

□

Lemma 10.2. *Let (X,d) be a complete metric space and $\{T_n : X \to P_{CB}(X) : n \in \mathbb{N}\}$ be a sequence of multivalued mappings uniformly convergent to a multivalued mapping $T : X \to P_{CB}(X)$. If T_n satisfies the condition (10.14) of Theorem 10.3 for every $n \in \mathbb{N}$, then T also satisfies the condition (10.14).*

Proof. As T_n satisfies the axiom (10.14) of Theorem 10.3 for every $n \in \mathbb{N}$, we have for $(x,y) \in A_i \times A_{i+1}$, $i = 1,2,\ldots,p$ and $A_{p+1} = A_1$ such that $H(T_nx,T_ny) > 0$ implies

$$F(H(T_nx,T_ny)) + \tau(M_{T_n}(x,y)) \leq F(M_{T_n}(x,y)),$$

where $F \in \mathcal{F}$, $\tau : \mathbb{R}_+ \to \mathbb{R}_+$ is a continuous function such that $\liminf_{s\to t^+} \tau(s) \geq 0$ for all $t \geq 0$, and

$$M_{T_n}(x,y) = \max\{d(x,y), d(x,T_nx), d(y,T_ny), \frac{d(y,T_nx)+d(x,T_ny)}{2}\}.$$

Since the sequence $\{T_n\}$ is uniformly convergent to T, taking the limit $n \to \infty$ we get for $H(Tx,Ty) > 0$

$$F(H(Tx,Ty)) + \tau(M_T(x,y)) \leq F(M_T(x,y)),$$

where $F \in \mathcal{F}$, $\tau : \mathbb{R}_+ \to \mathbb{R}_+$ is a continuous function such that $\liminf_{s\to t^+} \tau(s) \geq 0$ for all $t \geq 0$, and

$$M_T(Tx,Ty) = \max\{d(x,y), d(x,Tx), d(y,Ty), \frac{d(y,Tx)+d(x,Ty)}{2}\}.$$

which shows that T also satisfies the (10.14). □

Theorem 10.4. *Let (X,d) be a complete metric space and $\{T_n : X \to P_{CB}(X) : n \in \mathbb{N}\}$ be a sequence of multivalued mappings uniformly convergent to $T : X \to P_{CB}(X)$. If T_n satisfies (10.14) for every $n \in \mathbb{N}$, then*

$$\lim_{n\to\infty} H(F(T_n),F(T)) = 0,$$

that is, the fixed point sets of T_n are stable.

Proof. By the above Lemma 10.2, T_n satisfies (10.14). Let $k_n = \sup_{x\in X} H(T_nx,Tx)$. Since the sequence $\{T_n\}$ is uniformly convergent to T on X,

$$\lim_{n\to\infty} k_n = \lim_{n\to\infty} \sup_{x\in X} H(T_nx,Tx) = 0.$$

By using Theorem 10.3, we have

$$H(F(T_n),F(T)) \leq k_n \text{ for every } n \in \mathbb{N}.$$

Taking limit on the above inequality implies

$$\lim_{n\to\infty} H(F(T_n),F(T)) \leq \lim_{n\to\infty} k_n = 0.$$

Thus, the fixed point sets of T_n are stable. □

10.5 MULTIVALUED MAPPINGS UNDER CYCLIC SIMULATION FUNCTION

In this section, we present the idea of fixed points for multivalued mappings under the cyclic simulation function. The surely understood Banach contraction principle empowers the existence and uniqueness of fixed point of contraction in complete metric space. Because of this principle, numerous authors generalized this principle by presenting distinctive contractions on a metric space. Thus in this manner in this work, the mapping $\zeta : [0,\infty)\times[0,\infty)\to\mathbb{R}$ called the simulation function, and the idea of $Z-$contraction concerning ζ. $Z-$contraction sums up the Banach contraction principle and bring together few known types of contractions including the combination of $d(Tx,Ty)$ and $d(x,y)$. We create a connection amongst cyclic and simulation function.

Recently, the authors in [8] introduced the class of simulation functions as follows.

Definition 10.7. *A mapping $\zeta : [0,\infty)\times[0,\infty)\to\mathbb{R}$ is said to be simulation function if*

ζ_1) $\zeta(0,0)=0$.
ζ_2) $\zeta(t,s)<s-t$ *for all* $t,s>0$.
ζ_3) *if* $\{t_n\}$, $\{s_n\}$ *are sequences in* $(0,\infty)$ *such that* $\lim_{n\to\infty} t_n = \lim_{n\to\infty} s_n > 0$ *then*

$$\lim_{n\to\infty}\sup\zeta(t_n,s_n)<0.$$

We denote the set of all simulation functions by Z.

Definition 10.8. *For a metric space (X,d), let $A_1,A_2,\dots,A_p$ be nonempty closed subsets of X, where p be a positive integer, $Y=\cup_{i=1}^{p}A_i$. A multivalued mapping $T:Y\to P_{CB}(X)$ is called*

1. *cyclic $Z-$contractive mapping if:*
   (a) *Y has cyclic representation with respect to T.*
   (b) *For every $(w,v)\in A_i\times A_{i+1}$, $i=1,2,\dots,p$ and $A_{p+1}=A_1$, with $H(Tw,Tv)>0$ implies*

$$\zeta(H(Tw,Tv),M_1(w,v))\geq 0, \tag{10.19}$$

where

$$M_1(w,v)=\max\{d(w,v),d(w,Tw),d(v,Tv),\frac{d(w,Tv)+d(v,Tw)}{2}\}.$$

2. *Hardy Roger-type-cyclic $Z-$contractive mapping if*
   (a) *Y has cyclic representation with respect to T.*
   (b) *For every $(w,v)\in A_i\times A_{i+1}$, $i=1,2,\dots,p$ and $A_{p+1}=A_1$, with $H(Tw,Tv)>0$ implies*

$$\zeta(H(Tw,Tv),M_2(w,v)) \geq 0, \tag{10.20}$$

where

$$M_2(w,v) = a_1 d(w,v) + a_2 d(w,Tw) + a_3 d(v,Tv) + a_4 d(w,Tv) + a_5 d(w,Tv)$$

and $a_1,\ldots,a_5 \geq 0$ *with* $a_1+a_2+a_3+2a_4+a_5 < 1$.

Example 10.3. *Let* $X = \mathbb{R}$ *and* d *be a usual metric on* X. *Let* $A_1 = [-20,10]$ *and* $A_2 = [-10,30]$ *be any two closed subsets of* X, *such that* $Y = A_1 \cup A_2$ *and* $A_1 \cap A_2 = [-10,10]$. *Let* $T : Y \to P_{CB}(X)$ *be cyclic multivalued mapping such that*

$$T(x) = \begin{cases} [0, \frac{1-x}{40}] & if\ x \leq 1, \\ [0, \frac{1+x}{40}] & if\ x > 1. \end{cases}$$

Note that the set Y *has cyclic representation with respect to* T, *that is,* $T(A_1) \subseteq A_2$ *and* $T(A_2) \subseteq A_1$.
Let $\zeta : [0,\infty) \times [0,\infty) \to \mathbb{R}$ *be a simulation function in [8] defined by*

$$\zeta(t_1,t_2) = ct_2 - t_1 \text{ for all } t_1,t_2 \in [0,\infty),$$

where $c = \frac{9}{10} \in [0,1)$.

1. *Then for every* $(w,v) \in A_i \times A_{i+1}$, $i = 1,2$ *with* $H(Tw,Tv) > 0$ *implies*

$$\zeta(H(Tw,Tv),M_1(w,v)) \geq 0,$$

where

$$M_1(w,v) = \max\{d(w,v), d(w,Tw), d(v,Tv), \frac{d(w,Tv)+d(v,Tw)}{2}\}.$$

Thus, T is cyclic Z−contractive mapping.

2. *If we take* $a_1 = \frac{1}{4}$, $a_2 = a_3 = \frac{1}{8}, a_4 = \frac{1}{6}, a_5 = \frac{1}{4}$, *then for every* $(w,v) \in A_i \times A_{i+1}$, $i = 1,2$, *with* $H(Tw,Tv) > 0$ *implies*

$$\zeta(H(Tw,Tv),M_2(w,v)) \geq 0,$$

where

$$M_2(w,v) = a_1 d(w,v) + a_2 d(w,Tw) + a_3 d(v,Tv) + a_4 d(w,Tv) + a_5 d(w,Tv)$$

with $a_1+a_2+a_3+2a_4+a_5 < 1$. *Thus, T is Hardy Roger-type-cyclic Z−contractive mapping.*

10.6 FIXED POINT THEOREMS UNDER CYCLIC SIMULATION FUNCTION

In this section, we present fixed point results under cyclic simulation function. We start with the following result:

Theorem 10.5. *Let (X,d) be a complete metric space, $p \in \mathbb{N}$, $A_1, A_2, \ldots, A_p$ be nonempty closed subsets of X with $\cap_{i=1}^{p} A_i \neq \emptyset$ and $Y = \cup_{i=1}^{p} A_i$. If $T : Y \to P_{CB}(X)$ be a cyclic $Z-$contractive mapping with $\zeta \in Z$. Then, T has a fixed point in $\cap_{i=1}^{p} A_i$.*

Proof. Let $x_0 \in A_1$ and $x_1 \in T(x_0) \subset A_2$. Then, by Lemma 10.1, there exists $x_2 \in T(x_1) \subset A_3$, such that

$$d(x_1, x_2) \leq \lambda H(Tx_0, Tx_1). \tag{10.21}$$

Since $x_2 \in T(x_1) \subset A_3$. Then, again by Lemma 10.1, there exists $x_3 \in T(x_2) \subset A_4$ such that

$$d(x_2, x_3) \leq \lambda H(Tx_1, Tx_2). \tag{10.22}$$

And there exists $x_4 \in T(x_3) \subset A_5$ such that

$$d(x_3, x_4) \leq \lambda H(Tx_2, Tx_3), \tag{10.23}$$

$$\vdots$$

and there exists $x_{p-2} \in T(x_{p-3}) \subset A_{p-1}$ and $x_{p-1} \in T(x_{p-2}) \subset A_p$ such that

$$d(x_{p-2}, x_{p-1}) \leq \lambda H(Tx_{p-3}, Tx_{p-2}). \tag{10.24}$$

Similarly, there exists $x_{n+1} \in T(x_n) \subset A_{n+2}$ such that

$$d(x_n, x_{n+1}) \leq \lambda H(Tx_{n-1}, Tx_n). \tag{10.25}$$

We assume that $d(x_n, x_{n+1}) > 0$, for all $n \in \mathbb{N}$.
If not, then for some $k \in \mathbb{N}$, we have

$$d(x_k, x_{k+1}) = 0,$$

which implies $x_k = x_{k+1} \in Tx_k$, and the proof is finished. Therefore, we assume that $d(x_n, x_{n+1}) > 0$ for all $n \in \mathbb{N}$. Now

$$0 < d(x_{n+1}, x_{n+2}) \leq \lambda H(Tx_n, Tx_{n+1}),$$

where $\lambda \geq 1$, implies

$$H(Tx_n, Tx_{n+1}) > 0.$$

This implies that

$$\begin{aligned} 0 &\leq \zeta(H(Tx_n, Tx_{n+1}), M_1(x_n, x_{n+1})) \\ &< M_1(x_n, x_{n+1}) - H(Tx_n, Tx_{n+1}), \end{aligned}$$

which implies

$$\begin{aligned} H(Tx_n, Tx_{n+1}) &< M_1(x_n, x_{n+1}), \\ d(x_{n+1}, x_{n+2}) &< M_1(x_n, x_{n+1}), \end{aligned} \tag{10.26}$$

where

$$\begin{aligned} M_1(x_n, x_{n+1}) &= \max\{d(x_n, x_{n+1}), d(x_n, Tx_n), d(x_{n+1}, Tx_{n+1}), \\ &\quad \frac{d(x_n, Tx_{n+1}) + d(x_{n+1}, Tx_n)}{2}\} \\ &\leq \max\{d(x_n, x_{n+1}), d(x_n, x_{n+1}), d(x_{n+1}, x_{n+2}), \\ &\quad \frac{d(x_n, x_{n+2}) + d(x_{n+1}, x_{n+1})}{2}\} \\ &= \max\{d(x_n, x_{n+1}), d(x_{n+1}, x_{n+2})\}. \end{aligned}$$

Now, if $M_1(x_n, x_{n+1}) = d(x_{n+1}, x_{n+2})$, then (10.26) implies

$$d(x_{n+1}, x_{n+2}) < d(x_{n+1}, x_{n+2}),$$

a contradiction. Therefore, $M_1(x_n, x_{n+1}) = d(x_n, x_{n+1})$, which implies

$$d(x_{n+1}, x_{n+2}) < d(x_n, x_{n+1}) \text{ for all } n \in \mathbb{N}.$$

Thus, $\{d(x_{n+1}, x_{n+2})\}$ is a monotonically decreasing sequence of non-negative real numbers which is bounded below, and then, there exists $r \geq 0$ such that $\{d(x_{n+1}, x_{n+2})\}$ converges to r. we claim that $r = 0$. Assume on contrary $r > 0$. Obviously,

$$\lim_{n\to\infty} d(x_{n+1}, x_{n+2}) = \lim_{n\to\infty} d(x_n, x_{n+1}) = r.$$

It follows from the property (δ_3) of a simulation function that

$$0 \leq \lim_{n\to\infty} \sup \zeta(d(x_{n+1}, x_{n+2}), d(x_n, x_{n+1})) < 0.$$

Implies a contradiction, this shows that $r = 0$. Hence,

$$\lim_{n\to\infty} d(x_{n+1}, x_{n+2}) = 0.$$

Now to show $\{x_n\}$ is a Cauchy sequence in X, it is sufficient to show for Cauchyness that its subsequence $\{x_{2n}\}$ is also Cauchy in X. If it doesn't hold, then there exists $\varepsilon > 0$ such that we can find subsequences $\{x_{2m_k}\}$ and $\{x_{2n_k}\}$ of $\{x_{2n}\}$ with $n_k > m_k \geq k$ such that

$$d(x_{2m_k}, x_{2n_k}) \geq \varepsilon \text{ for all } k \in \mathbb{N}. \tag{10.27}$$

Without any loss of generality, we assume that for all $k \in \mathbb{N}$, n_k is the smallest positive integer greater than m_k for which this inequality holds, then

$$d(x_{2m_k}, x_{2n_k-2}) < \varepsilon \text{ for all } k \in \mathbb{N}. \tag{10.28}$$

Now by (10.27) and (10.28), we have

$$\begin{aligned}\varepsilon &\le d(x_{2m_k}, x_{2n_k})\\ &\le d(x_{2m_k}, x_{2n_k-2}) + d(x_{2n_k-2}, x_{2n_k-1}) + d(x_{2n_k-1}, x_{2n_k})\\ &< \varepsilon + d(x_{2n_k-2}, x_{2n_k-1}) + d(x_{2n_k-1}, x_{2n_k}),\end{aligned} \tag{10.29}$$

and by taking limit as $k \to \infty$ in the above inequality and using (10.28), we get that

$$\lim_{k\to\infty} d(x_{2m_k}, x_{2n_k}) = \varepsilon.$$

Similarly, we have

$$\begin{aligned}\varepsilon &\le d(x_{2m_k}, x_{2n_k}) \\ &\le d(x_{2m_k}, x_{2m_k+1}) + d(x_{2m_k+1}, x_{2n_k+1}) + d(x_{2n_k+1}, x_{2n_k}).\end{aligned} \tag{10.30}$$

Also,

$$d(x_{2m_{k+1}}, x_{2n_k+1}) \le d(x_{2m_{k+1}}, x_{2m_k}) + d(x_{2m_k}, x_{2n_k}) + d(x_{2n_k}, x_{2n_k+1}). \tag{10.31}$$

By taking limit as $k \to \infty$ on both sides of (10.30) and (10.31), we have

$$\lim_{k\to\infty} d(x_{2m_{k+1}}, x_{2n_k+1}) = \varepsilon.$$

As $\{x_{2m_k}\}$ and $\{x_{2n_k}\}$ are subsequences of $\{x_{2n}\}$ with $n_k > m_k \ge k$ and the property δ_3 of simulation function that

$$0 \le \limsup_{n\to\infty} \zeta(d(x_{2m_{k+1}}, x_{2n_k+1}), d(x_{2m_k}, x_{2n_k})) < 0,$$

a contradiction arises and hence $\{x_n\}$ is a Cauchy sequence in X. As A is a closed subset of a complete metric space X, there exists $x \in A$ such that $\lim_{n\to\infty} x_n = x$. Hence, x is a fixed point of T in $\cap_{i=1}^{p} A_i$. □

Example 10.4. *Let $X = \mathbb{R}$ and d be a usual metric on X. Let $A_1 = [-10, 0]$ and $A_2 = [0, 20]$ be any two closed subsets of X, such that $Y = A_1 \cup A_2$ and $A_1 \cap A_2 \neq \emptyset$. Consider a mapping $T : Y \to P_{CB}(X)$ defined as*

$$T(x) = \begin{cases} [0, \frac{-x}{15}] \text{ if } x \le 0 \\ \\ [\frac{-x}{15}, 0] \text{ if } x > 0. \end{cases}$$

Clearly, $T(A_1) \subseteq A_2$ and $T(A_2) \subseteq A_1$, so T is cyclic multivalued mapping. Let $\zeta : [0,\infty) \times [0,\infty) \to \mathbb{R}$ be a simulation function in [8] defined by

$$\zeta(t_1, t_2) = \kappa t_2 - t_1 \text{ for all } t_1, t_2 \in [0, \infty),$$

where $\kappa = \frac{4}{5} \in [0, 1)$.

To consider the contractive condition that for every $(w,v) \in A_i \times A_{i+1}$, $i = 1,2$ *with* $H(Tw,Tv) > 0$ *such that*

$$\zeta(H(Tw,Tv), M_1(w,v)) \geq 0,$$

where

$$M_1(w,v) = \max\{d(w,v), d(w,Tw), d(v,Tv), \frac{d(w,Tv)+d(v,Tw)}{2}\},$$

we have the following two cases:
Case (i): For $w \in A_1$, $v \in A_2$, $T(w) = [0, \frac{-w}{15}]$ *and* $T(v) = [\frac{-v}{15}, 0]$. *Then*

$$\begin{aligned} H(Tw,Tv) &= \max\{\sup_{u\in[0,\frac{-w}{15}]} d(u,[\frac{-v}{15},0]), \sup_{k\in[\frac{-v}{15},0]} d(k,[0,\frac{-w}{15}])\} \\ &= \max\{\frac{|w|}{15}, \frac{v}{15}\} = \frac{1}{15}\max\{|w|, v\} \end{aligned}$$

and

$$\begin{aligned} M_1(w,v) &= \max\{d(w,v), d(w,Tw), d(v,Tv), \frac{d(w,Tv)+d(v,Tw)}{2}\} \\ &= \max\{0, |w|, v, \frac{1}{30}(|w+15v| + |15w+v|)\} \\ &= \max\{|w|, v\}. \end{aligned}$$

So we have

$$\begin{aligned} \zeta(H(Tw,Tv), M_1(w,v)) &= \frac{4}{5}\max\{|w|, v\} - \frac{1}{15}\max\{|w|, v\} \\ &= \frac{11}{15}\max\{|w|, v\} \geq 0. \end{aligned}$$

Case (ii): When $w \in A_2$ *and* $v \in A_1$, $T(w) = [\frac{-w}{15}, 0]$ *and* $T(v) = [0, \frac{-v}{15}]$,

$$\begin{aligned} H(Tw,Tv) &= \max\{\sup_{u\in[0,\frac{-y}{15}]} d(u,[\frac{-w}{15},0]), \sup_{r\in[\frac{-x}{15},0]} d(r,[0,\frac{-y}{15}])\} \\ &= \max\{\frac{w}{15}, \frac{|v|}{15}\} = \frac{1}{15}\max\{w, -v\} \end{aligned}$$

and

$$\begin{aligned} M_1(w,v) &= \max\{d(w,v), d(w,Tw), d(v,Tv), \frac{d(w,Tv)+d(v,Tw)}{2}\} \\ &= \max\{0, w, |v|, \frac{1}{30}(|15w+v| + |w+15v|)\} \\ &= \max\{|w|, v\}. \end{aligned}$$

So we get

$$\zeta(H(Tw,Tv),M_1(w,v)) = \frac{4}{5}\max\{|w|,v\} - \frac{1}{15}\max\{|w|,v\}$$
$$= \frac{11}{15}\max\{|w|,v\} \geq 0.$$

Therefore, T is cyclic $Z-$contractive mapping. Thus, all the conditions of Theorem 10.5 are satisfied. Moreover, 0 is the fixed point of T.

Example 10.5. *Let $X = [0,10]$ and d be a usual metric on X. Let $A_1 = [0,5]$, $A_2 = [5,10]$ be any two nonempty closed subsets of X such that $X = A_1 \cup A_2$ and $A_1 \cap A_2 = \{5\}$. Let $T : X \to P_{CB}(X)$ be cyclic multivalued mapping defined by*

$$T(x) = \begin{cases} \{5\} & \text{if } x \in A_1, \\ \{0,5\} & \text{if } x \in A_2. \end{cases}$$

Let $\zeta : [0,\infty) \times [0,\infty) \to \mathbb{R}$ be a simulation function in [8] defined by

$$\zeta(t,s) = \begin{cases} 2(s-t) < s-t & \text{if } s < t, \\ ks - t < s - t & \text{otherwise where } k < 1. \end{cases}$$

Now, we consider the following cases:
Case (i): *If $x \in A_1$, $y \in A_2$ implies*

$$H(Tx,Ty) = H(\{5\},\{0,5\})$$
$$= \max(\{5\},\{5\}) = 5$$

and

$$M_1(x,y) = \max\left\{d(x,y), d(x,Tx), d(y,Ty), \frac{d(x,Ty)+d(y,Tx)}{2}\right\},$$

where

$$d(x,Tx) = \inf d([0,5],\{5\}) = \inf d(0,\{5\}) = 0$$
$$d(y,Ty) = \inf d([5,10],\{0,5\}) = \inf d(\{5\},0) = 0$$
$$\frac{d(x,Ty)+d(y,Tx)}{2} = \frac{5}{2}$$

implies that

$$\zeta(H(Tx,Ty),d(x,y)) = \zeta(5,d(x,y)) \geq 0.$$

Case (ii): *If $x \in A_2$, $y \in A_1$ implies*

$$H(Tx,Ty) = H(\{0,5\},\{5\})$$
$$= \max(\{5\},\{5\}) = 5$$

and

$$M_T(x,y) = \max\{d(x,y), d(x,Tx), d(y,Ty), \frac{d(x,Ty)+d(y,Tx)}{2}\},$$

where

$$d(x,Tx) = \inf d([5,10],\{0,5\}) = \inf d(\{5\}.0) = 0$$
$$d(y,Ty) = \inf d([0,5],\{5\}) = \inf d(0,\{5\}) = 0$$
$$\frac{d(x,Ty)+d(y,Tx)}{2} = \frac{5}{2}$$

implies that

$$\zeta(H(Tx,Ty),d(x,y)) = \zeta(5,d(x,y)) \geq 0.$$

Thus, mapping T satisfies the cyclic $Z-$contractive condition. Moreover, 5 is the fixed point of T in $A_1 \cap A_2$.

Theorem 10.6. *Let (X,d) be a complete metric space, $p \in \mathbb{N}$, $A_1, A_2, \ldots, A_p$ be nonempty closed subsets of X with $\cap_{i=1}^{p} A_i \neq \emptyset$ and $Y = \cup_{i=1}^{p} A_i$. If $T : Y \to P_{CB}(X)$ is Hardy Roger-type-cyclic $Z-$contractive mapping. Then, T has a fixed point in $\cap_{i=1}^{p} A_i$.*

Proof. Let $x_0 \in A_1$ and $x_1 \in T(x_0) \subset A_2$. Then, there exists $x_2 \in T(x_1) \subset A_3$, and by Lemma 10.1, we have

$$d(x_1,x_2) \leq \lambda H(Tx_0,Tx_1). \tag{10.32}$$

Since $x_2 \in T(x_1) \subset A_3$, again by Lemma 10.1, there exists $x_3 \in T(x_2) \subset A_4$ such that

$$d(x_2,x_3) \leq \lambda H(Tx_1,Tx_2). \tag{10.33}$$

And there exists $x_4 \in T(x_3) \subset A_5$ such that

$$d(x_3,x_4) \leq \lambda H(Tx_2,Tx_3). \tag{10.34}$$

And there exists $x_{p-2} \in T(x_{p-3}) \subset A_{p-1}$ and $x_{p-1} \in T(x_{p-2}) \subset A_p$ such that

$$d(x_{p-2},x_{p-1}) \leq \lambda H(Tx_{p-3},Tx_{p-2}). \tag{10.35}$$

By continuing in this way, we obtain a sequence $x_{n+1} \in T(x_n) \subset A_{n+2}$ such that

$$d(x_n,x_{n+1}) \leq \lambda H(Tx_{n-1},Tx_n). \tag{10.36}$$

We assume that $d(x_n,x_{n+1}) > 0$, for all $n \in \mathbb{N}$. If not, then for some $k \in \mathbb{N}$, we have

$$d(x_k,x_{k+1}) = 0,$$

which implies $x_k = x_{k+1} \in Tx_k$. The proof is finished. Therefore, $d(x_n,x_{n+1}) > 0$, for all $n \in \mathbb{N}$.

$$0 < d(x_{n+1},x_{n+2}) \leq \lambda H(Tx_n,Tx_{n+1}).$$

As $\lambda \geq 1$, so we have $H(Tx_n, Tx_{n+1}) > 0$, which implies

$$\begin{aligned} 0 &\leq \zeta(H(Tx_n, Tx_{n+1}), M_2(x_n, x_{n+1})) \\ &< M_2(x_n, x_{n+1}) - H(Tx_n, Tx_{n+1}), \end{aligned}$$

which implies

$$H(Tx_n, Tx_{n+1}) < M_2(x_n, x_{n+1}), \tag{10.37}$$

where

$$\begin{aligned} M_2(x_n, x_{n+1}) &= \alpha d(x_n, x_{n+1}) + \beta d(x_n, Tx_n) + \gamma d(x_{n+1}, Tx_{n+1}) \\ &\quad + \delta_1 d(x_n, Tx_{n+1}) + \delta_2 d(x_{n+1}, Tx_n) \\ &\leq \alpha d(x_n, x_{n+1}) + \beta d(x_n, x_{n+1}) + \gamma d(x_{n+1}, x_{n+2}) \\ &\quad + \delta_1 d(x_n, x_{n+2}) + \delta_2 d(x_{n+1}, x_{n+1}) \\ &\leq (\alpha + \beta + \delta_1) d(x_n, x_{n+1}) + (\gamma + \delta_1) d(x_{n+1}, x_{n+2}). \end{aligned}$$

Now, if $d(x_n, x_{n+1}) < d(x_{n+1}, x_{n+2})$, then

$$\begin{aligned} M_2(x_n, x_{n+1}) &\leq (\alpha + \beta + \gamma + 2\delta_1) d(x_{n+1}, x_{n+2}) \\ &< d(x_{n+1}, x_{n+2}) \end{aligned}$$

and (10.37) implies

$$H(Tx_n, Tx_{n+1}) < d(x_{n+1}, x_{n+2}),$$

that is,

$$d(x_{n+1}, x_{n+2}) < d(x_{n+1}, x_{n+2}),$$

a contradiction.

Therefore, $d(x_{n+1}, x_{n+2}) < d(x_n, x_{n+1})$ for all $n \in \mathbb{N}$. Thus, $\{d(x_{n+1}, x_{n+2})\}$ is a monotonically decreasing sequence of non-negative real numbers, bounded from below; then, there exists $r \geq 0$ such that $\{d(x_{n+1}, x_{n+2})\}$ converges to r. If we take $r = 0$. Assume on contrary that $r > 0$. Obviously,

$$\lim_{n \to \infty} d(x_{n+1}, x_{n+2}) = \lim_{n \to \infty} d(x_n, x_{n+1}) = r.$$

It follows from the property (δ_3) of a simulation function that

$$0 \leq \limsup_{n \to \infty} \zeta(d(x_{n+1}, x_{n+2}), d(x_n, x_{n+1})) < 0.$$

This shows that $r = 0$. Hence,

$$\lim_{n \to \infty} d(x_{n+1}, x_{n+2}) = 0.$$

Now to show $\{x_n\}$ is a Cauchy sequence in X, it suffices to show that $\{x_{2n}\}$ is a Cauchy sequence in X. If not, then there exists $\varepsilon > 0$ such that we can find subsequences $\{x_{2m_k}\}$ and $\{x_{2n_k}\}$ of $\{x_{2n}\}$ with $n_k > m_k \geq k$ such that

$$d(x_{2m_k}, x_{2n_k}) \geq \varepsilon \text{ for all } k \in \mathbb{N}. \tag{10.38}$$

Without any loss of generality, we assume that for all $k \in \mathbb{N}$, n_k is the smallest positive integer greater than m_k for which this inequality holds, then

$$d(x_{2m_k}, x_{2n_k-2}) < \varepsilon \text{ for all } k \in \mathbb{N}. \tag{10.39}$$

Now by (10.38) and (10.39), we have

$$\begin{aligned}\varepsilon &\leq d(x_{2m_k}, x_{2n_k}) \\ &\leq d(x_{2m_k}, x_{2n_k-2}) + d(x_{2n_k-2}, x_{2n_k-1}) + d(x_{2n_k-1}, x_{2n_k}) \\ &< \varepsilon + d(x_{2n_k-2}, x_{2n_k-1}) + d(x_{2n_k-1}, x_{2n_k}).\end{aligned} \tag{10.40}$$

By taking limit as $k \to \infty$ in the above inequality and using (10.39), we obtain that

$$\lim_{k\to\infty} d(x_{2m_k}, x_{2n_k}) = \varepsilon.$$

Similarly, we have

$$\begin{aligned}\varepsilon &\leq d(x_{2m_k}, x_{2n_k}) \\ &\leq d(x_{2m_k}, x_{2m_k+1}) + d(x_{2m_k+1}, x_{2n_k+1}) + d(x_{2n_k+1}, x_{2n_k}).\end{aligned} \tag{10.41}$$

Also,

$$d(x_{2m_{k+1}}, x_{2n_k+1}) \leq d(x_{2m_{k+1}}, x_{2m_k}) + d(x_{2m_k}, x_{2n_k}) + d(x_{2n_k}, x_{2n_k+1}). \tag{10.42}$$

By taking limit as $k \to \infty$ on both sides of (10.41) and (10.42), we have

$$\lim_{k\to\infty} d(x_{2m_{k+1}}, x_{2n_k+1}) = \varepsilon.$$

As $\{x_{2m_k}\}$ and $\{x_{2n_k}\}$ are subsequences of $\{x_{2n}\}$ with $n_k > m_k \geq k$, and the property δ_3 of simulation function that

$$0 \leq \lim_{n\to\infty} \sup \zeta(d(x_{2m_{k+1}}, x_{2n_k+1}), d(x_{2m_k}, x_{2n_k})) < 0,$$

a contradiction arises, and hence, Cauchyness of sequence is proved. As A is a closed subset of a complete metric space X, then there exists $x \in A$ such that x_n converges to x as $n \to \infty$. Hence x is a fixed point of T in $\cap_{i=1}^{p} A_i$. □

10.7 STABILITY OF FIXED POINT SETS UNDER CYCLIC SIMULATION FUNCTION

In this section, we present the stability results of cyclic simulation function.

Theorem 10.7. *Let (X,d) be a complete metric space, $p \in \mathbb{N}$ be a positive integer, $A_1, A_2, \ldots, A_p$ be nonempty closed subsets of X with $\cap_{i=1}^{p} A_i \neq \emptyset$ such that $Y = \cup_{i=1}^{p} A_i$ and $\{T_i\}_{i=1}^{m} : Y \to P_{CB}(X)$, $i = 1, 2, \ldots, m$ be family of multivalued mappings satisfying cyclic simulation function such that*

(i) $Y = \cup_{i=1}^{p} A_i$ *has cyclic representation of with respect to* T.
(ii) *For any* $(x,y) \in A_i \times A_{i+1}$, $i = 1,2,\ldots,p$ *and* $A_{p+1} = A_1$ *such that* $H(T_ix, T_iy) > 0$ *implies*

$$\zeta(H(T_ix,T_iy), M_{T_i}(x,y)) \geq 0, \tag{10.43}$$

where

$$M_{T_i}(x,y) = \max\{d(x,y), d(x,T_ix), d(y,T_iy), \frac{d(x,T_iy)+d(y,T_ix)}{2}\}.$$

Then,

$$H(F(T_i), F(T_{i+1})) \leq k,$$

where $k = \sup_{x\in X} H(T_ix, T_{i+1}x)$.

Proof. From Theorem 10.5, the fixed point set of T_i is nonempty, that is, $F(T_i) \neq \emptyset$ for $i = 1,2,\ldots,m$. Let $y_0 \in A_1$, then for any $y_1 \in F(T_1)$, that is, $y_1 \in T_1y_1 \subset A_2$. By Lemma 10.1, there exists $y_2 \in T_2y_1 \subset A_3$ such that

$$d(y_1,y_2) \leq H(T_1y_1, T_2y_1). \tag{10.44}$$

Since $y_2 \in T_2y_1 \subset A_3$, again by Lemma 10.1 there exists $y_3 \in T_2y_2 \subset A_4$ such that

$$d(y_2,y_3) \leq H(T_2y_1, T_2y_2). \tag{10.45}$$

By continuing in this way, we obtain a sequence $y_{n+1} \in T_2y_n \subset A_{n+2}$ such that

$$d(y_n,y_{n+1}) \leq H(T_2y_{n-1}, T_2y_n). \tag{10.46}$$

Since $d(y_n,y_{n+1}) > 0$ for all $n \in \mathbb{N}$. Therefore, $H(T_2y_{n-1}, T_2y_n) > 0$ for all $n \in \mathbb{N}$. This implies that

$$\begin{aligned} 0 &\leq \zeta(H(T_2y_{n-1},\ T_2y_n), M_{T_2}(y_{n-1},y_n)) \\ &< M_{T_2}(y_{n-1},y_n) - H(T_2y_{n-1}, T_2y_n), \end{aligned}$$

that is,

$$H(T_2y_{n-1}, T_2y_n) < M_{T_2}(y_{n-1},y_n), \tag{10.47}$$

where

$$\begin{aligned} M_{T_2}(y_{n-1},y_n) &= \max\{d(y_{n-1},y_n), d(y_{n-1},T_2y_{n-1}), d(y_n,T_2y_n), \\ &\quad \frac{d(y_{n-1},T_2y_n)+d(y_n,T_2y_{n-1})}{2}\} \\ &\leq \max\{d(y_{n-1},y_n), d(y_{n-1},y_n), d(y_{n-1},y_{n+1}), \\ &\quad \frac{d(y_{n-1},y_{n+1})+d(y_n,y_n)}{2}\} \\ &\leq \max\{d(y_{n-1},y_n), d(y_n,y_{n+1})\}. \end{aligned}$$

As $\frac{d(y_{n-1},y_{n+1})}{2} \le \frac{d(y_{n-1},y_n)+d(y_n,y_{n+1})}{2} \le \max\{d(y_{n-1},y_n),d(y_n,y_{n+1})\}$.
If $M_{T_2}(y_{n-1},y_n) = d(y_n,y_{n+1})$, then (10.46) and (10.47) implies

$$d(y_n,y_{n+1}) < d(y_n,y_{n+1}),$$

a contradiction. Therefore, $M_{T_2}(y_{n-1},y_n) = d(y_{n-1},y_n)$, which implies

$$d(y_n,y_{n+1}) < d(y_{n-1},y_n).$$

Thus, $\{d(y_n,y_{n+1})\}$ is a monotonically decreasing sequence of non-negative real numbers bounded from below, and then, there exists $r \ge 0$ such that $\{d(y_n,y_{n+1})\}$ converges to r. We claim that $r = 0$. Assume on contrary that $r > 0$.

$$\lim_{n\to\infty} d(y_n,y_{n+1}) = \lim_{n\to\infty} d(y_{n-1},y_n) = r.$$

It follows from the property (δ_3) of a simulation function that

$$0 \le \limsup_{n\to\infty} \zeta(d(y_n,y_{n+1}),d(y_{n-1},y_n)) < 0.$$

This shows that $r = 0$. Hence,

$$\lim_{n\to\infty} d(y_n,y_{n+1}) = 0.$$

Now to show $\{y_n\}$ is a Cauchy sequence in X, it suffices to show that $\{y_{2n}\}$ is a Cauchy sequence in X. If not, then there exists $\varepsilon > 0$ such that we can find subsequences $\{y_{2m_k}\}$ and $\{y_{2n_k}\}$ of $\{y_{2n}\}$ with $n_k > m_k \ge k$ such that

$$d(y_{2m_k},y_{2n_k}) \ge \varepsilon \text{ for all } k \in \mathbb{N}. \tag{10.48}$$

Without any loss of generality, we assume that for all $k \in \mathbb{N}$, n_k is the smallest positive integer greater than m_k for which this inequality holds, then

$$d(y_{2m_k},y_{2n_k-2}) < \varepsilon \text{ for all } k \in \mathbb{N}. \tag{10.49}$$

Now by (10.48) and (10.49), we have

$$\begin{aligned}\varepsilon &\le d(y_{2m_k},y_{2n_k})\\ &\le d(y_{2m_k},y_{2n_k-2}) + d(y_{2n_k-2},y_{2n_k-1}) + d(y_{2n_k-1},y_{2n_k})\\ &< \varepsilon + d(y_{2n_k-2},y_{2n_k-1}) + d(y_{2n_k-1},y_{2n_k})\end{aligned} \tag{10.50}$$

and by taking limit as $k \to \infty$ on both sides of (10.50) and using (10.49), we have

$$\lim_{k\to\infty} d(y_{2m_k},y_{2n_k}) = \varepsilon. \tag{10.51}$$

Similarly, we have

$$\begin{aligned}\varepsilon &\le d(y_{2m_k},y_{2n_k})\\ &\le d(y_{2m_k},y_{2m_k+1}) + d(y_{2m_k+1},y_{2n_k+1}) + d(y_{2n_k+1},y_{2n_k}).\end{aligned} \tag{10.52}$$

Also,

$$d(y_{2m_{k+1}}, y_{2n_k+1}) \leq d(y_{2m_{k+1}}, y_{2m_k}) + d(y_{2m_k}, y_{2n_k}) + d(y_{2n_k}, y_{2n_k+1}). \tag{10.53}$$

By taking limit as $k \to \infty$ in the above inequality and using (10.51), we obtain that

$$\lim_{k\to\infty} d(y_{2m_{k+1}}, y_{2n_k+1}) = \varepsilon.$$

Since $\{y_{2m_k}\}$ and $\{y_{2n_k}\}$ are subsequences of $\{y_{2n}\}$ with $n_k > m_k \geq k$, and the property δ_3 of simulation function that

$$0 \leq \lim_{k\to\infty} \sup \zeta(d(y_{2m_{k+1}}, y_{2n_k+1}), d(y_{2m_k}, y_{2n_k})) < 0,$$

a contradiction and hence $\{y_n\}$ is a Cauchy sequence in X and there exists a $u \in X$ such that $y_n \to u$ as $n \to \infty$. Also u is a fixed point of T_2; that is, $u \in T_2 u$.
From (10.44) and the definition of k, it follows that

$$\begin{aligned} d(y_1, y_2) &\leq H(T_1 y_1,\ T_2 y_1) \\ &\leq k = \sup_{x\in X} H(T_1 x,\ T_2 x). \end{aligned}$$

Again by the triangle inequality and using the above inequality, we have

$$d(y_1, u) \leq \sum_{i=1}^{n} d(y_i, y_{i+1}) + d(y_{n+1}, u),$$

on taking limit as $n \to \infty$ we have

$$\begin{aligned} d(y_1, u) &\leq \sum_{i=1}^{\infty} d(y_i, y_{i+1}) \\ &\leq \varepsilon. \end{aligned}$$

Thus, given an arbitrary $y_1 \in F(T_1)$, we can find $u \in F(T_2)$ for which

$$d(y_1, u) \leq k.$$

Similarly for an arbitrary $z_0 \in F(T_2)$, there exists $w \in F(T_1)$ such that

$$d(z_0, w) \leq k.$$

Hence, we conclude that

$$H(F(T_1), F(T_2)) \leq k.$$

□

Lemma 10.3. *Let (X, d) be a complete metric space. Let $\{T_n : X \to P_{CB}(X) : n \in \mathbb{N}\}$ be a sequence of multivalued mappings uniformly convergent to a multivalued mapping $T : X \to P_{CB}(X)$. If T_n satisfies condition (10.43) of Theorem 10.7 for every $n \in \mathbb{N}$, then T also satisfies condition (10.43).*

Proof. As T_n satisfies axiom (10.43) of Theorem 10.7 for every $n \in \mathbb{N}$, we have for $(x,y) \in A_i \times A_{i+1}$, $i = 1,2,\ldots,p$ and $A_{p+1} = A_1$ such that $H(T_n x, T_n y) > 0$ implies

$$\zeta(H(T_n x, T_n y), M_{T_n}(x,y)) \geq 0,$$

where

$$M_{T_n}(x,y) = \max\{d(x,y), d(x,T_n x), d(y,T_n y), \frac{d(y,T_n x) + d(x,T_n y)}{2}\}.$$

Since the sequence $\{T_n\}$ is uniformly convergent to T, taking the limit $n \to \infty$ in the above inequality, we get for $H(Tx,Ty) > 0$

$$\zeta(H(Tx,Ty), M_T(x,y)) \geq 0,$$

where

$$M_T(Tx,Ty) = \max\{d(x,y), d(x,Tx), d(y,Ty), \frac{d(y,Tx) + d(x,Ty)}{2}\},$$

which shows that T also satisfies the (10.43). □

We now present the stability result of fixed point.

Theorem 10.8. *Let (X,d) be a complete metric space. Let $\{T_n : X \to P_{CB}(X) : n \in \mathbb{N}\}$ be a sequence of multivalued mappings uniformly convergent to $T : X \to P_{CB}(X)$. If T_n satisfies (10.43) for every $n \in \mathbb{N}$, then*

$$\lim_{n\to\infty} H(F(T_n), F(T)) = 0;$$

that is, the fixed point sets of T_n are stable.

Proof. By Lemma 10.3, T satisfies (10.43). Let $k_n = \sup_{x\in X} H(T_n x, Tx)$. Since the sequence $\{T_n\}$ is uniformly convergent to T on X, by taking limit as $n \to \infty$, we have

$$\lim_{n\to\infty} k_n = \lim_{n\to\infty} \sup_{x\in X} H(T_n x, Tx) = 0.$$

From Theorem 10.7, we have

$$H(F(T_n), F(T)) \leq k_n, \text{ for every } n \in \mathbb{N}.$$

By taking limit as $n \to \infty$, we have

$$\lim_{n\to\infty} H(F(T_n), F(T)) \leq \lim_{n\to\infty} k_n = 0,$$

that is,

$$\lim_{n\to\infty} H(F(T_n), F(T)) = 0.$$

Thus, the fixed point sets of T_n are stable. □

BIBLIOGRAPHY

[1] Abbas, M. & Rhoades B. E. (2009). Fixed point theorems for two new classes of multivalued mappings. *Appl. Math.* Lett. 22, 1364–1368.

[2] Acar, O., Durmaz & Minak G. (2014). Generalized multivalued F-contractions on complete metric spaces. *Bulletin of the Iranian Mathematical Society.* 40 (6), 1469–1478.

[3] Aubin, J.P. & Frankowska, H. (1990). Set-valued Analysis, Systems and Control. BirkhR auser, Basel.

[4] Beg, I. & Butt A. R. (2013). Fixed point of set-valued graph contractive mappings. *Journal of Inequalities and Applications.* 1–7.

[5] Berinde, M. & Berinde V. (2007). On a general class of multi-valued weakly Picard mappings. *Journal of Mathematical Analysis and Applications.* 326(2):772–782.

[6] Banach S. (1922). Sur les opérations dans les ensembles abstraits et leur application aux équations intégrales. *Fund. Math.* 3 (1), 133–181.

[7] Karapinar, E. (2011). Fixed point theory for cyclic weak Φ-contraction. *Appl. Math.* Lett. 24 (6), 822 825.

[8] Khojasteh, F., Shukla, S. & Radenovic, S. (2015). A new approach to the study of fixed point theory for simulation functions. *Filomat* 29 (6), 1189–1194.

[9] Khojasteh, F. & Rakočević V. (2012). Some new common fixed point results for generalized contractive multi-valued non-self-mappings. *Applied Mathematics Letters* 25 (3), 287–293.

[10] Kirk, W. A., Srinivasan, P. S. & Veeramani P. (2003). Fixed points for mappings satisfying cyclical contractive conditions. *Fixed Point Theory* 4 (1), 79–89.

[11] Klim, D. & Wardowsk, D. (2015). Fixed points of dynamic processes of set-valued F-contractions and application to functional equations. *Fixed Point Theory and Appl.* 22.

[12] Minak, G. Helvasi, A. & Altun I. Ciric type generalized F-contractions on complete metric spaces and fixed point results. *Filomat* 28 (6) (2014), 1143–1151.

[13] Mizoguchi, N. & Takahashi W. (1989). Fixed point theorems for multivalued mappings on complete metric spaces, *Journal of Mathematical Analysis and Applications*, 141 (1), 177–188.

[14] Nadler, S.B. (1969). Multivalued contraction mappings, *Pacific J. Math.* 20 (2), 457–488.

[15] Păcurar, M. (2011). Fixed point theory for cyclic Berinde operators, *Fixed Point Theory* 12 (2), 419–428.

[16] Păcurar, M. & Rus I. A. (2010). Fixed point theory for cyclic weak-contractions. *Nonlinear Anal.* 72, 1181–1187.

[17] Petric, M. A. (2010). Some results concerning cyclical contractive mappings. *Gen. Math.* 18 (4), 213–226.

[18] Rhoades, B. E.(1977). A comparison of various definitions of contractive mappings. *Transactions of the American Mathematical Society* 226, 257–290.

[19] Rus, I. A., Petruşel, A. & Sîntămărian A. (2003). Data dependence of the fixed point set of some multivalued weakly Picard operators. *Nonlinear Analysis: Theory, Methods & Applications.* 52 (8), 1947–1959.

[20] Sintunavarat W. & Kumam P. (2012). Common fixed point theorem for cyclic generalized multivalued mappings. *Appl. Math. Lett.* 25, 1849–1855.

[21] Wardowski, D. (2012). Fixed points of new type of contractive mappings in complete metric spaces. *Fixed Point Theory Appl.* 94, 6.

11 Significance and Relevances of Functional Equations in Various Fields

B. V. Senthil Kumar
Nizwa College of Technology

Hemen Dutta
Gauhati University

CONTENTS

11.1 INTRODUCTION

An equation that indicates a function in hidden type is a functional equation. A functional equation communicates the value of a function at a particular point with its values at other points. Since a function is a solution of a functional equation, we can determine the properties of such function which satisfies the equation. It is not so easy to reduce a functional equation to an algebraic equation.

The significant role of functional equations in other fields of mathematics such as abstract algebra, number theory, differential geometry, differential equations, iterations, and analytic functional equations shows the importance of application of functional equations. Hence, this theory of functional equations has gained its own behavior. By the motivation of outdated analytic methods to some level in many areas of mathematics, this theory of functional equations has attracted many mathematicians. This theory comprises simple elementary methods which are very easy when compared with the classical methods of mathematical analysis.

There are many noteworthy applications of functional equations in various branches of mathematics, such as measure theory, group theory, geometry, mechanics, statistics, and algebraic geometry. There are some other important applications of functional equations in various other fields such as economics, classical mechanics, game theory, dynamic programming, computer graphics, statistics, wireless sensor networks, astronomy, stochastic processes, cognitive science, fuzzy set theory, artificial intelligence, information theory, neural networks, coding theory, digital image processing, cluster analysis, interest formulae, Gaussian functions, population ethics, multivalued logic, electric circuits, financial management, iteration, psychometry, probability theory, scalar products, Gaussian functions, determinants and sum of powers, nondifferentiable functions, the Cobb–Douglas production function and quasilinearity, binomial expansion, and Chebyshev polynomials. The usage of functional equations is increasing day-by-day vigorously in the above mentioned fields to investigate many problems. Hence, it is not so easy to provide an explanatory survey of all the applications of functional equations. We mention some interesting and remarkable applications of functional equations.

1. The communication and networks systems consider the ensuing functional equation involving two variables of the form:

   $$K_1(u,v)P(u,v) = K_2(u,v)P(u,0) + K_3(u,v)P(0,v) + K_4(u,v)P(0,0)$$

   where $K_i(u,v); i = 1,2,3,4$ are given polynomials in two complex variables u, v to study the nature of the systems [5].
2. The Cauchy additive equation $\phi(u+v) = \phi(u) + \phi(v)$ arises in genetics to determine the combinatorial function $g_r(n) =$ the number of possible ways of choosing r objects at a time from n objects allowing repetitions, since this function produces the number of possibilities from a gene pool. For detailed information, one can refer [7].
3. By the application of logarithmic Cauchy equation $\ell(ab) = \ell(a) + \ell(b)$, we can show that $\int_1^y \frac{1}{u} du = \ln y$ [17].

Let us define $\ell : \mathbb{R}_+ \to \mathbb{R}$ by

$$\ell(a) = \int_1^y \frac{1}{u} du, \qquad a > 0.$$

Hence, in the case $a, b \in (1, \infty)$, we have

$$\begin{aligned} \ell(a) + \ell(b) &= \int_1^a \frac{1}{u} du + \int_1^b \frac{1}{u} du \\ &= \int_1^a \frac{1}{u} du + \int_a^{ab} \frac{1}{v} dv, \qquad \text{where} \quad v = ua \\ &= \int_1^{ab} \frac{1}{v} dv \quad \text{(additive property of the integral)} \\ &= \ell(ab). \end{aligned} \tag{11.1}$$

4. Murali et al. [10] employed the generalized additive functional equation

$$k[f(kx+y)+f(kx-y)]+f(x+ky)+f(x-ky) = f(x+y)+f(x-y)+2k^2 g(x)$$

with $k \neq 0, \pm 1$ to encrypt and decrypt the digital spatial image, and investigated its modular stability in [11].

Here, we illustrate some examples how functional equations are applied to solve some interesting problems in geometry, finance, information theory, wireless sensor networks, and electric circuits with parallel resistances.

11.2 APPLICATION OF FUNCTIONAL EQUATION IN GEOMETRY

Legendre used functional equations to find the area of a rectangle [17]. He considered a rectangle whose length is ℓ and breadth is b. Then, the area of the rectangle is $\phi(\ell, b)$.

Then, he divided the rectangle horizontally so that the rectangle is divided into two subrectangles with breadths b_1 and b_2 and the same length ℓ as in Figure 11.1. Then, the areas of subrectangles are $\phi(\ell, b_1)$ and $\phi(\ell, b_2)$, and the area of the full rectangle is $\phi(\ell, b_1 + b_2)$. Then, he arrived at

$$\phi(\ell, b_1 + b_2) = \phi(\ell, b_1) + \phi(\ell, b_2). \tag{11.2}$$

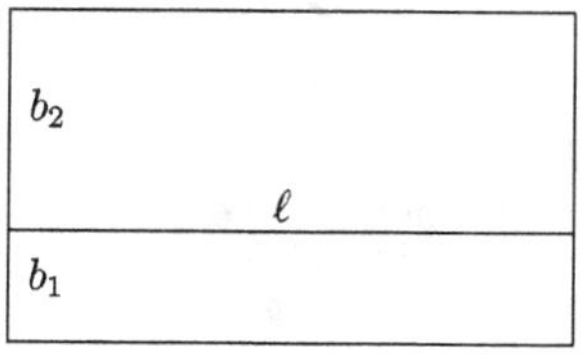

Figure 11.1 Rectnagle divided into two subrectangles along the breadth.

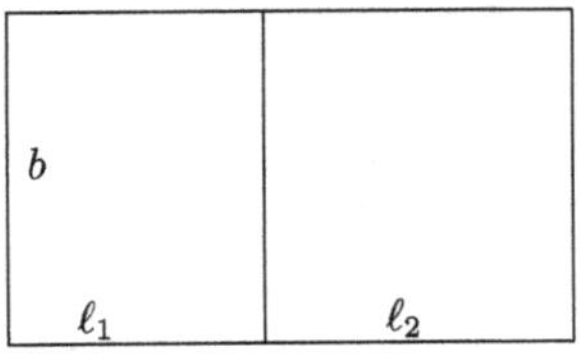

Figure 11.2 Rectangle divided into two subrectangles along the length.

In a similar manner, he divided the rectangle vertically with breadths ℓ_1 and ℓ_2 and the same length b as in Figure 11.2. Then, the resulting areas are $\phi(\ell_1,b)$ and $\phi(\ell_2,b)$ and $\phi(\ell_1+\ell_2,b)$. In the latter division of the rectangle, he obtained

$$\phi(\ell_1+\ell_2,b)=\phi(\ell_1,b)+\phi(\ell_2,b). \tag{11.3}$$

In equation (11.2), ℓ is a constant, and whereas in equation (11.3), b is a constant. Both the equations are similar to Cauchy's equation $\phi(u+v)=\phi(u)+\phi(v)$ whose solution is $\phi(u)=ku$. Therefore, he obtained the solution of (11.2) and (11.3) as

$$\phi(\ell,b)=k_2(\ell)b=k_1(b)\ell. \tag{11.4}$$

From (11.4)

$$\frac{k_1(b)}{b}=\frac{k_2(\ell)}{\ell}=k. \tag{11.5}$$

From (11.5)

$$k_1(b)=kb, \qquad k_2(\ell)=k\ell. \tag{11.6}$$

Substituting (11.6) in (11.4), he obtained

$$\phi(\ell,b)=k\ell b$$

where k is an arbitrary positive constant. He assumed the initial conditions, that is, when $\ell=1$, $b=1$, the area of the rectangle $=1$, to get the value of k as 1. Therefore, he proved $\phi(\ell,b)=\ell b$ which is the area of the rectangle.

11.3 APPLICATION OF FUNCTIONAL EQUATION IN FINANCIAL MANAGEMENT

In this section, the role of functional equation in calculating compound interest is explained in [17]. Suppose a person invests a principal of Rs.p at the rate of interest $r\%$ for the period of q years. Then, the final amount after q years is a function of p and q, that is, $C(p,q)$. Now, we have two cases:

i. The final amount will be the same, if we invest the principal amount p_1+p_2 together or principals p_1,p_2 are invested separately for q years. This can be expressed in the following functional equation:

$$g(p_1+p_2,q)=g(p_1,q)+g(p_2,q) \tag{11.7}$$

whose solution is

$$g(p,q)=K(q)p. \tag{11.8}$$

ii. The final amount is same, whether we invest the amount p for a period of q_1+q_2 years or p invested for q_1 years and then invest the resultant amount in q_2 years. It is expressed as

$$g(p,q_1+q_2)=g(g(p,q_1),q_2). \tag{11.9}$$

Applying (11.8) in (11.9), we get

$$\begin{aligned}K(q_1+q_2)p&=g(K(q_1)p,q_2)\\K(q_1+q_2)p&=K(q_2)K(q_1)p\\K(q_1+q_2)&=K(q_2)K(q_1).\end{aligned}$$

It is a multiplicative functional equation whose solution is

$$K(q)=a^q. \tag{11.10}$$

Substituting (11.10) in (11.8), we get

$$g(p,q)=a^q p.$$

For one year, $g(p,1)=p+$interest. Hence,

$$\begin{aligned}ap&=p+\frac{p\times 1\times r}{100}\\ap&=p\left(1+\frac{r}{100}\right)\\a&=\left(1+\frac{r}{100}\right).\end{aligned}$$

Therefore, $$g(p,q)=p\left(1+\frac{r}{100}\right)^q.$$

So, we arrive the compound interest formula $=p\left(1+\frac{r}{100}\right)^q$.

11.4 APPLICATION OF FUNCTIONAL EQUATION IN INFORMATION THEORY

Some functional equations of the following forms:

i. $h(u)+(1-u)^\alpha h\left(\frac{v}{1-u}\right)=h(v)+(1-v)^\alpha h\left(\frac{u}{1-v}\right)$
ii. $g(uv)+g((1-u)v)=g(v)\{m(u)+m(1-u)\}+m(v)\{g(u)+g(1-u)\}$
iii. $\sum_{\ell=1}^{n}\sum_{k=1}^{m}h(u_\ell v_k)=\sum_{\ell=1}^{n}h(u_\ell)+\sum_{k=1}^{m}h(v_k)$
iv. $\sum_{\ell=1}^{n}\sum_{k=1}^{m}s_{\ell k}(u_\ell v_k)=\sum_{\ell=1}^{n}p_\ell(u_\ell)+\sum_{k=1}^{m}q_k(v_k)+\sum_{\ell=1}^{n}r_\ell(u_\ell)\sum_{k=1}^{m}l_k(v_k)$

are applied in the information theory.

Definition 11.1. *Let $\Delta_m=\{p=(p_1,p_2,\ldots,p_m)|p_i\geq 0,\sum_j p_j=1\}$ be the set of all finite complete discrete probability distribution on a given partition of the sure event E into m events $E_1,E_2,\ldots,E_m$. In 1948, Shannon, in his paper [C.E. Shannon, A*

Mathematical theory of Communication, Bell System Tech. J. 27 (1948), 378-423 and 632-656], introduced the measure of information

$$H_m(p) = -\sum_{j=1}^{m} p_j log p_j, \qquad p \in \Delta_m$$

known as **Shannon's entropy**.

We have multiplicative functional equation

$$M(uv) = M(u) + M(v)$$

whose solution is $M(u) = \log u$, which is a function whose value on the product of probabilities of events is equal to the sum of its values on the probabilities of the individual events. Shannon used the above functional equation, in the information theory since with an intuitive notion that the information content of two independent events should be the sum of the information in each event.

In particular, the functional equation

$$f_1(u) + \alpha(1-u)f_2\left(\frac{v}{1-u}\right) = f_3(v) + \alpha(1-v)f_4\left(\frac{u}{1-v}\right)$$

for all $u, v \in [0,1]$ with $u+v \in [0,1]$. When $f_1 = f_2 = f_3 = f_4$ and $\alpha =$ the identity map, it is known as the fundamental equation of information. It has been extensively investigated by many authors. The general solution of the above equation is dealt by P.L. Kannappan [*Can. J. Math. Vol. XXXV, No. 5, 1983, pp. 862–872*].

The effect of the information communicated in a message can be measured by the changes in the probability concerning the receiver of the message. The effect of information will depend upon the expectation of receiver before and after receiving the message. Naturally, the information received can be taken as the ratio of the logarithm of two probabilities. Thus, the information received about the event E is given by

$$I(E) = \frac{\text{Probability concerning the receiver after receiving the information}}{\text{Probability concerning the receiver before receiving the information}}. \tag{11.11}$$

In case of noiseless channel, probability concerning the receiver after receiving the information equals to 1, as there will be no distortion of information during the process. The above equation becomes

$$\begin{aligned} I(E) &= \log\left[\frac{1}{\text{Probability concerning the receiver before receiving the information}}\right] \\ &= -\log[\text{Probability concerning the receiver before receiving the information}]. \end{aligned} \tag{11.12}$$

Shannon, with his intuitive idea, proposed a decreasing function $h(p)$, as a measure of the amount of information satisfying

$$h(p) = -\log p, \qquad 0 < p \leq 1. \tag{11.13}$$

The function $h(p)$ is called the information function, and it satisfies the additive property. Let A and B be any two events with $p(A) > 0$, $p(B) > 0$. Suppose that first we are informed that A has occurred and next that we are informed that B has occurred and if A and B are independent, then

$$-\log[p(A)] - \log[p(B)] = -\log[p(AB)].$$

If $p(A) = p_1$, $p(B) = p_2$ and $p(AB) = p_1 p_2$, then

$$-\log(p_1) - \log(p_2) = -\log(p_1 p_2).$$

Hence, $h(p_1) + h(p_2) = h(p_1 p_2)$, which shows that the information function h satisfies the Cauchy's functional equation:

$$f(xy) = f(x) + f(y). \tag{11.14}$$

11.5 APPLICATION OF FUNCTIONAL EQUATION IN WIRELESS SENSOR NETWORKS

The applications of functional equation in wireless sensor networks are discussed in [14]. Wireless sensor networks (WSNs) consist of small nodes with sensing, computation, and wireless communications capabilities. Many routing, power management, and data dissemination protocols have been specifically designed for WSNs where energy awareness is an essential design issue.

Due to recent technological advances, the manufacturing of small and low-cost sensors became technically and economically feasible. The sensing electronics measure ambient conditions related to the environment surrounding the sensor and transforms them into an electric signal. Processing such a signal reveals some properties about objects located and/or events happening in the vicinity of the sensor. A large number of these disposable sensors can be networked in many applications that require unattended operations. A WSN contains hundreds or thousands of these sensor nodes. These sensors have the ability to communicate either among each other or directly to an external base station (BS). A greater number of sensors allow for sensing over larger geographical regions with greater accuracy. Basically, each sensor node comprises sensing, processing, transmission, mobilizer, position finding system, and power units (some of these components are optional like the mobilizer). Sensor nodes are usually scattered in a sensor field, which is an area where the sensor nodes are deployed. Sensor nodes coordinate among themselves to produce high-quality information about the physical environment. Each sensor node bases its decisions on its mission, the information it currently has, and its knowledge of its computing, communication, and energy resources. Each of these scattered sensor nodes has the capability to collect and route data either to other sensors or back to an external base station(s).

Networking unattended sensor nodes may have profound effect on the efficiency of many military and civil applications such as target field imaging, intrusion detection, weather monitoring, security and tactical surveillance, distributed computing, detecting ambient conditions such as temperature, movement, sound, light, or the presence of certain objects, inventory control, and disaster management. Deployment of a sensor network in these applications can be in random fashion (e.g., dropped from an airplane) or can be planted manually (e.g., fire alarm sensors in a facility). For example, in a disaster management application, a large number of sensors can be dropped from a helicopter. Networking these sensors can assist rescue operations by locating survivors, identifying risky areas, and making the rescue team more aware of the overall situation in the disaster area.

Routing is the process of selecting path in a network along which to send network traffic. Routing trees are typical structures used in WSN to deliver data to sink. To ensure robust data communication, efficient methods are required to choose routes across a network that can react quickly to communication link changes. Many algorithms have been proposed in literature to support the routing protocols of the network. In 1958, Richard Bellman [*Dynamic Programming, Princeton University Press, 1957*] applied the functional equation approach to devise an algorithm which converges to the solution at atmost $N-1$ steps for a network with N nodes.

It is stated as follows:

"Given a set of N cities, with every two cities linked by a road. The time required to travel from i to j is not directly proportional to the distance between i and j, due to road conditions and traffic. Given the matrix $T=(t_{ij})$ not necessarily symmetric, where t_{ij} is the time required to travel from i to j. We wish to determine the path from one given city to another given city which minimizes the travel time."

The functional equation technique of dynamic programming, combined with approximation in policy space, yields an iterative algorithm which converges after at most $(N-1)$ iterations.

Let us now introduce the functional equation technique of dynamic programming. Let $f_i =$ the time required to travel from i to N, $i=1,2,\dots,N-1$, using an optimal policy with $f_n=0$.

Employing the principle of optimality, we see that the f_i satisfies the non-linear system of equations

$$\begin{aligned} f_i &= \min[t_{ij}+f_i], \qquad i=1,2,\dots,N-1 \\ f_N &= 0. \end{aligned} \tag{11.15}$$

The equation (11.15) is a functional equation because functions appear on both sides. We try to obtain the solution of the system (11.15) by using method of successive approximations. Choose an initial sequence $\left\{f_i^{(0)}\right\}$, and then, proceed iteratively, setting

$$f_i^{(k+1)} = \min_{i \neq j} \left(t_{ij} + f_i^{(k)} \right), \qquad i = 1, 2, \ldots, N-1$$
$$f_N^{(k+1)} = 0, \text{ for } k = 0, 1, 2, \ldots \qquad (11.16)$$

The sequence in (11.16) converges to the solution after $(N-1)$ iterations by suitable algorithm. In this way, we can solve routing problem by functional equation.

11.6 APPLICATION OF RATIONAL FUNCTIONAL EQUATION

In this section, we deal with the following new rational functional equation:

$$r(m+n) = \frac{r(m)r(n)}{r(m)+r(n)}. \qquad (11.17)$$

It is easy to see the function $r(m) = \frac{k}{m}$ is a solution of the equation (11.17), and hence, it is said to be a reciprocal functional equation. For additional facts of (11.17), one can refer [12,13].

11.6.1 GEOMETRICAL INTERPRETATION OF EQUATION (11.17)

Consider a right-angled triangle ABC with sides "a" and "b" as shown in Figure 11.3.

Construct a square BDEF inside the triangle ABC as shown in Figure 11.3 with side "p". Then, AD $= a - p$, FC $= b - p$. Now,

$$\text{Area of triangle ABC} = \frac{1}{2}ab. \qquad (11.18)$$

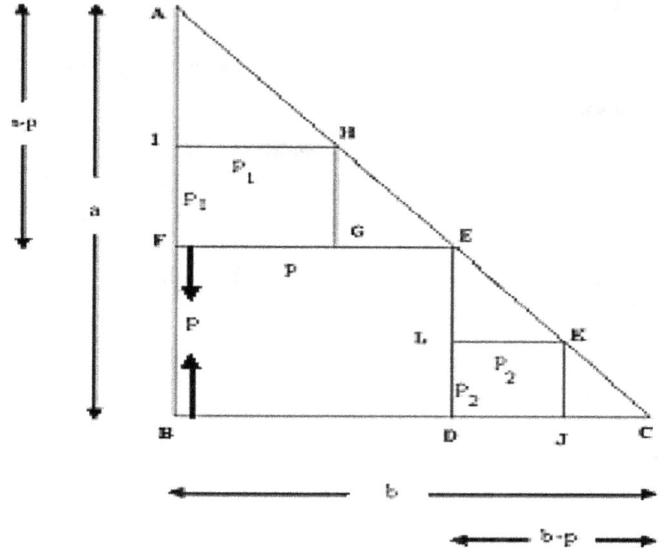

Figure 11.3 Geometrical interpretation of (11.17).

From the equation (11.18), it is easy to show that $p = \frac{ab}{a+b}$. Now, construct another two squares FGHI and DJKL with sides "p_1" and "p_2", respectively, as shown in Figure 11.3. Then $p_1 = \frac{p(a-p)}{a} = \frac{a^2b}{(a+b)^2}$ and $p_2 = \frac{p(b-p)}{b} = \frac{ab^2}{(a+b)^2}$. Now, $p_1 + p_2 = \frac{ab}{(a+b)^2}(a+b) = \frac{ab}{a+b} = p$. Therefore,

$$p_1 + p_2 = p. \tag{11.19}$$

In this construction, if we take $a = \frac{1}{m}$, $b = \frac{1}{n}$, then

$$p_1 = \frac{n}{(m+n)^2}, \tag{11.20}$$

$$p_2 = \frac{m}{(m+n)^2} \tag{11.21}$$

and

$$p = \frac{\frac{1}{m}\frac{1}{n}}{\frac{1}{m}+\frac{1}{n}} \tag{11.22}$$

Substituting the relations (11.20), (11.21), and (11.22) in (11.19), we get

$$\frac{1}{m+n} = \frac{\frac{1}{m}\frac{1}{n}}{\frac{1}{m}+\frac{1}{n}}.$$

Since $r(m) = \frac{k}{m}$ is a solution of the functional equation (11.17), the property (11.19) is satisfied by the functional equation (11.17). Hence, the functional equation (11.17) holds good in the above geometric construction.

11.6.2 AN APPLICATION OF EQUATION (11.17) TO RESISTANCES CONNECTED IN PARALLEL

We know the conductance of any material is reciprocal of its resistance. Consider two resistors R_1 and R_2 with resistances $\frac{1}{x}$ and $\frac{1}{y}$, respectively, connected in parallel as shown in Figure 11.4. Then, x and y are the conductances of the resistors R_1 and R_2, respectively, and hence, $x+y$ is the total circuit conductance. To find the equivalent resistance of loads wired in parallel, we use a mathematical formula known as the **"reciprocal formula"** which is given by

$$\frac{1}{\text{Total circuit conductance}} = \text{Total equivalent resistance of the parallel circuit.} \tag{11.23}$$

The reciprocal formula (11.23) satisfies the following algebraic identity:

$$\frac{1}{x+y} = \frac{\frac{1}{x}\frac{1}{y}}{\frac{1}{x}+\frac{1}{y}}$$

and hence, the functional equation (11.17) holds good in the above circuit.

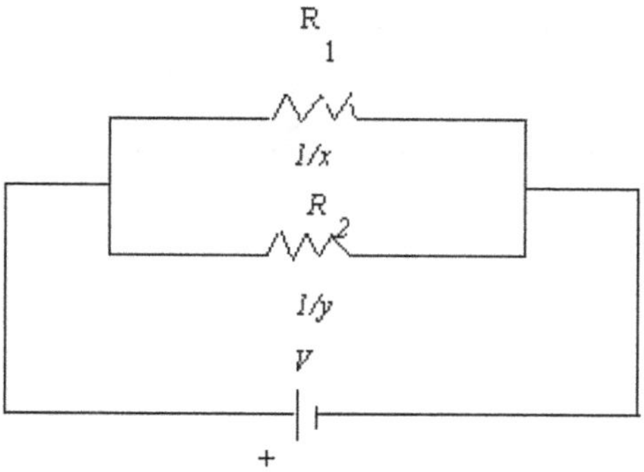

Figure 11.4 Couple of resistors connected in parallel.

11.7 APPLICATION OF RQD AND RQA FUNCTIONAL EQUATIONS

In this section, we introduce the following functional equation in several variables:

$$F\left(\frac{r_1+r_2}{2}\right) - F(r_1+r_2) = \frac{3F(r_1)F(r_2)}{F(r_1)+F(r_2)+2\sqrt{F(r_1)F(r_2)}}. \tag{11.24}$$

and

$$F\left(\frac{r_1+r_2}{2}\right) + F(r_1+r_2) = \frac{5F(r_1)F(r_2)}{F(r_1)+F(r_2)+2\sqrt{F(r_1)F(r_2)}}. \tag{11.25}$$

We establish the geometrical descriptions of functional equations (11.24) and (11.25) using Newton's law of gravitation.

Newton's law of universal gravitation states that a particle attracts every other particle in the universe with a force which is directly proportional to the product of their masses and inversely proportional to the square of the distance between their centers.

Let F be the force between the masses m_1 and m_2 denoted in Figure 11.5. Let G be the gravitational constant and r be the distance between the centres of the masses m_1 and m_2. Then, the force of attraction F between m_1 and m_2 is

$$F = G\frac{m_1 m_2}{r^2}.$$

Suppose both the above two objects are of unit mass, then the force of attraction between them is

$$F = \frac{G}{r^2}.$$

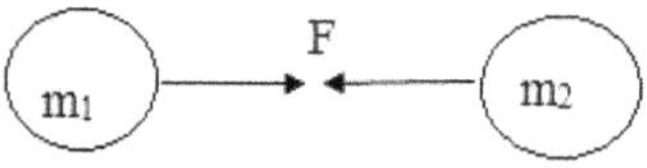

Figure 11.5 Force between two masses m_1 and m_2.

Now, consider three objects of unit mass with the following situation. Let r_1 be the distance between the objects 1 and 2. Let r_2 be the distance between the objects 2 and 3 (Figure 11.6).

The force of attraction between the objects 1 and 2 is

$$F(r_1) = \frac{G}{r_1^2}. \tag{11.26}$$

The force of attraction between the objects 2 and 3 is

$$F(r_2) = \frac{G}{r_2^2}. \tag{11.27}$$

The force of attraction between the objects 1 and 3 is

$$F(r_1+r_2) = \frac{G}{(r_1+r_2)^2}. \tag{11.28}$$

Using (11.26), (11.27), and (11.28), we have a relation as follows:

$$\begin{aligned} F(r_1+r_2) &= \frac{G}{r_1^2+r_2^2+2r_1r_2} \\ &= \frac{G}{r_1^2r_2^2\left(\frac{1}{r_1^2}+\frac{1}{r_2^2}+\frac{2}{r_1r_2}\right)} \\ &= \frac{\frac{G}{r_1^2r_2^2}}{\frac{1}{r_1^2}+\frac{1}{r_2^2}+\frac{2}{r_1r_2}} \\ &= \frac{F(r_1)F(r_2)}{F(r_1)+F(r_2)+2\sqrt{F(r_1)F(r_2)}}. \end{aligned} \tag{11.29}$$

It is not hard to verify that the reciprocal-quadratic function $F(r) = \frac{k}{r^2}$, where k is a constant, is a solution of (11.29). Thus, the reciprocal-quadratic functional equation (11.29) arises in the above physical phenomena.

If the distance between object 1 and object 3 is halved, that is, $\frac{r_1+r_2}{2}$, then the force of attraction between them becomes

$$F\left(\frac{r_1+r_2}{2}\right) = \frac{G}{\left(\frac{r_1+r_2}{2}\right)^2} = 4\frac{G}{(r_1+r_2)^2}.$$

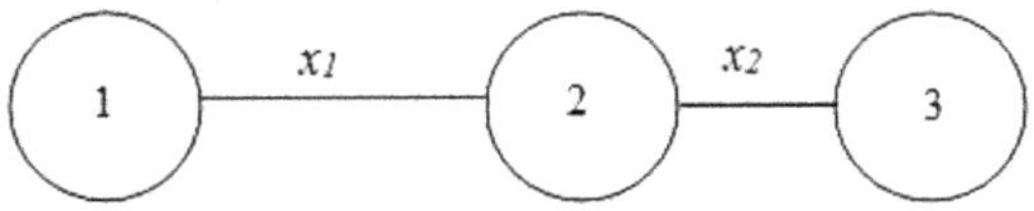

Figure 11.6 Forces when three unit masses in the same line.

The relation between the difference of the forces $F\left(\frac{r_1+r_2}{2}\right)$ and $F(r_1+r_2)$ and the forces between object 1 and object 2; and object 2 and object 3 can be modeled as a functional equation as follows:

$$F\left(\frac{r_1+r_2}{2}\right)-F(r_1+r_2)=\frac{3G}{(r_1+r_2)^2}$$
$$=\frac{3F(r_1)F(r_2)}{F(r_1)+F(r_2)+2\sqrt{F(r_1)F(r_2)}}. \quad (11.30)$$

We call the above functional equation (11.30) as reciprocal-quadratic difference functional (RQDF) equation in two variables r_1 and r_2.

Similarly, the relation between the sum of the forces $F\left(\frac{r_1+r_2}{2}\right)$ and $F(r_1+r_2)$ and the forces between the objects 1 and 2, and the objects 2 and 3 can be modeled as a functional equation as follows:

$$F\left(\frac{r_1+r_2}{2}\right)+F(r_1+r_2)=\frac{5G}{(r_1+r_2)^2}$$
$$=\frac{5F(r_1)F(r_2)}{F(r_1)+F(r_2)+2\sqrt{F(r_1)F(r_2)}}. \quad (11.31)$$

The above functional equation (11.31) is said to be reciprocal-quadratic adjoint functional (RQAF) Equation with a couple of variables r_1 and r_2.

11.8 APPLICATION OF OTHER MULTIPLICATIVE INVERSE FUNCTIONAL EQUATIONS

In this section, we discuss about various other multiplicative inverse functional equations and their applications.

11.8.1 MULTIPLICATIVE INVERSE SECOND POWER DIFFERENCE AND ADJOINT FUNCTIONAL EQUATIONS

Here, we deal with the following multiplicative inverse second power difference functional equation in several variables

$$m_i\left(\frac{1}{p}\sum_{\ell=1}^{p}v_\ell\right)-m_i\left(\sum_{\ell=1}^{p}v_\ell\right)=\frac{(p^2-1)\prod_{\ell=1}^{p}m_i(v_\ell)}{\left[\sqrt{\sum_{\ell=1}^{p}m_i(v_\ell)\prod_{m=1,j\neq\ell}^{p}m_i(v_m)}\right]^2} \quad (11.32)$$

and multiplicative inverse second power adjoint functional equation in several variables

$$m_i\left(\frac{1}{p}\sum_{\ell=1}^{p}v_\ell\right)+m_i\left(\sum_{\ell=1}^{p}v_\ell\right)=\frac{(p^2+1)\prod_{\ell=1}^{p}m_i(v_i)}{\left[\sqrt{\sum_{\ell=1}^{p}m_i(v_\ell)\prod_{m=1,m\neq\ell}^{p}r(v_m)}\right]^2}. \quad (11.33)$$

When $p = 2$, the above equations (11.32) and (11.33) are interpreted using Newton's law of gravitation. The left-hand side of (11.32) represents the difference of forces of attraction between two objects O_1 and O_2 and the force of attraction between the those objects when the distance is halved with an object in between them. The right-hand side of (11.32) relates with the force of attraction between the first two objects and second two objects. For detailed information, one can refer [3]. Other forms of multiplicative inverse second power functional equations are dealt in [1,16].

11.8.2 MULTIPLICATIVE INVERSE THIRD POWER FUNCTIONAL EQUATION

We consider the following multiplicative inverse third power functional equation:

$$R_c(u_1+u_2) = \frac{R_c(u_1)R_c(u_2)}{\left(R_c(u_1)^{\frac{1}{3}}+R_c(u_2)^{\frac{1}{3}}\right)^3}. \tag{11.34}$$

It is easy to check that the multiplicative inverse third power function $R_c(u) = \frac{c}{u^3}$ is a solution of (11.34). Equation (11.34) is associated with the well-known inverse cube law occuring in electromagnetism. The left-hand side of (11.34) denotes the magnetic field strength due to a dipole at a distance u_1+u_2. The right-hand side of (11.34) involves the magnetic fields strengths at the distances u_1 and u_2.

11.8.3 MULTIPLICATIVE INVERSE FOURTH POWER FUNCTIONAL EQUATION

We consider the following multiplicative inverse fourth power functional equation:

$$m_q(u+v) = \frac{m_q(u)m_q(v)}{\left[m_q(u)^{\frac{1}{4}}+m_q(v)^{\frac{1}{4}}\right]^4}. \tag{11.35}$$

One can easily verify that the multiplicative inverse fourth power function $m_q(u) = \frac{1}{u^4}$ is a solution of (11.35). The solution of (11.35) is associated with the intensity of scattered light and decay of radar energy in physics. According to Rayleigh's law of scattering, the intensity of scattered light is inversely proportional to the fourth power of its wavelength, that is, $I \propto \frac{1}{\lambda^4}$ or $I = \frac{c}{\lambda^4}$, where I is the intensity, λ is wavelength of the scattered light, and c is a constant. The energy of radar decays in inverse proportion to the fourth power of the distances to the targets. Hence, the intensity of scattered light and decay of radar energy could be studied through modeling equation (11.2).

11.8.4 MULTIPLICATIVE INVERSE QUINTIC FUNCTIONAL EQUATION

Now, we introduce a new multiplicative inverse quintic functional equation:

$$m_Q(x_1+x_2) = \frac{m_Q(x_1)m_Q(x_2)}{\left[m_Q(x_1)^{\frac{1}{5}}+m_Q(x_2)^{\frac{1}{5}}\right]^5}. \tag{11.36}$$

It is easy to substantiate that the multiplicative inverse quintic function $m_Q(x) = \frac{1}{x^5}$ is a solution of (11.36). The solution of (11.36) is linked with various hypotheses arising in physics, chemistry, and mechanics.

1. If an object moves on a circular path under the impact of an attractive central force which is directing the object towards a point on the circle, then this central force is inversely proportional to the fifth power of distance. Hence, the central force is a function of reciprocal fifth power of the distance.
2. In wave spectra theory, the rate of energy decay for the high-frequency part is inversely proportional to the fifth power of frequency. In this case, the rate of energy decay is a function of reciprocal fifth power of frequency.
3. In kinetic theory of gases, molecules are conceived as portions of matter moving as single bodies, repelling by a force inversely proportional to the fifth power of the distance between their centres of gravity. The inverse fifth power law is consistent with gas viscosity in terms of the time relaxation of stresses in the gas, a concept related to the macroscopic elastic properties of the gas.
4. Turbulent flow occurs in the aortic arch and in the large airways of the tracheobronchial tree down to the fifth bronchial division. The pressure drop is inversely proportional to the fifth power of radius of the airway.

11.8.5 MULTIPLICATIVE INVERSE FUNCTIONAL EQUATION INVOLVING TWO VARIABLES

We study about a multiplicative inverse functional equation involving two variables of the form:

$$r(u_1+u_2, v_1+v_2) = \frac{\prod_{j=1}^{2}\prod_{k=1}^{2} r(u_j, v_k)}{\sum_{j=1}^{2}\sum_{k=1}^{2}\left(\frac{1}{r(u_j,v_k)}\prod_{m=1}^{2}\prod_{n=1}^{2} r(u_m, v_n)\right)}. \tag{11.37}$$

It can be verified that the function $r(u,v) = \frac{1}{uv}$ is a solution of (11.37). Some system designs occurring in physics, such as to find the rigidity moduli in several situations and pressures of material in various occasions, are interpreted through (11.37). More details about (11.37) are available in [21].

11.8.6 SYSTEM OF MULTIPLICATIVE INVERSE FUNCTIONAL EQUATIONS WITH THREE VARIABLES

We discuss about the following system of multiplicative inverse functional equations involving three variables:

$$\left.\begin{aligned} T_r(u_1+u_2,v,w) &= \frac{T_r(u_1,v,w)T_r(u_2,v,w)}{T_r(u_1,v,w)+T_r(u_2,v,w)},\\ T_r(u,v_1+v_2,w) &= \frac{T_r(u,v_1,w)T_r(u,v_2,w)}{T_r(u,v_1,w)+T_r(u,v_2,w)},\\ T_r(u,v,w_1+w_2) &= \frac{T_r(u,v,w_1)T_r(u,v,w_2)}{T_r(u,v,w_1)+T_r(u,v,w_2)}. \end{aligned}\right\} \tag{11.38}$$

It can be verified that the multiplicative inverse function $T_r(u,v,w) = \frac{c}{uvw}$ is a solution of the system (11.38), where c is a constant. The basic properties like density and volume of a cuboid shaped sponge are studied through this system (11.38). The density of a cuboid shaped sponge represents left hand side of any member of the system (11.38) which is equal the half of the harmonic mean of the volumes of the two sub cuboid shaped sponges when original sponge is divided along its length. Further features of (11.38) are dealt in [20].

11.9 APPLICATIONS OF FUNCTIONAL EQUATIONS IN OTHER FIELDS

- The parallelogram law of forces is obtained using the functional equation $h(u+v)+h(u-v) = 2h(u)h(v)$. Further, this equation is used to describe circular functions and to illustrate their relation to spherical trigonometry.
- Functional equations are used by Einstein in the theory of relativity to demonstrate the light signaling process in the space-time coordinates. [Bulletin of the American Mathematical Society Vol. XXVI, Oct 1919 to July 1920, pp. 26–34].
- Ramanujan adopted functional equations to describe Bernoulli's number of negative index.
- The system of functional equations

  $$f(u+v) = \frac{f(u)+\{[g(u)]^2-[f(u)]^2\}f(v)}{1-f(u)f(v)}; \qquad f(u+v) = \frac{f(u)f(v)}{1-g(u)g(v)}$$

  has significant role in the optics and in the probability theory.
- The system of functional equations

  i. $\phi(u) = \sup_{v\in D} H(x,y,f(T(u,v)))$, for all $u \in S$;

  ii. $\phi(u) = \sup_{v\in D}\{h(u,v)+G(u,v,g(T(u,v)))\}$, for all $u \in S$;

  iii. $\varphi(u) = \sup_{v\in D}\{h(u,v)+F(u,v,\phi(T(u,v)))\}$, for all $u \in S$

  is used in dynamic programming of multistage decision processes.

- Several other types of multiplicative inverse functional equations were considered to obtain their solution, to prove their fundamental stability results associated with Ulam stability theory and their role in various fields. More information are available in [2,4,6,8,9,15,18,19,22–26].

11.10 OPEN PROBLEMS

1. Is it possible to form a mixed-type functional equation arising from additive and multiplicative inverse functions? If possible, form such an equation.
2. Form a functional equation whose solution is a rational function of the form $\frac{x}{x^2+1}$.
3. Find the solution of $f(x)f(y) = f(xy) + f\left(\frac{x}{y}\right)$.

BIBLIOGRAPHY

[1] A. Bodaghi, P. Narasimman, J. M. Rassias and K. Ravi, Ulam stability of the reciprocal functional equation in non-Archimedean fields, *Acta Mathematica Universitatis Comenianae,* LXXXV, 1 (2016), 113–124.

[2] A. Bodaghi and B. V. Senthil Kumar, Estimation of inexact reciprocal-quintic and reciprocal-sextic functional equations, *Mathematica,* 59 (82), No 1–2 (2017), 3–14.

[3] H. Dutta and B. V. Senthil Kumar, Geometrical elucidations and approximation of some functional equations in numerous variables, *Proc. Indian Natn. Sci. Acad.* 85, No. 3, (2019), 603–611.

[4] H. Dutta and B. V. Senthil Kumar, Classical stabilities of an inverse fourth power functional equation, *J. Interdisciplinary Math.* 22 (7) (2019), 1061–1070.

[5] E.-S. El-Hady, W. Forg-Rob and M. Mahmoudi, On a two-variable functional equation arising from databases, *WSEAS Transactions on Mathematics,* 14 (2015), 265–270.

[6] E. Gupta, R. Chugh and B. V. Senthil Kumar, Non-Archimedean approximation of MIQD and MIQA functional equations, *Int. J. Appl. Engg. Research,* 14(6) (2019), 1313–1318.

[7] P. Kannappan, Application of Cauchy's equation in combinatorics and genetics, *Mathware & Soft Computing,* 8 (2001), 61–64.

[8] S. O. Kim, B. V. Senthil Kumar and A. Bodaghi, Approximation on the reciprocal-cubic and reciprocal-quartic functional equations in non-Archimedean fields, *Advances in Difference Equ.* 77 (2017), 1–12.

[9] S. Merlin and B. V. Senthil Kumar, Approximations and applications of a reciprocal fifth power mapping, *Int. J. Engg. Advanced Tech.* 9(1S4) (2019), 983–986.

[10] R. Murali, P. Divyakumari and Sandra Pinelas, Generalized additive functional equation in digital spatial image crypto techniques system, *IOSR J. Math.,* 15 (4) (2019), 69–76.

[11] R. Murali, P. Divyakumari and H. Dutta, Functional equation and its modular stability with and without Δ_p-condition, *Filomat,* 34(3) (2020), 1–12.

[12] P. Narasimman, K. Ravi and S. Pinelas, Stability of Pythagorean mean functional equation, *Global J. Math.* 4 (1) (2015), 398–411.

[13] K. Ravi and B. V. Senthil Kumar, Stability and geometrical interpretation of reciprocal type functional equation, *Asian Journal of Current Engineering and Maths,* 1 (2012), 300–304.

[14] K. Ravi, R. Jamuna and R. Dhinesh Kumar, Functional equations in wireless sensor networks, *Bull. Marathwada Math. Soc.* 14 (2) (2013), 86–93.

[15] K. Ravi, J. M. Rassias and B. V. Senthil Kumar, Stability of reciprocal difference and adjoint functional equations in paranormed spaces: Direct and fixed point methods, *Functional Anal. Approx. Comp.* 5(1) (2013), 57–72.

[16] K. Ravi, J. M.Rassias, S. Pinelas and P. Narasimman, The Stability of a generalized radical reciprocal quadratic functional equation in Felbin's space, *PanAmerican Math. J.* 24 (1) (2014), 75–92.

[17] P. K. Sahoo and P. Kannappan, *Introduction to Functional Equations*, CRC Press, Taylor & Francis Group, 2011.

[18] B. V. Senthil Kumar, Ashish Kumar and G. Suresh, Functional equations related to spatial filtering in image enhancement, *Int. J. Control Theory Appl.* 9(28) (2016), 555–564.

[19] B. V. Senthil Kumar and H. Dutta, Non-Archimedean stability of a generalized reciprocal-quadratic functional equation in several variables by direct and fixed point methods, *Filomat,* 32(9) (2018), 3199–3209.

[20] B. V. Senthil Kumar, H. Dutta and S. Sabarinathan, Approximation of a system of rational functional equations of three variables, *Int. J. Appl. Comput. Math.* 5 (2019),1–16.

[21] B. V. Senthil Kumar and H. Dutta, Fuzzy stability of a rational functional equation and its relevance to system design, *Int. J. General Syst.,* 48, No. 2 (2019), 157–169.

[22] B. V. Senthil Kumar and H. Dutta, Approximation of multiplicative inverse undecic and duodecic functional equations, *Math. Meth. Appl. Sci.* 42 (2019), 1073–1081.

[23] B. V. Senthil Kumar, J. M. Rassias and S. Sabarinathan, Stabilities of various multiplicative inverse functional equations, *Tbilisi Math. J.* 12(4) (2019), 15–28.

[24] B. V. Senthil Kumar, H. Dutta and K. Al-Shaqsi, On a functional equation arising from subcontrary mean and its pertinences, *Advances in Intelligent Systems and Computing 1111, 4th International Conference on Computational Mathematics and Engineering Sciences (CMES2019)*, 2020, 241–247.

[25] B. V. Senthil Kumar, H. Dutta and S. Sabarinathan, Fuzzy approximations of a multiplicative inverse cubic functional equation, *Soft Comput.,* 24 (2020), 13285–13292, https://doi.org/10.1007/s00500-020-04741-x.

[26] B.V. Senthil Kumar and A. Bodaghi, A fixed point approach to generalized Hyers-Ulam stability of Jensen type reciprocal functional equation, *Boletim da Sociedade Paranaense de Matematica,* 38(3) (2020), 125–132.

12 Unified-Type Nondifferentiable Second-Order Symmetric Duality Results over Arbitrary Cones

Ramu Dubey
J.C. Bose University of Science and Technology

Vishnu Narayan Mishra
Indira Gandhi National Tribal University

CONTENTS

12.1 INTRODUCTION

Multiobjective-type programming problems are common in mathematical modeling of realistic phenomenon for a greatly spectrum of utilization. Optimality conditions given by Pareto that noticeable to be the physical evolvement of the optimization of a single purpose to the contemplation of several purposes. In his perusal, Hanson [9] has mentioned an illustration which exhibits the application of the second-order duality in certain apart frame of mind. Mathematical programming is the branch of operational research, which deals with the method of solving problems by findings the extrema of functions (defined by linear or nonlinear constraints) in finite dimensional vector space. In the development of mathematical programming, concept of optimality and concept of duality are very important for solving the problems. It has found its way into all branches of science and engineering due to its wide range of applications.

Duality plays an important role in nonlinear programming. Indeed, when the solution of a problem poses some difficulties, we shall see the solution of its dual problem provides some valuable information about the original problem. Symmetric duality in nonlinear programming deals with the situation where dual of the dual is primal duality in mathematical programming, which is not only used in many theoretical and computational developments in mathematical programming itself but also used in economics, control theory, business problems, and other diverse fields. Mangasarian [17] introduced the concept of second- and higher-order dualities for nonlinear problems. The study of higher-order duality is significant due to the computational advantage over the first-order duality as it provides tighter bounds for the value of the objective function when approximations are used.

In recent past, several definitions such as nonsmooth univex, nonsmooth quasiunivex, and nonsmooth pseudoinvex functions have been introduced by Xianjun [29]. By introducing these new concepts, sufficient optimality conditions for a nonsmooth multiobjective problem were obtained, and, a fortiori, weak and strong duality results were established for a Mond–Weir-type multiobjective dual program. Mond and Zhang [18] obtained duality results for various higher-order dual problems under higher-order invexity assumptions. Chandra et al. [2] and Yang et al. [30] discussed a mixed symmetric dual formulation for a nonlinear programming problem and for a class of nondifferentiable nonlinear programming problems, respectively. Later on, Chen [5] studied duality relations for Mond–Weir-type multiobjective higher-order symmetric dual programs under F-convexity assumptions.

In this chapter, we have formulated a new mixed-type second-order nondifferentiable symmetric duality in scalar-objective programming problem. In literature, we have discussed the results either Wolfe- or Mond–Weir-type dual or separately, whereas in this, we have combined result over one model over arbitrary cones. The duality theorems are proved for these programs over arbitrary cones under bonvexity and pseudobonvexity with respect to η assumptions.

12.2 LITERATURE REVIEW

The following convention for vector inequalities will be used: If a, b $\in R^n$, then
$a \geqq b \Leftrightarrow a_i \geqq b_i, i = 1,2,...,n;$
$a \geqslant b \Leftrightarrow a \geqq b$ and $a \neq b;$
$a > b \Leftrightarrow a_i > b_i, i = 1,2,...,n.$

A problem in which the dual of the dual is again the primal problem is called symmetric problem. Linear programming problems are always symmetric, but it is not in general true for nonlinear problems. This concept in nonlinear problems was first introduced by Dorn [8] for quadratic functions. Later on, Dantzig et al. [6] extended this notion for a general nonlinear problem and derived duality results under convexity/concavity assumptions. The problem considered in Dantzig et al. [6] is as follows:

(PP1) Minimize $F = k(x,y) - y^T \nabla_y k(x,y)$
subject to $\nabla_y k(x,y) \leqq 0,$
$x \geqq 0,\ y \geqq 0.$

(DP1) Minimize $G = k(x,y) - x^T \nabla_x k(x,y)$
subject to $\nabla_x k(x,y) \geqq 0,$
$x \geqq 0,\ y \geqq 0.$

Mond and Weir [19] considered the following primal-dual pair and established duality theorems under pseudoconvexity/pseudoconcavity assumptions:

(PP2) Minimize $H = k(x,y)$
subject to $\nabla_y k(x,y) \leqq 0,$
$y^T \nabla_y k(x,y) \geqq 0,$
$x \geqq 0.$

(DP2) Minimize $T = k(x,y)$
subject to $\nabla_x k(x,y) \geqq 0,$
$x^T \nabla_x k(x,y) \leqq 0,$
$y \geqq 0.$

Bazaraa and Goode [1] further extended the problem of Dantzig et al. [6] over arbitrary cones and proved duality relations under convexity assumptions. Later on, Chandra et al. [3] constructed a pair of nondifferentiable symmetric dual problems and discussed duality theorems. After that, Mond and Schecter [20] studied the following nondifferentiable symmetric dual programs:

(PP3) Minimize $\varphi(x,y) + S(x|C) - y^T \nabla_y \varphi(x,y)$
subject to $\nabla_y \varphi(x,y) - z \leqq 0,$
$x \geqq 0,\ z \in D.$

(DP3) Maximize $\varphi(u,v) - S(v|D) - u^T \nabla_x \varphi(u,v)$
subject to $\nabla_x \varphi(u,v) + \omega \geqq 0,$
$v \geqq 0,\ \omega \in C,$

where $\varphi : R^n \times R^m \to R$ is a differentiable function, and C and D are compact convex sets in R^n and R^m, respectively. Further, they obtained the following duality results:

Theorem 12.1 (Weak duality). Suppose that φ is convex in x for fixed y and is concave in y for fixed x. Then, inf (PP3)$\geqq$ sup (DP3).

Theorem 12.2 (Strong duality). Suppose that φ is twice continuously differentiable and that $(\bar{x}, \bar{y}, \bar{z})$ is optimal for problem (PP3). Suppose also that the Hessian

$(\nabla_{xx})\varphi(\bar{x},\bar{y})$ is nonsingular. Then, $\exists\ \alpha$ such that $(u,v,w)=(\bar{x},\bar{y},\alpha)$ is feasible for problem (DP3), and the objective function of problem (DP3) has the value min (PP3) at this feasible solution.
Motivated by [1,19], Chandra and Kumar [4] considered the following Mond–Weir-type symmetric model over arbitrary cones:

(PP4) Minimize $\Psi = \psi(x,y)$

$$\begin{aligned} \text{subject to} \quad & \nabla_y \psi(x,y) \in C_2^*, \\ & y^T \nabla_y \psi(x,y) \geqq 0, \\ & x \in C_1, \end{aligned}$$

(DP4) Maximize $\Omega = \psi(u,v)$

$$\begin{aligned} \text{subject to} \quad & -\nabla_x \psi(u,v) \in C_1^*, \\ & u^T \nabla_x \psi(u,v) \leqq 0, \\ & v \in C_2, \end{aligned}$$

where $\phi \neq C_i \subseteq R^n$ and $\phi \neq C_2 \subseteq R^m$ are closed convex cones. For $i=1,2$, C_i^* is the positive polar of C_i. $S_1 \subseteq R^n$ and $S_2 \subseteq R^m$ are open sets such that $C_1 \times C_2 \subset S_1 \times S_2$ and $H : S_1 \times S_2 \to R$ is a twice differentiable function.

Further, they derived duality results under generalized convexity assumptions. Recently, Mandal and Nahak [15] introduced a new type of invex functions called $(p,r)-\rho-(\eta,\theta)$-invex and proved weak, strong, and converse duality relations for the following multiobjective symmetric primal-dual problems:

(PP5) Minimize $f(x,y)$

$$\begin{aligned} \text{subject to} \quad & \nabla_y f(x,y) \in C_2^*, \\ & y^T \nabla_y f(x,y) \geqq 0, \\ & x \in C_1. \end{aligned}$$

(DP5) Minimize $f(u,v)$

$$\begin{aligned} \text{subject to} \quad & -\nabla_x f(u,v) \in C_1^*, \\ & u^T \nabla_x f(u,v) \leqq 0, \\ & v \in C_2, \end{aligned}$$

where f is a twice differentiable function. Further, they also established duality results for multiobjective symmetric programming problem over arbitrary cones under $(p,r)-\rho-(\eta,\theta)$-invexity assumptions. Many researchers have worked related to the symmetric problems over arbitrary cones [23,25,27]. Recently, Gulati et al. [14] considered the multiobjective nondifferentiable symmetric dual problems over arbitrary cones and derived duality results under K-preinvexity/K- convexity/K-pseudoinvexity assumptions.

The main importance of the second-order duality over the first-order duality is that it gives more closer bounds whenever numerical calculations/approximations are involved. This will also help to provide suitable stopping technique for a numerical algorithm. Mangasarian [16] was the first who has formulated the second-order dual for a general nonlinear program and then obtained duality theorems under generalized convexity assumptions. Kim and Yun [24] considered the following Wolfe-type multiobjective symmetric dual pair:

Primal Problem (PP6)
minimize $\Psi_P(x,y,\lambda,p) = \phi(x,y) - (y^T\nabla_y(\lambda^T\phi)(x,y))e - (y^T\nabla_{yy}(\lambda^T\phi)(x,y)p)e$

subject to

$$\nabla_y(\lambda^T\phi)(x,y) + \nabla_{yy}(\lambda^T\phi)(x,y)p \leqq 0,\ x,p \geqq 0,$$
$$\lambda > 0,\ \lambda^T e = 1.$$

Dual Problem (DP6)
maximize $\Psi_D(u,v,\lambda,q) = \phi(u,v) - (u^T\nabla_x(\lambda^T\phi)(u,v))e - (u^T\nabla_{xx}(\lambda^T\phi)(u,v)p)e$

subject to

$$-\nabla_x(\lambda^T\phi)(u,v) - \nabla_{xx}(\lambda^T\phi)(u,v)p \geqq 0,\ v,q \geqq 0,$$
$$\lambda > 0,\ \lambda^T e = 1,$$

and established duality theorems under convexity/concavity assumptions. Later on, Devi [7] constructed a pair of second-order symmetric dual, and corresponding duality results are established under η-bonvexity/η-pseudobonvexity conditions. Hou and Yang [10] formulated a nondifferentiable Mond–Weir-type symmetric model. After that, Suneja et al. [28] constructed a pair of Mond–Weir-type multiobjective second-order symmetric dual program and established duality theorems under η-bonvexity/η-pseudobonvexity assumptions. After that, Yang et al. [32] formulated a pair of Mond–Weir-type nondifferentiable second-order symmetric dual program and proved appropriate duality theorems under F-convexity assumptions. The work in Ref. [10] was further extended to multiobjective in Yang et al. [31] under generalized convexity assumptions. Kim et al. [27] constructed and proved usual duality theorems under second-order invexity assumptions for the following two multiobjective nondifferentiable symmetric pairs over arbitrary cones $C_1 \subseteq R^n$ and $C_2 \subseteq R^m$:

Primal Problem (PP7)
minimize $\Phi(x,y,\lambda,z,p) = \phi(x,y) + S(x|D) - (y^T w)e - \frac{1}{2}(p^T\nabla_{yy}(\lambda^T\phi)(x,y)p)e$

subject to

$$-[\nabla_y(\lambda^T\phi)(x,y) - w + \nabla_{yy}(\lambda^T\phi)(x,y)p] \in C_2^*,$$
$$y^T[\nabla_y(\lambda^T\phi)(x,y) - w + \nabla_{yy}(\lambda^T\phi)(x,y)p] \leqq 0,$$
$$x \in C_1,\ w \in E_i,\ \lambda \in K^*,\ \lambda^T e = 1,\ e \in \text{int}K.$$

Dual Problem (DP7)
minimize $\Psi(u,v,\lambda,z,r) = \phi(u,v) - S(v|E) + (u^T z)e - \frac{1}{2}(r^T\nabla_{xx}(\lambda^T\phi)(u,v)r)e$

subject to

$$[\nabla_x(\lambda^T\phi)(x,y) + z + \nabla_{xx}(\lambda^T\phi)(u,v)r] \in C_1^*,$$
$$u^T[\nabla_x(\lambda^T\phi)(u,v) + z + \nabla_{xx}(\lambda^T\phi)(u,v)r] \geqq 0,$$
$$v \in C_2,\ z \in D_i,\ \lambda \in K^*,\ \lambda^T e = 1,\ e \in \text{int}K.$$

Primal Problem (PP8)

minimize $\Psi_P(x,y,\lambda,z,p) = \phi(x,y) - (y^T\nabla_x(\lambda^T\phi)(x,y))e - (y^T\nabla_x(\lambda^T\phi)(x,y)p)e$

$$-\frac{1}{2}(p^T\nabla_{yy}(\lambda^T\phi)(x,y)p)e$$

subject to

$$-[\nabla_y(\lambda^T\phi)(x,y) - w + \nabla_{yy}(\lambda^T\phi)(x,y)p] \in C_2^*,$$

$$x \in C_1,\ w \in E_i,\ \lambda \in K^*,\ \lambda^T e = 1,\ e \in \text{int}K.$$

Dual Problem (DP8)

maximize $\Psi_D(u,v,\lambda,z,r) = \phi(u,v) - (u^T\nabla_x(\lambda^T\phi)(u,v))e - (u^T\nabla_x(\lambda^T\phi)(x,y)r)e$

$$-\frac{1}{2}(u^T\nabla_{xx}(\lambda^T\phi)(u,v)r)e$$

subject to

$$[\nabla_x(\lambda^T\phi)(u,v) + z + \nabla_{xx}(\lambda^T\phi)(u,v)r] \in C_1^*,$$

$$v \in C_2,\ z \in D_i,\ \lambda \in K^*,\ \lambda^T e = 1,\ e \in \text{int}K,$$

where $\phi : R^n \times R^m \to R^k$ is a thrice differentiable function. K is a closed convex cone in R^k such that int$K \neq \phi$ and $R^k_+ \subseteq K$. $r,z \in R^n,\ p,w \in R^m,\ e = (1,\ldots,1)^T \in R^k,\ D_i$, and $E_i,\ i \in K$ are compact and convex sets in R^n and R^m, respectively.

Gulati et al. [11] formulated the second-order multiobjective Mond–Weir-type second-order symmetric duality model over arbitrary cones and discussed duality theorems under η-bonvexity/η-pseudobonvexity assumptions. Gulati and Geeta [14] studied nondifferentiable multiobjective Mond–Weir-type symmetric duality model in which the objective function is optimized over cones and derived duality relations under second-order K-F-convexity/K-η-bonvexity/K-η-pseudobonvexity assumptions.

The concept of higher-order duality was first introduced by Mangasarian [16]. Higher-order duality is more significant over second-order duality since it gives more tighter bounds due to the presence of more parameters. It may provide a lower bound to the infimum of a primal optimization problem when it is difficult to find a feasible solution for the first-order dual. This makes the study of second- and higher-order dual interesting and useful. Chen [5] presented the Mond–Weir-type higher-order symmetric duality for multiobjective nondifferentiable programming problems and derived duality results under higher-order F-convexity. Further, Gulati and Gupta [12] formulated and discussed the results for the following Wolfe-type nondifferentiable symmetric dual pair:

Primal Problem (PP9)

Minimize $\Omega(x,y,p) = f(x,y) + S(x|C) + h(x,y,p) - p^T\nabla_p h(x,y,p) - y^T\nabla_y f(x,y)$
$\qquad - y^T\nabla_p h(x,y,p)$

subject to

$$\nabla_y f(x,y) - z + \nabla_p h(x,y,p) \leq 0.$$

$$z \in D.$$

Dual Problem (DP9)

Maximize $\Upsilon(u,v,r) = f(u,v) - S(v|D) + g(u,v,r) - r^T\nabla_r g(u,v,r) - u^T\nabla_x f(u,v) - u^T\nabla_r g(u,v,r)$

subject to

$$\nabla_x f(u,v) + w + \nabla_r g(u,v,r) \geq 0,$$
$$w \in C,$$

where $f : R^n \times R^m \mapsto R$, $g : R^n \times R^m \times R^n \mapsto R$ and $h : R^n \times R^m \times R^m \to R$ are differentiable functions.

Gulati and Gupta [13] proved duality results for Mond–Weir- and Wolfe-type models under generalized existing convexity assumptions over arbitrary cones constraints. After that, Kassem [21] formulated a higher-order symmetric primal-dual problem and discussed duality results under cone-invexity assumptions. Later on, Kassem [22] considered the symmetric dual problems over arbitrary cones and established duality relations under pseudoinvexity assumptions. Ying [33] introduced the following new class of higher-order multiobjective Mond–Weir-type dual pair over arbitrary cones:

(PP10) K-minimize $\left(f(x,y) + h(x,y,p) - p^T\nabla_p h(x,y,p)\right)$

subject to

$$-\sum_{i=1}^{k}\lambda_i[\nabla_y f_i(x,y) + \nabla_p h_i(x,y,p)] \in C_2^*,$$

$$y^T\sum_{i=1}^{k}\lambda_i[\nabla_y f_i(x,y) + \nabla_p h_i(x,y,p)] \geqq 0,$$

$$\lambda \in \text{int}K^*,\ x \in C_1.$$

(DP10) K-maximize $\left(f(u,v) + g(u,v,q) - q^T\nabla_q g(u,v,q)\right)$

subject to

$$\sum_{i=1}^{k}\lambda_i[\nabla_x f_i(u,v) + \nabla_q g_i(u,v,q)] \in C_1^*,$$

$$u^T\sum_{i=1}^{k}\lambda_i[\nabla_x f_i(u,v) + w_i + \nabla_q g_i(u,v,q)] \leqq 0,$$

$$\lambda \in \text{int}K^*,\ v \in C_2,$$

where $f : R^n \times R^m \to R^k$, $h : R^n \times R^m \times R^m \to R^k$ and $g : R^n \times R^m \times R^n \to R^k$ are twice differentiable functions and established duality theorems under K-pseudoinvex/strongly K-pseudoinvex. Recently, Padhan and Nahak [26] discussed a higher-order symmetric duality programming problem and derived duality results under higher-order generalized invexity assumptions.

12.3 PRELIMINARIES AND DEFINITIONS

We consider the following scalar-objective programming problem:

(P) Minimize $F(x)$,

$$x \in X$$

where $X \subseteq R^{n+m}$ and $F : X \to R$.

12.3.1 DEFINITION

Let C be a nonempty compact convex set in R^n. The support function $s(x|C)$ of C is defined by

$$s(x|C) = \max\{x^T y : y \in C\}.$$

The subdifferential of $s(x|C)$ is given by

$$\partial s(x|C) = \{z \in C : z^T x = s(x|C)\}.$$

For any convex set $S \subset R^n$, the normal cone to S at a point $x \in S$ is defined by

$$N_S(x) = \{y \in R^n : y^T(z - x) \leqq 0 \text{ for all } z \in S\}.$$

It is readily verified that for a compact convex set E, y is in $N_E(x)$ if and only if

$$s(y|E) = x^T y.$$

Remark 12.1. A support function $\pi_A : R^n \mapsto R$ of a nonempty closed convex set A in R^n is given by

$$\pi_A(x) = \sup\{x \,.\, a : a \in A\}, \quad x \in R^n.$$

Its interpretation is most intuitive when x is a unit vector: by definition, A is contained in the closed half space

$$\{y \in R^n : y \,.\, x \leqq \pi_A(x)\},$$

and there is at least one point of A in the boundary

$$H(x) = \{y \in R^n : y \,.\, x = \pi_A(x)\}$$

of this half space. The hyperplane $H(x)$ is therefore called a supporting hyperplane with exterior (or outer) unit normal vector x. The word exterior is important here, as the orientation of x plays a role and the set $H(x)$ is in general different from $H(-x)$. Now π_A is the (signed) distance of $H(x)$ from the origin.

Example 12.1. If A is a line segment through the origin with end points $-\alpha$ and α, then $\pi_A(x) = |x \,.\, \alpha|$.

12.3.2 DEFINITION

The positive polar cone P^* of a cone P is defined by

$$P^* = \{y \in R^p : x^T y \geqq 0, \forall x \in P\}$$

12.3.3 DEFINITION

A function $\Phi : R^n \mapsto R$ is said to be bonvex at $u \in R^n$ with respect to $\eta : R^n \times R^n \mapsto R^n$, if for all $(x,p) \in R^n \times R^n$,

$$\Phi(x) - \Phi(u) + \frac{1}{2} p^T \nabla_{xx}\Phi(x)p \geqq \eta^T(x,u)\{\nabla_x\Phi(u) + \nabla_{xx}\Phi(x)p\}.$$

Example 12.2 Let $X = [0,2] \subseteq R$, $n = m = 1$ and $k = 1$. Consider the function $\Phi : X \to R$, which is given by

$$\Phi(x) = \left(\frac{e^x - e^{-x}}{2}\right)^3$$

and $\eta(x,u) = x^2u^2 + u$.
We have to claim that Φ is bonvex at $u \in X$ with respect to η. For this, it is sufficient to prove that the following expression is nonnegative, i.e.,

$$\Upsilon = \Phi(x) - \Phi(u) + \frac{1}{2}\nabla_{xx} p^T \Phi(x)p - \eta^T(x,u)\{\nabla_x\Phi(u) + \nabla_{xx}\Phi(x)p\}. \geqq 0.$$

Substituting the values of Φ and η in the above expression, we have

$$\begin{aligned}\Upsilon = & \left(\frac{e^x - e^{-x}}{2}\right)^3 - \left(\frac{e^u - e^{-u}}{2}\right)^3 + \frac{1}{2}p^T\left\{6\left(\frac{e^u - e^{-u}}{2}\right)\left(\frac{e^u + e^{-u}}{2}\right)^2\right. \\ & \left. + 3\left(\frac{e^u - e^{-u}}{2}\right)^3\right\}p - (x^2u^2 + u)\left[3\left(\frac{e^u - e^{-u}}{2}\right)^2\left(\frac{e^u + e^{-u}}{2}\right)\right. \\ & \left. + \left\{6\left(\frac{e^u - e^{-u}}{2}\right)\left(\frac{e^u + e^{-u}}{2}\right)^2 + 3\left(\frac{e^u - e^{-u}}{2}\right)^3\right\}p\right].\end{aligned}$$

Simplifying the above equation at the point $u = 0 \in X$, we obtain

$$\Upsilon = \left(\frac{e^x - e^{-x}}{2}\right)^3 \quad \forall\, x \in X.$$

It is clear that $\Upsilon \geqq 0$, $\forall\, x \in X$. Therefore, Φ is bonvex at $u = 0 \in X$ with respect to η.

12.3.4 DEFINITION

A function $\Phi : R^n \mapsto R$ is pseudobonvex at $u \in R^n$ with respect to $\eta : R^n \times R^n \mapsto R^n$, if for all $(x,p) \in R^n \times R^n$,

$$\eta^T(x,u)\{\nabla_x\Phi(u) + \nabla_{xx}\Phi(x)p\} \geqq 0 \Rightarrow \Phi(x) - \Phi(u) + \frac{1}{2}p^T\nabla_{xx}\Phi(x)p \geqq 0.$$

12.4 NONDIFFERENTIABLE SECOND-ORDER MIXED-TYPE SYMMETRIC DUALITY MODEL OVER ARBITRARY CONES

For N = $\{1,2,3, \ldots,n\}$ and M = $\{1,2,3, \ldots,m\}$, let us assume $J_1 \subset N$, $K_1 \subset M$ and $J_2 = N \setminus J_1$ and $K_2 = M \setminus K_1$, where $|J_1|$ denotes the number of elements in the set J_1. The other numbers $|J_2|$, $|K_1|$ and $|K_2|$ are defined similarly. Notice that if $J_1 = \varnothing$, then $J_2 = N$, then $R^{|J_1|}$ is zero-dimensional Euclidean space and $R^{|J_2|}$ is n-dimensional Euclidean space. It is clear that any $x \in R^n$ can be written as $x = \{x_1, x_2\}$, $x_1 \in R^{|J_1|}$, $x_2 \in R^{|J_2|}$. Similarly, any $y \in R^m$ can be written as $y = \{y_1, y_2\}$, $y_1 \in R^{|K_1|}$, $y_2 \in R^{|K_2|}$. Let

i. $f_1 : R^{|J_1|} \times R^{|K_1|} \to R$,
ii $f_2 : R^{|J_2|} \times R^{|K_2|} \to R$.

In this section, we introduce the following pair of nondifferentiable second-order symmetric duality model over arbitrary cones and derive duality theorems.

Primal Problem (MNHP):
Minimize $L(x,y,z,p) = f_1(x_1,y_1) + s(x_1|E_1) + f_2(x_2,y_2) + s(x_2|E_2) - y_1^T z_1$

$$-\frac{1}{2}p_1^T \nabla_{y_1y_1} f_1(x_1,y_1)p_1 - \frac{1}{2}p_2^T \nabla_{y_2y_2} f_2(x_2,y_2)p_2 - y_2^T[\nabla_{y_2} f_2(x_2,y_2) + \nabla_{y_2y_2} f_2(x_2,y_2)p_2]$$

subject to

$$-\Big(\nabla_{y_1} f_1(x_1,y_1) - z_1 + \nabla_{y_1y_1} f_1(x_1,y_1)p_1\Big) \in C_1^*, \tag{12.1}$$

$$-\Big(\nabla_{y_2} f_2(x_2,y_2) - z_2 + \nabla_{y_2y_2} f_2(x_2,y_2)p_2\Big) \in C_2^*, \tag{12.2}$$

$$y_1^T[\nabla_{y_1} f_1(x_1,y_1) - z_1 + \nabla_{y_1y_1} f_1(x_1,y_1)p_1] \geqq 0, \tag{12.3}$$

$$p_1^T[\nabla_{y_1} f_1(x_1,y_1) - z_1 + \nabla_{y_1y_1} f_1(x_1,y_1)p_1] \geqq 0, \tag{12.4}$$

$$p_2^T[\nabla_{y_2} f_2(x_2,y_2) - z_2 + \nabla_{y_2y_2} f_2(x_2,y_2)p_2] \geqq 0, \tag{12.5}$$

$$x_1, x_2, y_2 \geqq 0, \tag{12.6}$$

$$z_1 \in D_1, z_2 \in D_2, x_1 \in C_3, x_2 \in C_4. \tag{12.7}$$

Dual Problem (MNHD):
Minimize $M(u,v,w,r) = f_1(u_1,v_1) - s(v_1|D_1) + f_2(u_2,v_2) - s(v_2|D_2) + u_1^T w_1$

$$-\frac{1}{2}r_1 \nabla_{u_1u_1} f_1(u_1,v_1)r_1 - \frac{1}{2}r_2 \nabla_{u_2u_2} f_2(u_2,v_2)r_2 - u_2^T[\nabla_{u_2} f_2(u_2,v_2) + \nabla_{u_2u_2} f_2(u_2,v_2)r_2]$$

subject to

$$\left(\nabla_{u_1} f_1(u_1,v_1)+w_1+\nabla_{u_1u_1} f_1(u_1,v_1)r_1\right) \in C_3^*, \tag{12.8}$$

$$\left(\nabla_{u_2} f_2(u_2,v_2)+w_2+\nabla_{u_2u_2} f_2(u_2,v_2)r_2\right) \in C_4^*, \tag{12.9}$$

$$u_1^T[\nabla_{u_1} f_1(u_1,v_1)+w_1+\nabla_{u_1u_1} f_1(u_1,v_1)r_1] \leqq 0, \tag{12.10}$$

$$r_1^T[\nabla_{u_1} f_1(u_1,v_1)+w_1+\nabla_{u_1u_1} f_1(u_1,v_1)r_1] \leqq 0, \tag{12.11}$$

$$r_2^T[\nabla_{u_2} f_2(u_2,v_2)+w_2+\nabla_{u_2u_2} f_2(u_2,v_2)r_2] \leqq 0, \tag{12.12}$$

$$v_1, v_2, u_2 \geqq 0, \tag{12.13}$$

$$w_1 \in E_1, w_2 \in E_2, v_1 \in C_1, v_2 \in C_2, \tag{12.14}$$

where $p_1 \in R^{|K_1|}$, $p_2 \in R^{|K_2|}$, $r_1 \in R^{|J_1|}$ and $r_2 \in R^{|J_2|}$ and E_1, E_2, D_1 and D_2 are compact convex sets in $R^{|J_1|}, R^{|J_2|}, R^{|K_1|}$ and $R^{|K_2|}$, respectively.

12.4.1 REMARKS

For $C_i = R_+,\ i = 1,2,3,4.$

i. If $|K_1| = 0$ and $|J_1| = 0$, then our model becomes Wolfe-type dual model.
ii. If $|K_2| = 0$ and $|J_2| = 0$, then our model reduces to Mond–Weir-type dual model.

In the next section, we have discussed the duality theorems under generalized assumptions. Let P^0 and Q^0 be the set of feasible solution of primal problem (MNHP) and dual problem (MNHD), respectively.

12.5 DUALITY THEOREMS

Theorem 12.1. *(Weak Duality) Let* $(x_1,x_2,y_1,y_2,z_1,z_2,p_1,p_2) \in P^0$ *and* $(u_1,u_2,v_1,v_2,w_1,w_2,r_1,r_2) \in Q^0$. *Let*

i. $f_1(.,v_1)+(.)^T w_1$ *be pseudobonvex at* u_1 *with respect to* η_1,
ii. $-f_1(x_1,.)+(.)^T z_1$ *be pseudobonvex at* y_1 *with respect to* η_2,
iii. $f_2(.,v_2)+(.)^T w_2$ *be bonvex at* u_2 *with respect to* η_3,
iv. $-f_2(x_2,.)+(.)^T z_2$ *be bonvex at* y_2 *with respect to* η_4,
v. $\eta_1(x_1,u_1)+u_1+r_1 \in C_1$,
vi. $\eta_2(v_1,y_1)+y_1+p_1 \in C_2$,
vii. $\eta_3(x_2,u_2)+u_2+r_2 \in C_3$,
viii. $\eta_4(v_2,y_2)+y_2+p_2 \in C_4$.

Then,

$$L(x_1,x_2,y_1,y_2,z_1,p_1,p_2) \nleq M(u_1,u_2,v_1,v_2,w_1,r_1,r_2). \tag{12.15}$$

Proof: By hypotheses (*iii*) and (*iv*), we get

$$f_2(x_2,v_2)+x_2^T w_2 - f_2(u_2,v_2) - u_2^T w_2 + \frac{1}{2} r_2^T \nabla_{u_2u_2} f_2(u_2,v_2) r_2$$
$$\geqq \eta_3(x_2,u_2)[\nabla_{x_2} f_2(u_2,v_2) + w_2 + \nabla_{u_2u_2} f_2(u_2,v_2) r_2], \tag{12.16}$$

and

$$f_2(x_2,y_2) - y_2^T z_2 - f_2(x_2,v_2) + v_2^T z_2 + \frac{1}{2} p_2^T \nabla_{x_2x_2} f_2(x_2,v_2) p_2$$
$$\geqq \eta_4(v_2,y_2)[-\nabla_{y_2} f_2(x_2,y_2) + z_2 - \nabla_{x_2x_2} f_2(x_2,v_2) p_2]. \tag{12.17}$$

Using hypotheses (*vii*)*and* (*viii*) and dual constraints (12.2) and (12.9), we have

$$(\eta_3(x_2,u_2)+u_2+r_2)[\nabla_{x_2} f_2(u_2,v_2) + w_2 + \nabla_{u_2u_2} f_2(u_2,v_2) r_2] \geqq 0,$$

and

$$(\eta_4(v_2,y_2)+y_2+p_2)[-\nabla_{y_2} f_2(x_2,y_2) + z_2 - \nabla_{x_2x_2} f_2(x_2,v_2) p_2] \geqq 0.$$

Above inequalities follows that

$$\eta_3(x_2,u_2)[\nabla_{x_2} f_2(u_2,v_2) + w_2 + \nabla_{u_2u_2} f_2(u_2,v_2) p_2] + u_2[\nabla_{x_2} f_2(u_2,v_2) + w_2$$
$$+ \nabla_{u_2u_2} f_2(u_2,v_2) r_2] \geqq -r_2[\nabla_{x_2} f_2(u_2,v_2) + w_2 + \nabla_{u_2u_2} f_2(u_2,v_2) r_2],$$

and

$$\eta_4(v_2,y_2)[-\nabla_{y_2} f_2(x_2,y_2) + z_2 - \nabla_{y_2y_2} f_2(x_2,y_2) p_2] + y_2[-\nabla_{y_2} f(x_2,y_2) + z_2$$
$$-\nabla_{p_2} h_2(x_2,y_2,p_2))] \geqq p_2[\nabla_{y_2} f_2(x_2,y_2) - z_2 + \nabla_{y_2y_2} f_2(x_2,y_2) p_2].$$

Using inequalities (12.5) and (12.12) gives that

$$\eta_3(x_2,u_2)[\nabla_{x_2} f_2(u_2,v_2) + w_2 + \nabla_{u_2u_2} f_2(u_2,v_2) r_2]$$
$$\geqq -u_2[\nabla_{x_2} f_2(u_2,v_2) + w_2 + \nabla_{u_2u_2} f_2(u_2,v_2) r_2],$$

$$\eta_4(v_2,y_2)[-\nabla_{y_2} f_2(x_2,y_2) + z_2 - \nabla_{y_2y_2} h_2(x_2,y_2) p_2]$$
$$\geqq y_2[\nabla_{y_2} f_2(x_2,y_2) + z_2 + \nabla_{y_2y_2} f_2(x_2,y_2) p_2].$$

Further, from inequalities (12.16) and (12.17), we obtain

$$f_2(x_2,v_2) + x_2^T w_2 - f_2(u_2,v_2) - u_2^T w_2 + r_2^T \nabla_{u_2u_2} f_2(u_2,v_2) p_2$$
$$\geqq -u_2[\nabla_{x_2} f_2(u_2,v_2) + w_2 + \nabla_{u_2u_2} f_2(u_2,v_2) r_2],$$

$$f_2(x_2,y_2) - y_2^T z_2 - f_2(x_2,v_2) + v_2^T z_2 + p_2^T \nabla_{y_2y_2} f_2(x_2,v_2) p_2$$

$$\geqq y_2[\nabla_{y_2} f_2(x_2,y_2) + z_2 + \nabla_{y_2y_2} f_2(x_2,y_2) p_2].$$

Adding the above inequalities, we have

$$f_2(x_2,y_2) + x_2^T w_2 - y_2^T z_2 - \tfrac{1}{2} p_2^T \nabla_{y_2y_2} f_2(x_2,y_2) p_2$$

$$-y_2[\nabla_{y_2} f_2(x_2,y_2) - z_2 + \nabla_{y_2y_2} f_2(x_2,y_2) p_2]$$

$$\geqq f_2(u_2,v_2) + u_2^T w_2 - v_2^T z_2 - r_2^T \nabla_{u_2u_2} f_2(u_2,v_2) p_2$$

$$-u_2[\nabla_{x_2} f_2(u_2,v_2) + w_2 + \nabla_{u_2u_2} f_2(u_2,v_2) r_2].$$

Now, using $x_2^T w_2 \leqq s(x_2|E_2)$ and $v_2^T z_2 \leqq s(v_2|D_2)$, we obtain

$$f_2(x_2,y_2) + s(x_2|E_2) - \tfrac{1}{2} p_2^T \nabla_{y_2y_2} f_2(x_2,y_2) p_2$$

$$-y_2[\nabla_{y_2} f(x_2,y_2) + \nabla_{x_2x_2} f_2(x_2,y_2) p_2]$$

$$\geqq f_2(u_2,v_2) - s(v_2|D_2) - \tfrac{1}{2} r_2^T \nabla_{u_2u_2} f_2(u_2,v_2) p_2$$

$$-u_2[\nabla_{x_2} f_2(u_2,v_2) + \nabla_{u_2u_2} f_2(u_2,v_2) r_2]. \tag{12.18}$$

Similarly, using hypotheses $(v)-(vi)$, primal-dual constraints and the fact that $x_1^T w_1 \leqq s(x_1|E_1)$ and $v_1^T z_1 \leqq s(v_1|D_1)$, we get

$$\eta_1(x_1,u_1)[\nabla_{u_1} f_1(u_1,v_1) + w_1 + \nabla_{u_1u_1} f_1(u_1,v_1) r_1]$$

$$\geqq -(u_1 + r_1)[\nabla_{u_1} f_1(u_1,v_1) + w_1 + \nabla_{u_1u_1} f_1(u_1,v_1) r_1],$$

and

$$\eta_2(v_1,y_1)[-\nabla_{y_1} f_1(x_1,y_1) + z_1 - \nabla_{y_1y_1} f_1(x_1,y_1) p_1]$$

$$\geqq (y_1 + p_1)[\nabla_{y_1} f_1(x_1,y_1) - z_1 + \nabla_{y_1y_1} f_1(x_1,y_1) p_1].$$

Now inequalities (12.3), (12.4), (12.10), and (12.11) give

$$\eta_1(x_1,u_1)[\nabla_{u_1} f_1(u_1,v_1) + w_1 + \nabla_{u_1u_1} f_1(u_1,v_1) r_1] \geqq 0,$$

and

$$\eta_2(v_1,y_1)[-\nabla_{y_1} f_1(x_1,y_1) + z_1 - \nabla_{y_1y_1} f_1(x_1,y_1) p_1] \geqq 0,$$

which by hypothesis (i) and (ii) implies

$$f_1(x_1,v_1)+x_1^T w_1 - f_1(u_1,v_1) - v_1^T w_1 + \frac{1}{2} r_1^T \nabla_{u_1u_1} f_1(u_1,v_1) p_1 \geqq 0,$$

and

$$f_1(x_1,y_1)+v_1^T z_1 - f_1(x_1,v_1) - y_1^T z_1 - \frac{1}{2} p_1^T \nabla_{x_1x_1} f_1(x_1,v_1) p_1 \geqq 0.$$

Adding the above two inequalities, we get

$$f_1(x_1,y_1) - y_1^T z_1 + x_1^T w_1 - \tfrac{1}{2} p_1^T \nabla_{y_1y_1} f_1(x_1,y_1) p_1$$

$$\geqq f_1(u_1,v_1) + v_1^T w_1 - v_1^T z_1 - \tfrac{1}{2} r_1^T \nabla_{u_1u_1} f_1(u_1,v_1) p_1.$$

Using $x_1^T w_1 \leqq s(x_1|E_1)$ and $v_1^T z_1 \leqq s(v_1|D_1)$, we have

$$f_1(x_1,y_1) - y_1^T (y_1)^T z_1 + s(x_1|E_1) - \tfrac{1}{2} p_1^T \nabla_{y_1y_1} f_1(x_1,y_1) p_1$$

$$\geqq f_1(u_1,v_1) + v_1^T w_1 - s(v_1|D_1) + g_1(u_1,v_1,p_1) - r_1^T \nabla_{r_1} g_1(u_1,v_1,p_1). \quad (12.19)$$

Combining inequalities (12.18) and (12.19), we obtain

$$L(x_1,x_2,y_1,y_2,z_1,p_1,p_2) \geqq M(u_1,u_2,v_1,v_2,w_1,r_1,r_2).$$

This completes the proof.

Theorem 12.2. *(Strong Duality) Let $(\bar{x_1},\bar{x_2},\bar{y_1},\bar{y_2},\bar{z}_1,\bar{z}_2,\bar{p}_1,\bar{p}_2)$ be an optimal solution of (MNHP). Suppose that*

i. $\nabla_{y_1y_1} f_1(\bar{x_1},\bar{y_1})$ *is positive or negative definite and* $\nabla_{y_2y_2} f_2(\bar{x_2},\bar{y_2})$ *is negative definite,*
ii. $\nabla_{y_1} f_1(\bar{x_1},\bar{y_1}) - \bar{z}_1 + \nabla_{y_1y_1} f_1(\bar{x_1},\bar{y_1}) p_1 \neq 0$ *and*
$\nabla_{y_2} f_2(\bar{x_2},\bar{y_2}) - \bar{z}_2 + \nabla_{y_2y_2} f_2(\bar{x_2},\bar{y_2}) \bar{p}_2 \neq 0,$
iii. $(\bar{p}_1)^T [\nabla_{y_1} f_1(\bar{x_1},\bar{y_1}) - \bar{z}_1 + \nabla_{y_1y_1} f_1(\bar{x_1},\bar{y_1}) \bar{p}_1] = 0 \Rightarrow \bar{p}_1 = 0$ *and*
$y_2[\nabla_{y_2} h_2(\bar{x_2},\bar{y_2},\bar{p}_2) - \nabla_{p_2} h_2(\bar{x_2},\bar{y_2},\bar{p}_2) + \nabla_{y_2y_2} f_2(\bar{x_2},\bar{y_2}) \bar{p}_2] = 0 \Rightarrow \bar{p}_2 = 0.$

Then, their exist $\xi_1 \in E_1$, $\xi_2 \in E_2$ such that

i. $(\bar{x_1},\bar{x_2},\bar{y_1},\bar{y_2},\xi_1,\xi_2,\bar{r}_1 = 0,\bar{r}_2 = 0)$ *is feasible for (MNHD) and*

ii. $L(\bar{x_1},\bar{x_2},\bar{y_1},\bar{y_2},\bar{p}_1,\bar{p}_2) = M(\bar{x_1},\bar{x_2},\bar{y_1},\bar{y_2},\bar{r}_1,\bar{r}_2).$

Furthermore, if the hypotheses of Theorem 12.1 are satisfied for all feasible solutions of (MNHP) and (MNHD), then $(\bar{x_1},\bar{x_2},\bar{y_1},\bar{y_2},\xi_1,\xi_2,\bar{r}_1 = 0,\bar{r}_2 = 0)$ is an optimal solution for (MNHD).

Proof:
Since $(\bar{x}_1,\bar{x}_2,\bar{y}_1,\bar{y}_2,\bar{p}_1,\bar{p}_2)$ is a n optimal solution of (MNHP), by the Fritz John necessary optimality conditions [17], there exist $\alpha,\gamma \in R_+, \delta_1 \in C_3, \delta_2 \in C_4$, $\beta_1 \in R^{|K_1|}$, $\beta_2,\zeta \in R^{|K_2|}$, $\xi_1 \in R^{|J_1|}$, $\xi_2 \in R^{|J_2|}$ such that the following conditions are satisfied at $(\bar{x}_1,\bar{x}_2,\bar{y}_1,\bar{y}_2,\bar{p}_1,\bar{p}_2)$:

$$(x_1-\bar{x}_1)^T\Big(\alpha[\nabla_{x_1}f_1(\bar{x}_1,\bar{y}_1)+\xi_1-\tfrac{1}{2}\nabla_{x_1}\{\bar{p}_1\nabla_{yy}f_1(\bar{x}_1,\bar{y}_1)\bar{p}_1\}+[\nabla_{y_1x_1}f_1(\bar{x}_1,\bar{y}_1)$$

$$+\nabla_{x_1}\{\nabla_{y_1y_1}f_1(\bar{x}_1,\bar{y}_1)\bar{p}_1\}(\beta_1-\gamma\bar{y}_1-\delta_1\bar{p}_1)\Big)\geqq 0,\ \forall\, x_1\in C_1, \tag{12.20}$$

$$(x_2-\bar{x}_2)^T\Big(\alpha[\nabla_{x_2}f_2(\bar{x}_2,\bar{y}_2)+\xi_2-\tfrac{1}{2}\nabla_{x_2}\{\bar{p}_2\nabla_{y_2y_2}f_2(\bar{x}_2,\bar{y}_2)\bar{p}_2](\beta_2-\alpha\bar{y}_2$$

$$-\alpha\bar{p}_2-\delta_2\bar{p}_2)+\{\nabla_{y_2x_2}f_2(\bar{x}_2,\bar{y}_2)\}(\beta_2-\alpha\bar{y}_2-\delta_2\bar{p}_2)\geqq 0,\ \forall\, x_2\in C_2, \tag{12.21}$$

$$\alpha[\nabla_{y_1}f_1(\bar{x}_1,\bar{y}_1)-\bar{z}_1-\tfrac{1}{2}\nabla_{y_1}(\nabla_{y_1y_1}f_1(\bar{x}_1,\bar{y}_1)\bar{p}_1)]+(\nabla_{y_1y_1}f_1(\bar{x}_1,\bar{y}_1)$$

$$+\nabla_{y_1}(\nabla_{y_1y_1}h_1(\bar{x}_1,\bar{y}_1)\bar{p}_1)(\beta_1-\gamma\bar{y}_1-\delta_1\bar{p}_1)$$

$$-\gamma[\nabla_{y_1}f_1(\bar{x}_1,\bar{y}_1)-\bar{z}_1+\nabla_{y_1y_1}f_1(\bar{x}_1,\bar{y}_1)\bar{p}_1]=0, \tag{12.22}$$

$$\{\nabla_{y_2}(\nabla_{y_2y_2}f_2(\bar{x}_2,\bar{y}_2)\bar{p}_2)\}(\beta_2-\alpha\bar{y}_2-\alpha\bar{p}_2-\delta_2\bar{p}_2)-\tfrac{1}{2}\alpha[\nabla_{y_2}(\nabla_{y_2y_2}f_2(\bar{x}_2,\bar{y}_2)\bar{p}_2)]$$

$$+\{\nabla_{y_2y_2}f_2(\bar{x}_2,\bar{y}_2)\}(\beta_2-\alpha\bar{y}_2-\delta_2\bar{p}_2)-\zeta=0, \tag{12.23}$$

$$\{\nabla_{y_1y_1}f_1(\bar{x}_1,\bar{y}_1)\}(\beta_1-\alpha\bar{p}_1-\gamma\bar{y}_1-\delta_1\bar{p}_1)$$

$$-\delta_1[\nabla_{y_1}f_1(\bar{x}_1,\bar{y}_1)-\bar{z}_1+\nabla_{y_1y_1}f_1(\bar{x}_1,\bar{y}_1)\bar{p}_1]=0, \tag{12.24}$$

$$\{\nabla_{y_2y_2}f_2(\bar{x}_2,\bar{y}_2)\}(\beta_2-\alpha\bar{y}_2-\alpha\bar{p}_2-\delta_2\bar{p}_2)$$

$$-\delta_2[\nabla_{y_2}f_2(\bar{x}_2,\bar{y}_2)-\bar{z}_2+\nabla_{y_2y_2}f_2(\bar{x}_2,\bar{y}_2)\bar{p}_2]=0, \tag{12.25}$$

$$\beta_1[\nabla_{y_1}f_1(\bar{x}_1,\bar{y}_1)-\bar{z}_1+\nabla_{y_1y_1}h_1(\bar{x}_1,\bar{y}_1)\bar{p}_1]=0, \tag{12.26}$$

$$\beta_2[\nabla_{y_2}f_2(\bar{x}_2,\bar{y}_2)-\bar{z}_2+\nabla_{y_2y_2}f_2(\bar{x}_2,\bar{y}_2)\bar{p}_2]=0, \tag{12.27}$$

$$\gamma\bar{y}_1[\nabla_{y_1}f_1(\bar{x}_1,\bar{y}_1)-\bar{z}_1+\nabla_{y_1y_1}f_1(\bar{x}_1,\bar{y}_1)\bar{p}_1]=0, \tag{12.28}$$

$$\delta_1\bar{p}_1[\nabla_{y_1}f_1(\bar{x}_1,\bar{y}_1)-\bar{z}_1+\nabla_{y_1y_1}f_1(\bar{x}_1,\bar{y}_1)\bar{p}_1]=0, \tag{12.29}$$

$$\delta_2\bar{p}_2[\nabla_{y_2}f_2(\bar{x}_2,\bar{y}_2)-\bar{z}_2+\nabla_{y_2y_2}f_2(\bar{x}_2,\bar{y}_2)\bar{p}_2]=0, \tag{12.30}$$

$$(\alpha-\gamma^1)y_1+\beta_1-\delta_1 p_1\in N_{D_1}(\bar{z}_1), \tag{12.31}$$

$$\beta_2-\delta_2 p_2\in N_{D_2}(\bar{z}_2), \tag{12.32}$$

$${\xi_1}^T\bar{x}_1=s(x_1|E_1),\xi_1\in E_1, \tag{12.33}$$

$$\xi_2{}^T\bar{x}_2 = s(x_2|E_2), \xi_2 \in E_2, \tag{12.34}$$

$$\mu_1\bar{x}_1 = 0, \tag{12.35}$$

$$\mu_2\bar{x}_2 = 0, \tag{12.36}$$

$$\eta^1\bar{y}_1 = 0, \tag{12.37}$$

$$\zeta\bar{y}_2 = 0, \tag{12.38}$$

$$(\alpha, \beta_1, \beta_2, \gamma, \delta_1, \delta_2, \zeta) \neq 0, \tag{12.39}$$

$$(\alpha, \beta_1, \beta_2, \gamma, \delta_1, \delta_2, \zeta) \geqq 0, \tag{12.40}$$

Premultiplying equations (12.24) and (12.25) by $(\beta_1 - \alpha\bar{p}_1 - \gamma\bar{y}_1 - \delta_1\bar{p}_1)$ and $(\beta_2 - \alpha\bar{p}_2 - \alpha\bar{y}_2 - \delta_2\bar{p}_2)$, respectively, and then, using equations (12.26)–(12.30), we get

$$(\beta_1 - \alpha\bar{p}_1 - \gamma\bar{y}_1 - \delta_1\bar{p}_1)^T\nabla_{y_1y_1}h(\bar{x}_1,\bar{y}_1)(\beta_1 - \alpha\bar{p}_1 - \gamma\bar{y}_1 - \delta_1\bar{p}_1) = 0,$$

and

$$(\beta_2 - \alpha\bar{p}_2 - \alpha\bar{y}_2 - \delta_2\bar{p}_2)\nabla_{y_2y_2}f_2(\bar{x}_2,\bar{y}_2)(\beta_2 - \alpha\bar{p}_2 - \alpha\bar{y}_2 - \delta_2\bar{p}_2)$$
$$= -\alpha\delta\bar{y}_2[\nabla_{y_2}f_2(\bar{x}_2,\bar{y}_2) + \nabla_{y_2y_2}f_2(\bar{x}_2,\bar{y}_2)\bar{p}_2].$$

Using hypothesis (i), we get

$$\beta_1 = \alpha\bar{p}_1 + \gamma\bar{y}_1 + \delta_1\bar{p}_1. \tag{12.41}$$

Further using inequality (12.1), (12.6), and (12.40), we obtain

$$(\beta_2 - \alpha\bar{p}_2 - \alpha\bar{y}_2 - \delta_2\bar{p}_2)\nabla_{y_2y_2}f_2(\bar{x}_2,\bar{y}_2)(\beta_2 - \alpha\bar{p}_2 - \alpha\bar{y}_2 - \delta_2\bar{p}_2) \geqq 0,$$

which on using hypothesis (i)

$$\beta_2 = \alpha\bar{p}_2 + \alpha\bar{y}_2 + \delta_2\bar{p}_2. \tag{12.42}$$

From equations (12.24) and (12.25), and hypothesis (ii), we obtain

$$\delta_1 = 0, \tag{12.43}$$

and

$$\delta_2 = 0. \tag{12.44}$$

Now suppose, $\alpha = 0$. Then, equations (12.42) and (12.44) imply $\beta_2 = 0$ and (12.23), (12.42), (12.44) gives $\zeta = 0$. From equation (12.22) and hypothesis (ii)

yield $\gamma = 0$, which along with equations (12.41) and (12.43) gives $\beta_1 = 0$. Thus, $(\alpha, \beta_1, \beta_2, \gamma, \delta_1, \delta_2, \zeta) = 0$, a contradiction to equation (12.39). Hence, from (12.40),

$$\alpha > 0. \tag{12.45}$$

Using equations (12.26), (12.28), and (12.29), we have

$$(\beta_1 - \gamma\bar{y}_1 - \delta_1\bar{p}_1)^T[\nabla_{y_1} f_1(\bar{x}_1, \bar{y}_1) - \bar{z}_1 + \nabla_{y_1y_1} f_1(\bar{x}_1, \bar{y}_1)\bar{p}_1] = 0,$$

and now using equation (12.41) in the above equation gives

$$\alpha\bar{p}_1{}^T[\nabla_{y_1} f_1(\bar{x}_1, \bar{y}_1) - \bar{z}_1 + \nabla_{y_1y_1} f_1(\bar{x}_1, \bar{y}_1)\bar{p}_1] = 0. \tag{12.46}$$

which along with hypothesis (iii) yields

$$\bar{p}_1 = 0. \tag{12.47}$$

Further, from equations (12.23) and (12.42), we get

$$\alpha[\nabla_{y_2}(\nabla_{y_2y_2} f_2(\bar{x}_2, \bar{y}_2)\bar{p}_2) - \nabla_{y_2y_2} f_2(\bar{x}_2, \bar{y}_2)\bar{p}_2 + \nabla_{y_2y_2} f_2(\bar{x}_2, \bar{y}_2)\bar{p}_2] = 0. \tag{12.48}$$

Now from hypothesis (iii), we obtain

$$\bar{p}_2 = 0. \tag{12.49}$$

Therefore, equations (12.41) and (12.42) reduce to

$$\beta_1 = \gamma\bar{y}_1, \tag{12.50}$$

and

$$\beta_2 = \alpha\bar{y}_2. \tag{12.51}$$

Also, it follows from equations (12.22), (12.41), (12.47) and hypothesis (ii) that

$$\alpha - \gamma = 0$$

As $\alpha > 0$, we get

$$\alpha = \gamma > 0. \tag{12.52}$$

So equation (12.50) implies

$$\bar{y}_1 = \frac{\beta_1}{\gamma} \geqq 0. \tag{12.53}$$

Similarly,

$$\bar{y}_2 = \frac{\beta_2}{\alpha} \geqq 0. \tag{12.54}$$

Moreover, equations (12.20) and (12.21) along with (12.41), (12.42), and (12.49) yield

$$(x_1-\bar{x}_1)^T\left(\nabla_{x_1}f_1(\bar{x}_1,\bar{y}_1)-\bar{z}_1+\nabla_{x_1x_1}f_1(\bar{x}_1,\bar{y}_1)r_1\right)\geqq 0,\ x_1\in C_1, \tag{12.55}$$

$$(x_2-\bar{x}_2)^T\left(\nabla_{x_2}f_2(\bar{x}_2,\bar{y}_2)-\bar{z}_2+\nabla_{x_2x_2}f_2(\bar{x}_2,\bar{y}_2)r_2\right)\geqq 0,\ \ x_2\in C_2. \tag{12.56}$$

Let $x_1 \in C_1$. Then, $x_1+\bar{x}_1 \in C_1$, as C_1 is a closed convex cone. By substituting $x_1+\bar{x}_1$ in place of x_1 in (12.55), we get

$$x_1^T[\nabla_{x_1}f_1(\bar{x}_1,\bar{y}_1)-\bar{z}_1+\nabla_{x_1x_1}f_1(\bar{x}_1,\bar{y}_1)r_1]\geqq 0, \tag{12.57}$$

which in turn implies that for all $x_1 \in C_1$, we have

$$[\nabla_{x_1}f_1(\bar{x}_1,\bar{y}_1)-\bar{z}_1+\nabla_{x_1x_1}f_1(\bar{x}_1,\bar{y}_1)r_1]\in C_1^*, \tag{12.58}$$

Also, substituting $x_1 = 0$ and $x_1 = 2\bar{x}_1$ simultaneously in (12.55) yields

$$\bar{x}_1^T[\nabla_{x_1}f_1(\bar{x}_1,\bar{y}_1)-\bar{z}_1+\nabla_{x_1x_1}f_1(\bar{x}_1,\bar{y}_1)r_1]\geqq 0. \tag{12.59}$$

Using (12.53), we get

$$\bar{y}_1=\frac{\beta_1}{\gamma}\in C_3. \tag{12.60}$$

Let $x_2 \in C_2$. Then, $x_2+\bar{x}_2 \in C_2$, as C_2 is a closed convex cone. By substituting $x_2+\bar{x}_2$ in place of x_2 in (12.56), we obtain

$$x_2^T[\nabla_{x_2}f_2(\bar{x}_2,\bar{y}_2)-\bar{z}_2+\nabla_{x_2x_2}f_2(\bar{x}_2,\bar{y}_2)r_2]\geqq 0, \tag{12.61}$$

which in turn implies that for all $x_2 \in C_2$, we have

$$[\nabla_{x_2}f_2(\bar{x}_2,\bar{y}_2)-\bar{z}_2+\nabla_{x_2x_2}f_2(\bar{x}_2,\bar{y}_2)r_2]\in C_2^*. \tag{12.62}$$

Also, substituting $x_2 = 0$ and $x_2 = 2\bar{x}_2$, simultaneously, in (12.56) yields

$$\bar{x}_2^T[\nabla_{x_2}f_1(\bar{x}_2,\bar{y}_2)-\bar{z}_2+\nabla_{x_2x_2}f_2(\bar{x}_2,\bar{y}_2)r_2]\geqq 0. \tag{12.63}$$

Using (12.54), we get

$$\bar{y}_2=\frac{\beta_2}{\gamma}\in C_4. \tag{12.64}$$

$$\bar{x}_2^T[\nabla_{x_2}f_2(\bar{x}_2,\bar{y}_2)-\bar{z}_2+\nabla_{x_2x_2}f_2(\bar{x}_2,\bar{y}_2)r_2]\geqq 0 \text{ (using equation (12.36))}. \tag{12.65}$$

Thus, $(\bar{x}_1,\bar{x}_2,\bar{y}_1,\bar{y}_2,\bar{r}_1=0,\bar{r}_2=0)$ satisfies the dual constraints (12.8)–(12.13). Now using equations (12.27), (12.50), and (12.52), we obtain

$$\bar{y}_1^T\nabla_{y_1}f_1(\bar{x}_1,\bar{y}_1)=\bar{y}_1^T\bar{z}_1, \tag{12.66}$$

and from (12.30), (12.47), (12.50), and (12.52), we get

$$\bar{y}_2^T \bar{z}_2 = s(\bar{y}_2|D_2). \tag{12.67}$$

Moreover, since $\beta_2 = \alpha \bar{y}_2$ and $\alpha > 0$ then from (12.35) and (12.49), we obtain $\bar{y}_1 \in N_{D_1}(\bar{z}_1)$. Also D_1 is a compact convex set in $R^{|K_1|}$:

$$\bar{y}_1^T \bar{z}_1 = s(\bar{y}_1|D_1). \tag{12.68}$$

Thus, after using (12.33), (12.34), (12.47), (12.49), (12.66)–(12.68), the values of the objective functions of (MNHP) and (MNHD) at $(\bar{x}_1, \bar{x}_2, \bar{y}_1, \bar{y}_2, \bar{p}_1 = 0, \bar{p}_2 = 0)$ and $(\bar{x}_1, \bar{x}_2, \bar{y}_1, \bar{y}_2, \bar{r}_1 = 0, \bar{r}_2 = 0)$ are equal. Using weak duality, it has easily shown that $(\bar{x}_1, \bar{x}_2, \bar{y}_1, \bar{y}_2, \bar{r}_1 = 0, \bar{r}_2 = 0)$ is an optimal solution for (MNHD).

Theorem 12.3. *(Converse Duality) Let $(\bar{u}_1, \bar{u}_2, \bar{v}_1, \bar{v}_2, \bar{w}_1, \bar{w}_2, \bar{r}_1, \bar{r}_2)$ be an optimal solution of (MNHD). Suppose that*

i. $\nabla_{u_1u_1} f_1(\bar{u}_1, \bar{v}_1)$ *is positive or negative definite, and* $\nabla_{r_2r_2} g_2(\bar{u}_2, \bar{v}_2, \bar{r}_2)$ *is negative definite,*
ii. $\nabla_{u_1} f_1(\bar{u}_1, \bar{v}_1) + \bar{w}_1 + \nabla_{u_1u_1} f_1(\bar{u}_1, \bar{v}_1) \neq 0$ *and* $v_2[\nabla_{y_2} f_2(\bar{x}_2, \bar{y}_2) + \bar{w}_2 + \nabla_{y_2y_2} f_2(\bar{x}_2, \bar{y}_2)] \neq 0,$
iii. $\bar{r}_1^T [\nabla_{u_1} f_1(\bar{u}_1, \bar{v}_1) + \bar{w}_1 + \nabla_{u_1u_1} f_1(\bar{u}_1, \bar{v}_1)] = 0 \Rightarrow \bar{r}_1 = 0$ *and* $u_2[\nabla_{u_2} g_2(\bar{u}_2, \bar{v}_2, \bar{r}_2) - \nabla_{r_2} g_2(\bar{u}_2, \bar{v}_2, \bar{r}_2) + \nabla_{u_2u_2} f_2(\bar{u}_2, \bar{v}_2)\bar{r}_2] = 0 \Rightarrow \bar{r}_2 = 0,$

Then, their exist $\vartheta_1 \in D_1$, $\vartheta_2 \in D_2$ *such that*

i. $(\bar{u}_1, \bar{u}_2, \bar{v}_1, \bar{v}_2, \vartheta_1, \vartheta_2, \bar{r}_1 = 0, \bar{r}_2 = 0)$ *is feasible for (MNHP)*
ii. $L(\bar{u}_1, \bar{u}_2, \bar{v}_1, \bar{v}_2, \bar{p}_1, \bar{p}_2) = M(\bar{u}_1, \bar{u}_2, \bar{v}_1, \bar{v}_2, \bar{r}_1, \bar{r}_2).$

Furthermore, if the hypothesis of Theorem 12.1 is satisfied for all feasible solutions of (MNHP) and (MNHD), then $(\bar{x}_1, \bar{x}_2, \bar{y}_1, \bar{y}_2, \bar{r}_1 = 0, \bar{r}_2 = 0)$ *is an optimal solution for (MNHP).*

Proof. Follows on the line of Theorem 12.2.

12.6 SELF-DUALITY

In general, $(MNHP)$ and (MNHD) are not self-duals without some added restrictions on f. If we assume $D_1 = D_2, E_1 = E_2, C_1 = C_2 = C_3 = C_4 = C$, $f_1 : R^{|J_1|} \times R^{|K_1|} \to R$, $f_2 : R^{|J_2|} \times R^{|K_2|} \to R$ are skew symmetric, that is,

$$f_i(u_1, v_1) = -f_i(v_1, u_1), i = 1, 2, ...k,$$

then, we shall show that (MNHP) and (MNHD) are self-duals. By recasting the dual problem (MNHD) as a minimization problem, we have

Minimize $M(u,v,r) =$
$-\{f_1(u_1,v_1) - s(v_1|D_1) + f_2(u_2,v_2) - s(v_2|D_2) + u_1^T w_1$

$-\frac{1}{2}r_1^T \nabla_{u_1u_1} f_1(u_1,v_1,r_1) - \frac{1}{2}r_2^T \nabla_{u_2u_2} f_2(u_2,v_2,r_2) - u_2^T [\nabla_{u_2} f_2(u_2,v_2) + \nabla_{u_2u_2} f_2(u_2,v_2) r_2]\}$

subject to

$$\nabla_{u_1} f_1(u_1,v_1) + w_1 + \nabla_{u_1u_1} f_1(u_1,v_1) r_1 \in C_3^*,$$
$$\nabla_{u_2} f_2(u_2,v_2) + w_2 + \nabla_{u_2u_2} f_2(u_2,v_2) r_2 \in C_4^*,$$
$$u_1^T [\nabla_{u_1} f_1(u_1,v_1) + w_1 + \nabla_{u_1u_1} f_1(u_1,v_1) r_1] \leqq 0,$$
$$r_1^T [\nabla_{u_1} f_1(u_1,v_1) + w_1 + \nabla_{u_1u_1} f_1(u_1,v_1) r_1] \leqq 0,$$
$$r_2^T [\nabla_{u_2} f_2(u_2,v_2) + w_2 + \nabla_{u_2u_2} f_2(u_2,v_2) r_2] \leqq 0,$$
$$v_1, v_2, u_2 \geqq 0.$$

Since f are skew symmetric,
$\nabla_{u_1} f_1(u_1,v_1) = -\nabla_{u_1} f_1(v_1,u_1)$ and $\nabla_{u_1} f_2(u_2,v_2) = -\nabla_{u_1} f_2(v_2,u_2)$,

now the above problem becomes
Minimize $M(u,v,r) =$
$f_1(v_1,u_1) + s(v_1|D_1) + f_2(v_2,u_2) + s(v_2|D_2) - u_1^T w_1$

$-\frac{1}{2}r_1^T \nabla_{u_1u_1} f_1(v_1,u_1) r_1 - \frac{1}{2}r_2^T \nabla_{u_2u_2} f_2(u_2,v_2) r_2 - u_2^T [\nabla_{u_2} f_2(v_2,u_2) + \nabla_{u_2u_2} f_2(v_2,u_2) r_2]$

subject to

$$-\left(\nabla_{u_1} f_1(v_1,u_1) + w_1 + \nabla_{u_1u_1} f_1(v_1,u_1) r_1\right) \in C_1^*,$$
$$-\left(\nabla_{u_2} f_2(v_2,u_2) + w_2 + \nabla_{u_2u_2} f_2(v_2,u_2) r_2\right) \in C_2^*,$$
$$u_1^T [\nabla_{u_1} f_1(v_1,u_1) + w_1 + \nabla_{u_1u_1} f_1(v_1,u_1) r_1] \geqq 0,$$
$$r_1^T [\nabla_{u_1} f_1(v_1,u_1) + w_1 + \nabla_{u_1u_1} f_1(v_1,u_1) r_1] \geqq 0,$$
$$r_2^T [\nabla_{u_2} f_2(v_2,u_2) + w_2 + \nabla_{u_2u_2} f_2(v_2,u_2) r_2] \geqq 0,$$
$$v_1, v_2, u_2 \geqq 0,$$

which shows that $M(u,v,p)$ is identical to $L(x,y,p)$; that is, the objective and the constraint functions are equal. Thus, the problem $L(x,y,p)$ becomes self-dual.

It is obvious that the feasibility of $(x_1,x_2,y_1,y_2,p_1,p_2)$ for $L(x,y,p)$ implies the feasibility of $(y_1,y_2,x_1,x_2,p_1,p_2)$ $(MNHP)$ implies the feasibility of for $(MNHD)$ and conversely.

12.7 CONCLUSION

A pair of nondifferentiable mixed-type second-order symmetric dual programs has been formulated over arbitrary constraints by considering the optimization under the assumptions of bonvexity and pseudobonvexity with respect to η assumptions. While discussing the result of Wolfe-type higher-order dual model over arbitrary cones, we can see the result of duality over Mond–Weir also. It may be noted that the symmetric duality between $(MNHP)$ and $(MNHD)$ can be utilized to establish mixed symmetric duality in integer for scalar and multiobjective programming over cone and other related programming problems. The question arises as to whether the duality results developed in this chapter hold for Wolfe-type higher-order nondifferentiable multiobjective optimization problems under arbitrary cones. This may be the future direction for the researchers working in this area.

BIBLIOGRAPHY

[1] Bazaraa, M. S. and Goode, J. J.: On symmetric duality in nonlinear programming, *Operations Research,* 21 (1973), 1–9.

[2] Chandra, S., Husain, I. and Abha, On mixed symmetric duality in mathematical programming, *Opsearch,* 36 (1999), 165–171. View at Zentralblatt MATH.

[3] Chandra, S., Craven, B. D. and Mond, B.: Generalized concavity and duality with a square root term, *Optimization,* 16 (1985), 653–662.

[4] Chandra, S. and Kumar, V.: A note on pseudoinvexity and symmetric duality, *European Journal of Operational Research,* 105 (1998), 626–629.

[5] Chen, X.: Higher-order symmetric duality in nondifferentiable multiobjective programming problems, *Journal of Mathematical Analysis and Applications,* 290 (2004), 423–435.

[6] Dantzig, G. B., Eisenberg E. and Cottle R. W.: Symmetric dual nonlinear programming, *Pacific Journal of Mathematics,* 15 (1965), 809–812.

[7] Devi, G.: Symmetric duality for nonlinear programming problems involving η-bonvex functions, *European Journal of Operational Research,* 104 (1998), 615–621.

[8] Dorn, W. S.: A symmetric dual theorems for quadratic programs, *Journal of the Operations Research Society of Japan,* 2 (1960), 93–97.

[9] Hanson, M.A.: Second-order invexity and duality in mathematical programming, *Opsearch,* 30 (1989), 313–320.

[10] Hou, S. H. and Yang, X. M.: On second-order symmetric duality in nondifferentiable programming, *Journal of Mathematical Analysis and Applications,* 255 (2001), 491–498.

[11] Gulati, T. R., Himani, S. and Gupta, S. K.: 2010, Second-order multiobjective symmetric duality with cone constraints, *European Journal of Operational Research,* 205 (2005), 247–252.

[12] Gulati, T. R. and Gupta, S. K.: Higher-order nondifferentiable symmetric duality with generalized F-convexity, *Journal of Mathematical Analysis and Applications,* 329 (2007), 229–237.

[13] Gulati, T. R. and Gupta, S. K.: Higher-order symmetric duality with cone constraints, *Applied Mathematics Letters,* 22 (2009), 776–781.

[14] Gulati, T. R. and Mehndiratta, G.: Nondifferentiable multiobjective Mond-Weir type second-order symmetric duality over cones, *Optimization Letters,* 4 (2010), 293–309.

[15] Mandal, P. and Nahak, C.: Symmetric duality with $(p,r)-\rho-(\eta,\theta)$-invexity, *International Journal of Pure and Applied Mathematics,* 217 (2011), 8141–8148.

[16] Mangasarian, O. L.: *Nonlinear Programming,* McGraw-Hill, New York, (1969).

[17] Mangasarian, O. L.: Second and higher-order duality in nonlinear programming, *Journal of Mathematical Analysis and Applications,* 51 (1975), 607–620.

[18] Mond, B. and Zhang, J.: Higher-order invexity and duality in mathematical programming, in: J. P. Crouzeix, et al. (Eds.), *Generalized Convexity, Generalized Monotonicity: Recent Results,* Kluwer Academic, Dordrecht, (1998), 357–372.

[19] Mond, B. and Weir, T.: Generalized concavity and duality in Generalized Concavity in Optimization and Economics (S. Schaible and W. T. Ziemba Eds.), Academic Press, New York, (1981), 263–280.

[20] Mond, B. and Schechter, M.: Nondifferentiable symmetric duality, *Bulletin of the Australian Mathematical Society,* 53 (1996), 177–188.

[21] Kassem, M.: Higher-order symmetric duality in vector optimization problem involving generalized cone-invex functions, *Applied Mathematics and Computation,* 209 (2009), 405–409.

[22] Kassem, M.: Multiobjective nonlinear symmetric duality involving generalized pseudoconvexity, *Applied Mathematics and Computation,* 2 (2011), 1236–1242.

[23] Khurana, S.: Symmetric duality in multiobjective programming involving generalized cone-invex functions, *European Journal of Operational Research,* 165 (2005), 592–597.

[24] Kim, D. S. and Yun, Y. B.: Second-order symmetric and self duality in multiobjective programming, *Applied Mathematics Letters,* 10 (1997), 17–22.

[25] Kim, D. S., Kang, H. S. and Lee, Y. J.: Second-order symmetric duality for nondifferentiable multiobjective programming involving cones, *Taiwanese Journal of Mathematics,* 12 (2008), 1347–1363.

[26] Padhan, S. K. and Nahak, C.: Higher-order symmetric duality with higherorder generalized invexity, *Journal of Applied Mathematics and Computing,* 48 (2015), 407–420.

[27] Suneja, S. K., Sunila, A. and Sonia, D.: Multiobjective symmetric duality involving cones, *European Journal of Operational Research,* 141 (2002), 471–479.

[28] Suneja, S. K., Lalitha, C. S. and Khurana, S.: Second order symmetric duality in multiobjective programming, *European Journal of Operational Research,* 144 (2003), 492–500.

[29] Xianjun, L.: Sufficiency and duality for nonsmooth multiobjective programming problems involving generalized univex functions, *Journal of Systems Science and Complexity,* 26 (2013), 1002–1018.

[30] Yang, X. M., Teo, K. L. and Yang, X. Q.: Mixed symmetric duality in nondifferentiable mathematical programming, *Indian Journal of Pure and Applied Mathematics,* 34 (2003), 805–815. View at Zentralblatt MATH.

[31] Yang, X. M., Yang, X. Q., Teo, K. L. and Hou, S. H.: Second order symmetric duality in non-differentiable multiobjective programming with F -convexity, *European Journal of Operational Research,* 164 (2005), 406–416.

[32] Yang, X. M., Yang, X. Q., Teo, K. L. and Hou, S. H.: Multiobjective second-order symmetric duality with F-convexity, *European Journal of Operational Research,* 165 (2005), 585–591.

[33] Ying, G.: Higher-order symmetric duality in multiobjective programming problems, *Acta Mathematicae Applicatae Sinica, English Series,* 32 (2016), 485–494.

Index

For Product Safety Concerns and Information please contact our EU representative GPSR@taylorandfrancis.com
Taylor & Francis Verlag GmbH, Kaufingerstraße 24, 80331 München, Germany

www.ingramcontent.com/pod-product-compliance
Lightning Source LLC
Chambersburg PA
CBHW060812220326
41598CB00022B/2598
* 9 7 8 0 3 6 7 5 3 2 6 8 0 *